PROTEIN-BASED
SURFACTANTS

SURFACTANT SCIENCE SERIES

ADDITIONAL VOLUMES IN PREPARATION

PROTEIN-BASED SURFACTANTS

Synthesis, Physicochemical Properties, and Applications

edited by
Ifendu A. Nnanna
Proliant, Inc.
Ames, Iowa

Jiding Xia
Wuxi University of Light Industry
Wuxi, China

CRC Press
Taylor & Francis Group
Boca Raton London New York

CRC Press is an imprint of the
Taylor & Francis Group, an **informa** business

CRC Press
Taylor & Francis Group
6000 Broken Sound Parkway NW, Suite 300
Boca Raton, FL 33487-2742

First issued in paperback 2019

© 2001 by Taylor & Francis Group, LLC
CRC Press is an imprint of Taylor & Francis Group, an Informa business

No claim to original U.S. Government works

ISBN-13: 978-0-8247-0004-1 (hbk)
ISBN-13: 978-0-367-39729-6 (pbk)

Visit the Taylor & Francis Web site at
http://www.taylorandfrancis.com

and the CRC Press Web site at
http://www.crcpress.com

Preface

Surfactants have broad applications in many areas. For example, they are used either as essential additives or as processing aids in production and processing of materials that surround us. Surfactants are major constituents of laundry detergents, industrial cleaners, and cosmetics. Some surfactants are used as wetting, emulsifying, defoaming, or solubilizing agents in the pharmaceutical, textile, food, and petroleum industries.

In recent years, environmental concerns and regulatory pressure have provided the driving force to replace petrochemical-based surfactants partly with those based on naturally occurring renewable sources. There is a growing interest in the synthesis and formulation applications of surfactants from natural biopolymers. The notion that such surfactants would be biodegradable and/or biocompatible has provided strong incentive for research pursuit in this area. For the aforementioned reasons, surfactants such as protein-based surfactants (PBS) have attracted research attention. However, scientific information for this group of surfactants is scanty. This book intends to narrow the knowledge gap by presenting recent developments in this area.

Protein-based surfactants are usually synthesized with amino acids/peptides and fatty acids as building blocks. They are mainly of two types: peptide and amino acid surfactants. Both are interesting compounds that contain an amino or a peptide as the hydrophilic part and a long hydrocarbon chain as the hydrophobic portion. The hydrocarbon chain can be introduced through acyl, ester, amide, alkyl, or ether linkage. Protein-based surfactants are usually considered biodegradable, nontoxic, and nonirritating and, in some cases, they have antimicrobial properties. Information on the preparation, structure, and properties of PBS is found mostly in the patent literature. Relatively few commercial PBS are available.

The lack of a book devoted to PBS emphasizes the timeliness of and provides impetus for this book. The potential benefits of PBS are enormous and will be highlighted in this volume to stimulate more research in the area. The book provides insights into how underutilized protein and oil sources can be converted to new surfactants of potentially high value. It suggests potential large-volume new applications for waste and/or underutilized proteins in the surfactant and detergent industries. Furthermore, it presents viewpoints on chemical and enzymatic approaches in the synthesis of PBS. In general, the book examines synthesis approaches and the physical chemistry of PBS, application potentials, and the overall state of knowledge in the subject area.

This volume is intended for chemists, formulation chemists, physical chemists, biologists, food scientists, and research scientists involved in the synthesis, technology, properties, and application of new surfactants. Also, it is expected to serve as a reference source for graduate students or researchers in the fields of colloid and surface chemistry or chemical technology and engineering.

The book brings together fundamental and applied aspects of PBS, drawing perspectives from diversified researchers to provide comprehensive information on the subject. It provides viewpoints on synthesis (chemical, enzymatic, or chemoenzymatic processes), properties (safety, antimicrobial, biodegradability, etc.), and other special features of PBS. The first of the ten chapters provides an overview of surfactant properties, technology, application, environmental issues, and new concepts in the field. Chapter 2 takes a look at the natural raw materials and agricultural by-products for PBS synthesis. Chapter 3 discusses protein interactions at interfaces, reviewing enzymatic reactions at the interfaces and naturally occurring protein surfactants. Chapter 4 discusses amino acid surfactants, their chemical synthesis and physicochemical properties. Chapters 5 and 6 discuss enzymatic approaches to synthesis of PBS. Chapter 7 describes the preparation and properties of ionic and nonionic surfactants containing peptides or amino acids with perhydrogenated or perfluorinated chains. Chapter 8 provides insights on potential interactions of PBS with other compounds. Chapter 9 discusses potential applications of PBS including formulation, detergency, emulsion, foaming, cosmetics, personal cares, biotechnology, and other industrial applications. Chapter 10 examines the current market development and trends. Extensive references are given at the end of each chapter so that the reader can obtain further details elsewhere.

It is hoped that this volume will stimulate the interest of more researchers in the academic world and the surfactant industry to intensify work in the area of PBS and to facilitate the rapid movement of developments from laboratory to pilot level, to generate novel surfactants of unique properties.

Ifendu A. Nnanna
Jiding Xia

Contents

A. Pinazo   Technology of Surfactants, Departamento Tecnología de Tensioactivos, Instituto de Investigaciones Químicas y Ambientales de Barcelona, CSIC, Barcelona, Spain

Ludwig Rodehüser   LCPOC UMR 7565, CNRS–UHP, Université Henri Poincaré—Nancy I, Vandoeuvre-les-Nancy, France

Kazutami Sakamoto   Applied Research Department, AminoScience Laboratories, Ajinomoto Company, Inc., Kanagawa, Japan

J. Seguer   Laboratorios Miret, S.A. (LAMIRSA), Terrassa, Spain

Claude Selve   LCPOC UMR 7565, CNRS–UHP, Université Henri Poincaré—Nancy I, Vandoeuvre-les-Nancy, France

P. Vinardell   Departamento Fisiología, Facultad de Farmacia UB, Barcelona, Spain

Jiding Xia   Division of Surfactant Science and Technology, Department of Chemical Engineering, Wuxi University of Light Industry, Wuxi, China

Yun-Peng Zhu   Crompton Corporation, Dublin, Ohio

PROTEIN-BASED SURFACTANTS

1

An Overview of the Basis, Technology, and Surface Phenomena of Protein-Based Surfactants

JIDING XIA Wuxi University of Light Industry, Wuxi, China

IFENDU A. NNANNA Proliant, Inc., Ames, Iowa

I. INTRODUCTION

Surfactants (short for surface-active agents) consist of a hydrophilic (water-compatible) head group and a hydrophobic (water-repellent) hydrocarbon tail. They affect all aspects of our daily life, either directly via household detergents and personal care products or indirectly via the production and processing of the materials that surround us. Generally, they are important industrial chemicals widely used in the manufacture of household cleaning, personal care, agricultural, and food products. Most commercial surfactants in the market are synthesized by chemical process. Manufacturers of surfactants use a combination of petrochemicals and natural feedstock. Cost, performance, and availability considerations make the petrochemicals the raw materials of choice among manufacturers. The petrochemicals used include paraffin, benzene, olefin, fatty alcohol, fatty acid, fatty amine, ethanolamine, ethylene oxide, propyl oxide, betaine, and imidazoline and their polymerized products. However, some chemical intermediates from the preceding chemical groups are used directly, for example, alkylbenzene from the condensation of paraffin derivatives and benzene for manufacturing sulfonates through sulfonation with SO_3 in a falling-film reactor. These are mainly anionic surfactants, such as tetrapropylene benzene sulfonates.

Synthetic surfactants are usually produced in a series of strong and severe conditions such as high pressure (hydrogenation for fatty alcohol process), nitrogen protection (ethoxylation), and blended catalysts (alkyla-

1

tion, oxidation, ethoxylation, condensation, and polymerization). Due to their complex raw materials and a series of chemical treatments, synthetic surfactants prepared from petrochemicals often lead to by-products or isomers of the main products (e.g., ortho- or meta- in linear alkyl sulfonates, disulfonates in petrosulfonates or in methyl ester sulfonates), causing some quality, environmental, or ecological problems [1]. They may be examined easily by component and life cycle analysis [2,3]. Also, some minor harmful products are known to occur during chemical synthesis of many of the petrochemical-based surfactants. For example, the cancer-promoting substance dioxane is formed during sulfonation of polyoxyethylenated alcohol to make alcohol ether sulfate (AES), sultone that occurs during the abnormal oxidative bleaching of olefin sulfonates (optional), or sulfamine in the reaction of ethanolamine-like compounds with sulfoxidation. These potential environmental and biological problems associated with petrochemical-based surfactants are prompting the surfactant industries to seek natural alternatives.

The trend is toward mild and biodegradable surfactant products. The biodegradability issue is expected to be a major consideration in new product and market development in the 21st century. Companies are looking into new science and chemistry for developing mild, low-cost, biodegradable, and multifunctional surfactants.

The commercial prospects of protein-based surfactants (PBS) is expected to be huge, especially at the more expensive end of the market, such as pharmaceutical formulations and personal care products, where a broad range of functionality (e.g., safety, mildness to skin, high surface activity, antimicrobial activity, biodegradability) is desired.

According to Myers [4], *biodegradability* may be defined as the removal or destruction of chemical compounds through the biological action of living organisms. Such degradation may be divided into two stages: (1) primary degradation, leading to modification of the chemical structure of the material sufficient to eliminate any surface-active properties, and (2) ultimate degradation, in which the material is essentially completely removed from the environment as carbon dioxide, water, inorganic salts, or other materials that are the normal waste by-products of biological activity [4].

In this chapter we provide an overview of the basis, technology, and surface phenomena of PBS. It is hoped that this overview will stimulate interest in exploring strategies to develop PBS with desirable ecological and toxicological properties. It is expected that increased environmental awareness among consumers will drive the development of biological methods for the manufacture of industrial-scale PBS.

II. PROTEINS AS A BASE FOR SURFACTANT PREPARATION

Proteins are by nature amphipathic or amphiphilic molecules; that is, they contain both a hydrophobic (nonpolar) and a hydrophilic (polar) moiety. However, natural proteins per se are not used as commercial surfactants. Rather, proteins are modified by chemical or enzymatic means to products with surface-active properties. The use of modified proteins based on casein, soybean, albumen, collagen, or keratin is not new [5]. The Maywood Chemical Company introduced commercial protein-based surfactants (PBS) in the United States in 1937. They were primarily condensation products of fatty acids with hydrolyzed proteins [5]. Renewed interest in PBS has occurred not only as products based on renewable raw materials (i.e., proteins and fatty acids), but also as a solution to waste disposal for animal and vegetable protein by-products [5]. Among the commercial PBS, the following trade names have been active: Crotein, Lexein, Magpon Polypeptide, Protolate, Sol-U-Teins, and Super Pro.

Proteins are highly specific polypeptide polymers with three-dimensional structures that are formed by amino acids linked by peptide bonds in a set arrangement. The 3-D structure results from crosslinkage or interaction derived from amino acid sidechains and hydrogen bonds between peptides. Proteins differ from synthetic polymers in two major ways. In proteins, the number and variety of monomer units are considerably greater than those used in industrial polymer synthesis. Also, in a given protein species, the covalent structure or monomer order is essentially the same, as is the total number of monomer units. Thus, a given protein has a very narrow molecular weight range. However, proteins are similar to synthetic surfactants, because they are composed of both hydrophobic and hydrophilic amino acids that afford them a certain degree of surface activity. The main molecular properties of proteins responsible for their surface activity are size, charge, structural features, stability, amphipathicity, and lipophilicity [6]. The balance of polar, nonpolar, and charged amino acids determines the surface activity of proteins in a particular system [6]. This amphipathic nature of protein molecule allows it to bind with surfaces of different chemical nature [6].

Although various factors may affect the surface activity of proteins, a dominant parameter is hydrophobicity [7]. It influences adsorption and orientation of proteins at interfaces and correlates with surface activity in some instance (see the references in Ref. 6). However, the contribution of amino acids to the overall hydrophobicity of the protein is limited compared to the contribution of a hydrophobic tail in classic surfactants, such as ethoxylated fatty acids [7].

Hydrophobic modification of protein can be achieved either by chemical or enzymatic process to enhance the surface activity. The starting material in such processes may be a polypeptide, a peptide, or an amino acid as the hydrophilic part and a long hydrocarbon chain as the hydrophobic portion. The hydrocarbon can be introduced through acyl, ester, or alkyl linkage [8]. The following is a brief description of acylation and enzymatic modification strategies.

A.   Acylation Modification

Acylation is not just a widespread process among natural proteins but also the most important chemical means to modify protein functionality. Acylating agents include acyl anhydrides, acyl chlorides, and *N*-hydroxysuccinimide ester. All the acylating reagents can react with nucleophilic groups such as amino, phenolic (tyrosine), aliphatic hydroxyl (serine and threonine), and imidazole (histidine) groups, but the reactivity of these groups and the stability of their acyl derivatives differ respectively. Serine and threonine hydroxyl groups, which are weak nucleophiles, are not easily acylated in aqueous solutions, but succinyl derivatives are rather stable, whereas acylates of histidine and cysteine residues are unstable [9]. In most cases, the main groups involved are the α- and ε-NH$_2$, and to a lesser extent, the —SH and —OH groups. Essentially, the hydrophobicity and surface activity of the protein molecule can thus be tailored by controlling the molar ratio of the acyl anhydrides to proteins.

Acylated protein hydrolysates are known to be very mild surfactants. In the formulation of surfactants, the addition of small amounts of acylated protein hydrolysates to the more strongly irritating bulk surfactants results in a more than proportional improvement in their compatibility with the skin [10].

B.   Enzymatic Modification

Enzymatic techniques can be used to endow proteins with surface-active functionality. An enzymatic technique that has shown promise in enhancing surface properties of proteins is a modified version of the classical "plastein" reaction. The plastein reaction is known to be a protease-catalyzed reverse process in which a peptide–peptide condensation reaction [11,12] proceeds through the peptidyl-enzyme intermediate formation [13]. It is essentially a two-step process: enzymatic hydrolysis of a protein and plastein formation from the hydrolysate peptides. A novel one-step process was developed as a modified type of the plastein reaction by Yamashita et al. [14,15], which allowed papain-catalyzed incorporation of L-methionine and other amino acids directly into soy protein and flour. Although the novel one-step process of amino acid attachment was first designed for nutritional improvement, these researchers soon demonstrated its great potential for improving protein amphiphilicity and functionality as well [16]. It was hypothesized that by reacting a hydrophilic protein as the substrate with highly hydrophobic amino acid ester as the nucleophile, a product with amphiphilic properties would result from the localized regions of hydrophobicity [16]. To obtain adequately hydrophilic proteins as substrates, succinylation could be used to modify the proteins prior to their use as substrates for the one-step process [16]. To obtain adequately hydrophobic nucleophiles, amino acid esters with specific functional properties were produced [16]. For example, the application of the papain-catalyzed one-step process for L-norleucine n-dodecyl ester attachment to succinylate α-s_1-casein yielded a surface-active 20-kDa product with increased emulsifying activity compared to α-s_1-casein or succinylated α-s_1-casein [17]. Other enzymatic approaches used in modifying protein's surface properties or in novel surfactant synthesis are discussed in subsequent chapters.

III. CLASSIFICATION OF PROTEIN-BASED SURFACTANTS

Although this book is focused primarily on the synthesis and application of PBS from renewable sources or waste products, some naturally occurring PBS will also be discussed briefly. Some of these naturally occurring PBS have found applications in the pharmaceutical and personal care industries. Also, this book is primarily about amino acid– and peptide-based surfactants, two types of PBS that are discussed in some detail in the following sections.

A. Amino Acid Surfactants

Amino acid–based surfactants are derived from simple amino acids or mixed amino acids from synthesis or protein hydrolysates. They are composed of amino acid as the hydrophilic part and a long hydrocarbon chain as the hydrophobic part. The hydrophobic chain can be introduced through acyl, ester, amide, or alkyl linkage. Interest in amino acid surfactants is not new, as shown by early work in the area. In 1909, Bondi performed the first research on the introduction of a hydrophobic group to obtain N-acylglycine and N-acylalamine [18]. Subsequent work in this area focused on N-acylamino acids, as reported by Funk [19], Izar [20], Karrer [21], Staudinger and Becker [22],

Naudet [23], Tsubone [24], Heitmann [25], Kester [26], Fieser [27], Komatsu [28], Takehara [29], Imanaka [30], Seguer [31], Hatsutori [32], Abramzon [33], and others. Other derivatives of amino acids have also been prepared by reacting epoxidized fatty acid with ammonium or amines [34].

Some properties and uses of amino acid surfactants have been reviewed by Kariyoma [35]. A commercial amino acid surfactant, "Lamepon," derived from leather hydrolysates and a fatty acid acyl group has been reported. It was used as a mild detergent, an emulsifier, or dyeing additives. The surfactants of long-chain N^α-acyl amino acid derivatives from pure amino acids or protein hydrolysates have also been studied by many authors [36,37]. N-Acylsarcosinates (Medialan) is used extensively in the chemical industry as an intermediate in the Rapidogen series of fast cotton dyestuff [38]. N-Acyl sarcosinate salts are suitable for cosmetics, toothpaste, wound cleaners, personal cares, shampoo, bubble-bath pastes, aerosols and synthetic bars [39], flooding and reducing agents [40], and corrosion inhibitors [41]. Long-chain N-acylglutamates are surfactants derived from glutamic acid with less irritation on the skin than other conventional surfactants, such as SDS and LAS [42]. Acylglutamates generally showed weak acidity in an aqueous solution, the pH of which was almost equal to that of the human skin. Triethanolamine acylglutamates (AGTn) has shown strong stability to calcium ions and could be used with water having a hardness of 200–300 ppm calculated as $CaCO_3$ [42]. Hirofumi Yokota reported that the solubility of N^ε-acyllysine surfactant was improved by the introduction of N^ε-methyl group [43]. Ajinomoto Company (Japan) developed a series of amino acid emollients, including glycinates, glutamates, DL-pyrrolidone carboxylic acid salt of N^α-cocoyl-L-arginine ethyl ester, N^ε-lauroyl-L-lysine, and polyaspartates (Amilite GCK-12, Amisoft, CAE, Amihope, Aquadew SPA-30, and Eldew CL-301) [30]. A series of amphoteric surfactants, glycinates, sodium salts of N-(2-hytroxyethyl)-N-(2-hydroxyalkyl)-β-alanines (Na-HAA) was prepared by adding methyl acrylate to N-(hydroxyalkyl)-ethanolamine and subsequent saponification [44]. Pure N-acyl leucines of some structurally different and biologically active common fatty acids were synthesized; the N-acyl leucines exhibited greater activity in acid form than the methyl ester form and against gram positive bacteria than gram negative bacteria [45]. Sodium salts of long-chain N-alkyl-β-alanines, including N-(2-hydroxyalkyl-N-(2-hydroxyalkyl)-β-alanines, have been prepared that showed a wide pH range activity and less toxicity and irritation to human skin because of their structural similarity to amino acid [46–49]. Infante et al. synthesized a series of long-chain N^α-acyl-L-arginine, long-chain N^α-acyl-L-lysine, and a commercial cationic amino acid surfactant, N^ε-N^ε-N^εt-trimethyl N^α-lauroyl-L-lysine methyl (LLM$_q$) iodide [50].

Amino acid esters and amides are known to display excellent emulsifying characteristics and to possess strong antimicrobial properties [51–54], which makes them attractive as food additives [54]. Recently, Xia et al. [53] prepared and evaluated the structure–function relationships of acyl amino acid surfactants and their surface activity and antimicrobial properties. The amino acid surfactants had a general structure of α-amino-(*N*-acyl)-β-alkoxypropionate. A strong correlation existed between the critical micelle concentration (cmc) of amino acid surfactants and the chain length of the acyl group, as evident from Fig. 1, and also with their minimum inhibitory concentration (MIC) against *Escherichia coli, Pseudomonas aeruginosa, Aspergillus niger*, and *Pseudomonas cerevisiae* (see Table 1). Using methyl *p*-hydroxybenzoate as a control, the amino acid surfactants were shown to have 2–8, 64, and 4–8 times the activity against gram-negative, gram-positive bacteria, and fungi, respectively [53].

The foregoing is an example of many kinds of amino acid–based surfactants that are being produced either for investigative research or for commercialization. They have potential wide application in the cosmetic, personal care, food, and drug industries.

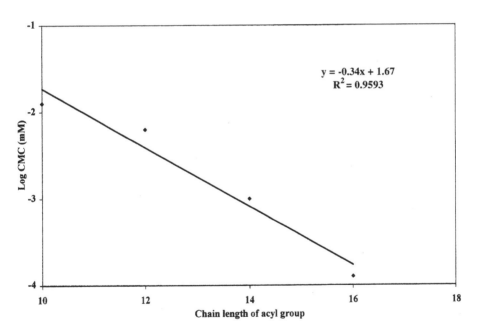

FIG. 1 Relationship of acyl group chain length of amino acid surfactants to critical micelle concentration (cmc). (From Ref. 53.)

TABLE 1   Correlation Between Log MIC for Microorganisms and Log CMC of Amino Acid Surfactants[a]

Microorganism	Regression[b]	Correlation coefficient (R^2)
Staphylococcus aureus	$Y = -1.4742 + 0.3611X$	0.080
Escherichia coli	$Y = -11.843 - 3.1745X$	0.878
Pseudomonas aeruginosa	$Y = 1.8777 + 1.9712X$	0.876
Aspergillus niger	$Y = 0.6358 + 1.0917X$	0.788
Saccharomyces cerevisiae	$Y = 1.4676 + 1.4123X$	0.940

MIC = minimum inhibitory concentration; CMC = critical micelle concentration.
[a] Amino acid surfactants of the general structure α-amino-(N-acyl)-β-alkoypropionate.
[b] $Y = \log$ MIC; $X = \log$ CMC.
Source: Ref. 53.

B. Peptide Surfactants

Peptide surfactants are derived from the condensation of dipeptides or tripeptides and hydrophobic chains such as fatty acids. Most in the literature have been synthesized chemically, although some have been produced biosynthetically. A number of useful reports on peptide surfactants have appeared in the literature. A few examples are cited here.

In a novel approach, Mhasker and Lakshminerayama [52] synthesized diethanolamides (DEA) of N-lauroyl dipeptides of various molecular structures. N-Lauroyl condensates of five amino acids were coupled with the corresponding amino acid methyl esters, and the resulting products were condensed with DEA in the presence of sodium methoxide to yield DEA of N-lauroyl dipeptides [52]. The physicochemical properties of the DEA-dipeptides showed that C12-glutamic-glutamic-DEA derivatives had a CMC of 0.13 wt% and a surface tension of 32.5 MN/m at 0.1 wt%; C12-glycine-glycine-DEA derivatives possessed good surface properties; C12-proline-proline-DEA had good wetting and emulsifying properties; C12-cysteine-cysteine-DEA was a good foaming agent. Their work showed correlation of the N-lauroyl dipeptides derivatives with surface activities and antimicrobial property [52]. For example, the leucine derivative showed maximum inhibition against both gram-positive and gram-negative bacteria. Also, whereas the DEA of glycine and phenylalanine were active against both bacteria, the cyclic structure of proline contributed to the activity only against gram-positive bacteria. The cysteine derivative did not show any inhibition against both bacteria. The authors [52] concluded that the thiol group in cysteine does not impart any antimicrobial activity to the derivatives. They further stated that the activity of the isobutyl sidechain in leucine in inhibiting the bacteria was more pronounced than the benzyl group in phenylalanine and the hydrogen atom of glycine [52]. The DEA derivatives of N-lauroyl dipeptides were nonionic, mild to the skin, and biodegradable surfactants.

Recently, Infante et al. [51] synthesized acidic and basic N^α-lauroyl arginine dipeptides, as methyl esters and free carboxylic acids, from pure amino acids. They found that amphoteric N^α-lauroyl-l-arginine dipeptide surfactants containing glutamic acid or lysine had no antimicrobial activities. However, the cationic versions of these peptides were antimicrobial, particularly the dipeptide derivatives containing lysine [51].

Also of interest are amphoteric perfluoroalkylated homologues of N-(2-hydroxy alkyl) amino acids prepared by the addition of 2-perfluoroalkyl-1,2-epoxy ethane to a starting (L,D or L) amino acid (glycine, alanine, β-alanine, serine, 2-amino butyric acid, norvaline, norleucine, methionne, sarcosine, aspartic acid, or glutamic acid). The presence of perfluoroalkyl groups produces a much more rigid and stable system, which in turn can lead to higher gel-to-liquid-crystal-phase transition temperatures [55,56]. Meanwhile, the presence of polyfluoro-alkyl groups in peptide surfactants lowers the surface tension markedly and could enhance the polarity of the compound [57]. Fluorinated peptide surfactants are described in more detail in a later chapter.

In addition to the preceding examples of protein-based surfactants, there are numerous biosurfactants that are produced by microorganisms as metabolic products. For example, surfactin is a lipopeptide-type biosurfactant produced by various strains of *Bacillus subtilis* and is one of the most powerful biosurfactant so far known [58]. Interest in biosurfactants is increasing, for two reasons: their diversity and biodegradability. There are also peptide-based surfactants that are of animal and plant origin. They are usually antimicrobial in function and are typically cationic (i.e., contain excess lysine and arginine residues) amphipathic molecules composed of 12–45 amino acid residues [59]. These antimicrobial peptides are described as exhibiting an α-helical structure or containing β-sheet elements with a α-helical domain, whereas others are usually rich in proline, tryptophan, or histidine residue [59].

IV.　SYNTHESIS OF PROTEIN-BASED SURFACTANTS

A.　PBS Building Blocks—Hydrophobic and Hydrophilic Groups

As indicated previously, as is the case with many surfactants, the molecules of PBS have the hydrophobic groups, have hydrophilic portions, and connect with intermediate linkage. The hydrophobic moieties may be a long hydrocarbon chain (generally C_8–C_{25}) in a linear, branched, or cyclic state. The hydrophobicity depends on the degree of chain branching, chain length, and chain distribution. Mostly the unsaturated or branched hydrocarbon chains have weak hydrophobicity. Hydrophilic moieties are the short polar or ionic portion, which can interact strongly with the water via dipole–dipole or ion–dipole interaction, such as the atoms O, N, P, S, and the groups —COO⁻, —SO₃⁻, —SO₃H, —PO₃²⁺, —N(CH₃)₂O⁻, —N(CH₃),⁺ —S(NH₃)₂⁺, —N(CH₃)₂CH₂CH₂O⁻, —(CH₂CH₂O)ₙH, —OH, —SH, —COOH, —NHCO—, —NH₂, —P≡, and so on.

The intermediate linkages are used as the connecting portion between hydrophobic and hydrophilic groups if they are not connected to each other directly. By inserting the weak functional groups in hydrophobic groups, the performance of surfactants may be improved and modified as a multifunctional surfactant so that the hydrophobicity and hydrophilicity within a molecule can be adjusted in a designed manner. Some practical intermediate linkages are as follows: —CH₂—, —CH₂—CH₂O—, —S—, —COO—, —CONH—, —NHCONH—. For example,

$$\text{RCO—N—CH}_2\text{CH}_2\text{COONa}$$
$$|$$
$$\text{CH}_3$$

—CON— and —CH₂—CH₂— are used as intermediate linkage between R and —COONa groups to increase the calcium ion tolerance and moderate mildness of RCOONa compound.

As has been indicated before in this chapter, it is possible to synthesize PBS by enzymatic techniques. More will be discussed on this subject in subsequent chapters. It is worth mentioning here selected work that has utilized the enzyme approach. Enzymatic acylation of the α-amino group of amino acid amides, using free fatty acids or their methyl esters, has been reported [60,54]. A range of N^ε-acyl amino acid amides was prepared with up to 50% yield [54]. The products were quantitatively converted into N-acyl-amino acids by means of a second enzyme, carboxypeptidase Y [54]. For more information on similar enzymatic approach, see the review by Sarney and Vulfson [54]. More recently, Izumi et al. [61] demonstrated enzymatic synthesis of N-acyl-amino acid homologous series in organic media. They performed enzymatic amidation of 3-amino-propionitrile and β-alanine ethyl ester using methyl laurate and immobilized lipase from *Candida antarctica*. This resulted in the formation of N-lauroyl-β-alanine ethyl ester and 3-N-lauroyl amino propionitrile, respectively, with the best yield (82%) attained using dioxane as solvent [61].

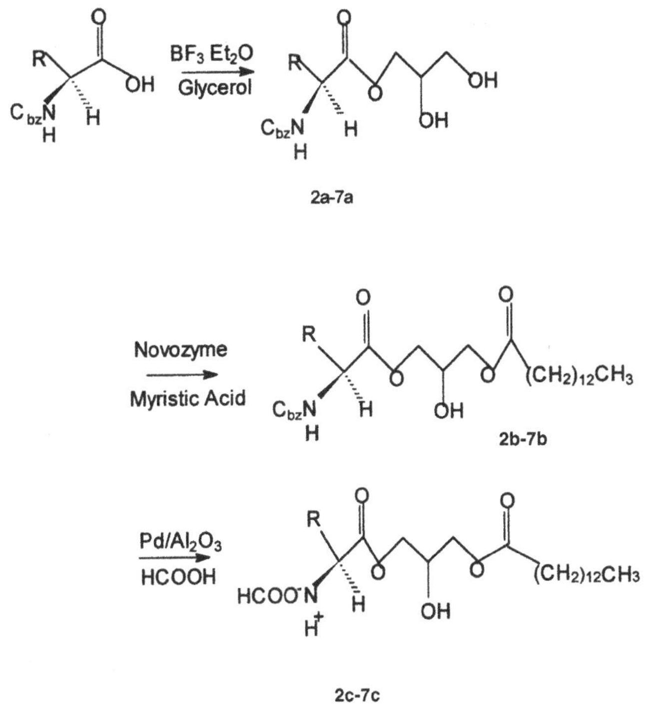

2a-7a

2b-7b

2c-7c

FIG. 2 Chemoenzymatic approach to the synthesis of 1-O-(L-aminoacyl)-3-O-myristoylglycerols. (From Ref. 62.)

Although both chemical and enzymatic synthetic methods have their uniqueness and limitations, a combination of the two approaches in the synthesis of PBS may be advantageous. The area of chemoenzymatic synthesis has been studied only a little, so not enough information is available to discuss its merits. Nevertheless, Rao et al. [62] described that glycerol-linked amino acid fatty acid esters can be obtained by a chemoenzymatic synthesis, as shown schematically in Fig. 2, with conversions of 50–90%.

V. SUMMARY AND CLOSING REMARKS

Several PBS have been studied, usually with amino acid/peptide and fatty acids as building blocks. Some are used as emulsifying agents in the cosmetic and pharmaceutical industries. They also play a role in biochemical research as detergents for the isolation and purification of membrane proteins. Information on the preparation, structure, and properties of PBS is found mostly in the patent literature. Protein-based proteins are usually considered biodegradable, nontoxic, nonirritating components of detergents and, in some cases, have antimicrobial properties.

Of the PBS, amino acid surfactants have been the subject of many studies, primarily on their applications as pharmaceuticals, biomedicals, cosmetics, household cleaners, and antimicrobial agents. On the other hand, except for few fragmentary reports, experimental work on peptide surfactants is relatively scanty. Both amino acid surfactants and peptide surfactants are interesting biocompatible compounds that contain amino acid or dipeptide as the hydrophilic part and a long hydrocarbon chain as the hydrophobic part. The hydrocarbon chain can be introduced through acyl, ester, amide, or alkyl linkage.

It is important to note renewable resources such as natural oils and proteins, especially those that are underutilized or ordinarily would go to waste, could afford the basis of new surfactant molecules. Surfactants derived from natural oils and proteins should be biodegradable and potentially of low toxicity.

The subsequent chapters will discuss the following areas in greater detail: structure–function of proteins, with specific reference to their surface property/interfacial behavior, amino acid surfactants, both chemically and enzymatically synthesized; peptide surfactants; and potential applications and the market assessment of PBS.

REFERENCES

1. P. A. Gilbert, Proceedings of 4th World Surfactants Congress, 1:57, Barcelona, 1996.
2. P. Avidsen, Proceedings of 4th World Surfactants Congress, 3:352, Barcelona, 1996.
3. J. L. Berna, Proceedings of 4th World Surfactants Congress, 3:321, Barcelona, 1996.
4. D. Myers, in Surfactant Science and Technology, VCH Publishers, New York, 1988, p 21.
5. R. D. Cowell and B. R. Bluestein, in Amphoteric Surfactants (B. R. Bluestein and C. L. Hilton, eds.), Marcel Dekker, New York, 1982, pp 265–279.
6. S. Magdassi and A. Kamyshny, in Surface Activity of Proteins (S. Magdassi, ed.), Marcel Dekker, New York, 1997, p 3.
7. S. Magdassi and O. Toledano, in Surface Activity of Proteins (S. Magdassi, ed.), Marcel Dekker, New York, 1997, pp 39–60.
8. J. Molinero, M. R. Julia, P. Erra, M. Robert, and M. R. Infante, J. Am. Oil Chem. Soc. 65:975 (1988).

9. A. P. Gounaris and G. E. Perlman, J. Biol. Chem. 242:2739 (1967).

10. A. Sander, Fett-Lipid 99(4):115 (1997).

11. A. I. Virtanen, Makromol. Chem. 6:94 (1951).

12. M. Yamashita, S. Arai, S. Tanimoto, and M. Fujimaki, Agric. Biol. Chem. (Japan) 37:95 (1973).

13. S. Tanimoto, M. Yamashita, S. Arai, and M. Fujimaki, Agric. Biol. Chem. (Japan) 36:1595 (1972).

14. M. Yamashita, S. Arai, Y. Imaizumi, and M. Fujimaki, J. Agric. Food Chem. 27:52 (1979).

15. M. Yamashita, S. Arai, Y. Ahrano, and M. Fujimaki, Agric. Biol. Chem. (Japan) 43:1065 (1979).

16. S. Nakai and E. Li-Chan, in Hydrophobic Interactions in Food Systems, CRC Press, Boca Raton, FL, 1988, pp 145–151.

17. S. Arai and M. Watanabe, Agric. Biol. Chem. (Japan) 44:1979 (1980).

18. S. Bondi, Z. Biochem. 17:543 (1909).

19. C. Funk, Z. Physiol. Chem. 65:61 (1910).

20. G. Izar, Biochem. 40:390 (1912).

21. P. Karrer, Chim. Acta 8:205 (1930).

22. H. Staudinger and H. V. Becker, Ber. Deut. Chem. 70:889 (1937).

23. M. Naudet, Bull. Soc. Chim. France 358 (1950).

24. T. Tsubone, J. Japan. Biochem. Soc. 35:67 (1963).

25. P. Heitmann, Eur. J. Biochem. Soc. 3:346 (1968).

26. E. B. Kester, U.S. Patent 2,463,779 (1949).

27. M. Fieser, J. Am. Chem. Soc. 78:2825 (1956).

28. S. Komatsu, Japan Patent 29,444 (1964).

29. M. Takehara, J. Am. Oil Chem. Soc. 49:157 (1972).

30. T. Imanaka, Jpn. Kokao Tokkyo Koho, JP05140057, 1993.

31. J. Seguer, New J. Chem. 18, 6:765–774 (1994).

32. M. Hatsutori, Jpn. Kokai Tokkyo Koho, JP0761952, 1995.

33. A. A. Abramzon, Zh. Prikl Khim, 68, 7:1178–1181 (1995) (Russian).

34. A. H. Cook and S. F. Coox, J. Chem. Soc. 2334 (1949).

35. Kariyoma, Yushi 45(4):75–82 (1992); 45(5):83–89 (1992).

36. R. K. Yoshida, J. Jpn. Oil Chem. Soc. 25:404 (1976).

37. H. Goldschmidt and J. Lubowe, Amer. Cosm. Perf. 87:35 (1972).

38. A. M. Schwartz and J. W. Perry, Surface Active Agents, Their Properties and Technology, R. E. Krieyer, Huntington, NY, 1978, p 35.

39. Y. T. Yakoto Yuichi, JP. Patent 02,237,911; CA, 114:128, 807h (1991).

40. D. Bradly, U.S. Patent 3,402,990 (1968).

41. M. Ako and K. Hideo, JP. Patent 01,164,446 (1989).

42. M. Takehara, I. Yoshimura, and R. Yoshida, J. Am. Oil Chem. Soc. 51:419–423 (1974).

43. H. Yokota, J. Am. Oil Chem. Soc. 62:1716–1719 (1985).

44. Ajinomoto Inc. Report, Kanayama, Japan (1997).

45. S. Y. Mhaskar, J. Am. Oil Chem. Soc. 70:19 (1993).

46. C. Blustein, J. Am. Oil Chem. Soc. 50:532 (1973).
47. T. Okumura, Bull. Chem. Soc. Jpn. 47:1067 (1974).
48. C. D. Moor, Soc. Chem. Ind. (London) Monograph 19:193 (1965).
49. Takal, J. Am. Oil Chem. Soc. 54:537–541 (1979).
50. M. R. Infante, J. Molinero, P. Erra, M. R. Julid, and J. J. Garcia Demingues, Fette Seifen Anstrichmittel 87(8):309 (1985).
51. M. R. Infante, J. Mollnero, P. Bosch, M. R. Julia, and P. Erra, J. Am. Oil Chem. Soc. 66(12): 1835 (1989).
52. S. Y. Mhasker and G. Lakshminerayana, J. Am. Oil Chem. Soc. 69:643 (1992).
53. J. Xia, Y. Xia, and Nnanna. J. Agric. Food Chem. 43:867 (1995).
54. D. B. Sarney and E. N. Vulfson. Trends in Biotech. 13:164 (1995).
55. M. Nasreddine, S. Szonyi, and A. Cambon. J. Am. Oil Chem. Soc. 70:1 (1993).
56. M. Nasreddine, S. Szonyi, and A. Cambon. Synth. Commun. 22:1597 (1992).
57. Daikin Kogyo Co., European Patent 0043108 (1981).
58. Y-H. Wei and I-M Chu, Enzyme Microbial Technol. 22:724 (1998).
59. R. E. W. Hancock and R. Lehrer, Trends in Biotech. 16:82 (1998).
60. S. E. Godtfredsen and F. Bjorkling. World Patent No. 90/14429 (1990).
61. T. Izumi, Y. Yagimuma, and M. Haga. J. Am. Oil Chem. Soc. 74:875 (1997).
62. V. Rao, P. Jauregi, I. Gill, and E. Vulfson. J. Am. Oil Chem. Soc. 74:879 (1997).

2

Natural Raw Materials and Enzymatic Modification of Agricultural By-Products for Protein-Based Surfactants

XIAO-QING HAN   Kraft Research and Development Center, Glenview, Illinois

I. PHYSICOCHEMICAL PROPERTIES OF PROTEINS

Because proteins play key roles in nearly all biological processes, their physicochemical properties have attracted scientists worldwide and been the subject of extensive study for decades. It is, therefore, impossible to cover all physicochemical properties of proteins in this section. Physicochemical properties that govern a protein's surface activity include: protein size, shape, amino acid composition, net charge and charge distribution, amino acid sequence and their higher structure, surface hydrophobicity or hydrophilicity, and molecular flexibility and rigidity. All these properties are related to the amino acid composition and sequence of proteins. The structure and functions of proteins depend on their amino acid sequences and the dynamic behavior of protein conformations.

A. Three-Dimensional Structure

Proteins are macromolecules consisting of 20 different amino acid residues arranged in a highly sophisticated three-dimensional structure. The basic structural units of proteins are amino acids. All proteins are constructed from the same set of 20 amino acids in different combinations. These 20 building blocks differ in size, shape, polarity, charge, and chemical reactivity. Amino acids are joined by peptide linkage between the α-carboxyl group of one amino acid and the α-amino group of the next one. The sequential linkage of amino acids is the primary structure of proteins, which is absolutely crucial to a protein's identity. In addition, the amino acid sequence of proteins specifies their three-dimensional structures, which are critical to their biological functions.

The secondary structure of proteins includes helices, sheet, turns, loops, and random coil. The linear linkage of amino acids under certain sequences can fold into regularly repeating structures that fall into two classes: helices and sheet. Helices and sheet are termed "regular" structures because their backbone N—N and C=O groups are arranged in a periodic pattern of hydrogen bonding [1]. The α-helix, for example, is formed by the hydrogen bonding of the backbone carbonyl oxygen of each residue to the backbone NH of the fourth residue along. The backbone atoms pack closely and form favorable van der Waals interactions, while the sidechains project out from the helix. The amino acid residue, proline (without an NH group), interrupts the hydrogen bonding pattern and lead to a kink in a helix. β-Strands, ranging from three to ten residues long, are associated side by side into parallel or antiparallel sheets called β-sheet. Hydrogen bonds in β-sheet are formed between backbone CO and NH groups of adjacent strands.

For those "nonregular" structures with nonrepeating backbone torsion angles and, at most, one internal hydrogen bond (N—H—O=C), they are classified as *turns* [2]. To make a spherical fold for globular proteins, the residues between regular helices and strands need to make sharp turns. Turns, also called *reverse turns* or β-*turns*, were first recognized by Venkatachalam [3]. Glycine and proline residues are often involved in turns because of their ability to adopt unusual backbone torsion angles. The remaining residues are often classified as *random coil*, although they are neither random nor coil [4]. Recently, it has been found that most proteins consist of a class of structure termed *loop*. A loop can be described as a continuous chain segment that adopts a "loop-shaped" conformation in three-dimensional space, with a small distance between its segment termini [5]. Loops are almost always situated at the molecular surface, which often implicate in molecular functions.

Although the well-known secondary structures are usually used for describing protein structure, no obligatory relationship to the functional domains has appeared. To form a functional domain and express the functionality of proteins, the secondary structures have to be further folded into higher-level structures, tertiary and/or quaternary structures. The tertiary structure of proteins describes the pattern of folding of secondary structures into a compact, more sophisticated molecule that can carry out biological functions. The tertiary-structured proteins may further associate into a higher degree of structure—quaternary structure. Quaternary structure refers to the noncovalent association of two or more subordinate entities, subunits, which may or may not be identical.

B. Hydration

Protein–water interaction plays an important role in the determination and maintenance of the three-dimensional structure of proteins. Water modified the physicochemical properties of proteins. Therefore, protein–water interactions have been the subject of intensive study and have provided significant advances in our understanding of the involvement of water in protein functionality, stability, and dynamics [6]. The thermodynamics of protein–water interaction directly affects dispersibility, wettability, swelling, and solubility of proteins. Surface-active properties of proteins are simply the result of the thermodynamically unfavorable interaction of exposed nonpolar patches of proteins with solvent water.

Protein hydration is a process from the dry state of a protein sample to the formation of solution state, which occurs over a very wide water activity range. Water molecules bind to both polar and nonpolar groups in proteins via dipole–dipole, dipole-induced dipole, and charge–dipole interactions. The hydration of a protein, therefore, is related to its amino acid composition and is affected by solution conditions such as pH, temperature, and ionic strength.

Most work on protein hydration involved the use of protein powders that are brought to the required hydration level by isopiestic equilibration with a solution of a salt or sulfuric acid of known water activity [7]. Quite apart from its fundamental importance and interest of proteins and their derivatives is important in applications of surface-active proteins. For instance, lyophilized protein stored in a vacuum desiccator over phosphorus pentoxide can only reach the lowest hydration level down to about 0.01 g H_2O per gram protein (defined as "*H*"), which represents about 8 moles of water per mole of protein [8]. Removal of the last few water molecules is extremely difficult and requires high vacuum for several days. Consequently, the removal of the most tightly bound water molecules could cause hysteresis during a solvation or absorption process. That is, water sorption isotherms for highly dried proteins show pronounced hysteresis, depending on the extent of prior dehydration of the protein sample. Experimental evidence suggested that conformational changes occur in the region of low hydration, indicating that adsorption hysteresis is a molecular phenomenon related to conformational change in protein molecules [9]. In practice, one should realize that a complete removal of water molecules from dehydrated protein samples is very difficult and may not be necessary.

When the protein is dehydrated to a certain level, its conformational flexibility decreases, in order to maintain a local free-energy minimum. Therefore, the level of hydration significantly affects the biological properties (e.g., enzyme activity) of protein. A number of enzyme activities have been studied as a function of hydration. In general, enzymes require only a small amount of water to express their catalytic activity [10]. However, although much experimental evidence suggested that conformational changes could occur at hydration levels below $0.2H$, the extent of the structural changes is still a matter of debate. Information from proton-exchange studies shows that efficient exchange occurs in near-dry protein ($0.08H$) [11]. Therefore, in most partially dried agricultural by-products, enzymatic reactions continue during storage, although at a much reduced rate. Although most moisture has been removed during the process of dehydration, most enzymes retain their activity at a certain level, depending on their stability and other characteristics. Dehydration alone can hardly eliminate the activity of enzymes.

C. Solubility

The solubility of proteins is an important property that affects and predicts other functional properties. Therefore, the dispersion of protein molecules in continuous phase is essential for expressing their surface activity. However, the effect of solubility on the surface activity of proteins is complicated, and there is no direct correlation between them. In general, a full dispersion of proteins is necessary in order to form a stable protein film at the interface, because insoluble proteins may precipitate at the interface.

The solubility of proteins is a thermodynamic manifestation of the equilibrium between protein–solvent and protein–protein interactions. Solubility is an intrinsic property of proteins that depends on their amino acid composition and sequence. The intrinsic factors that influence the solubility of proteins include surface hydrophobicity, polarity, and net charges. Hydrophobic interactions promote protein intermolecular interactions, resulting in decreased solubility. Polar and charged amino acid residues, on the other hand, promote protein–water interactions, resulting in increased solubility. As a rule, proteins containing more polar and charged groups, globular in shape, relatively small in molecular weight, have better solubility. On the other hand, highly hydrophobic proteins, proteins with random structure, or highly aggregated protein polymers are generally insoluble or unstable in solution.

Many thermodynamic variables, such as temperature, pH, ionic strength, and other compositions in the solution system, affect protein solubility and compatibility with other macromolecule components of the system. At constant pH and ionic strength, the solubility of most proteins increases with temperature in a certain range. Further increases of temperature may cause the denaturation of proteins, resulting in decreased solubility. In addition, many proteins form precipitates when completely heat-denatured.

While pH value affects the net charge of proteins, ionic strength affects the solubility of proteins in two different ways, depending on the characteristics of the protein surface. The ionic strength (μ) of a salt solution is given as follows:

$$\mu = 0.5 \sum C_i Z_i^2 \tag{1}$$

where C_i is the concentration of an ion and Z_i is its valance. Ionic strength affects the ζ potential of the protein surface, causing salting-in and/or salting-out reactions. Salts at a certain concentration usually cause an increase in protein solubility (salting-in). A further increase in salt concentration, however, causes a reversed change of ζ potential, resulting in a precipitation of proteins (salting-out).

The solubility of a protein in a system containing different proteins or different types of polymers (such as whey protein concentrate, WPC) possesses complicated characteristics. Intermolecular interactions significantly affect the overall solubility of the system, especially when protein concentration is high. Processing equipment and stirring conditions also contribute to protein hydration and solubility. The added energy for stirring gives a high degree of deformation of dispersed particles with a low (related to the dispersion medium) viscosity in flow. In the shear field of a protein solution system, macromolecules may orient themselves, interact with each other more intensively, and self-associate or dissociate frequently. A weak association of protein molecules may break down in a shear field, thereby increasing the cosolubility of proteins in a multicomponent solution system. The stirring effect on the solution structure, cosolubility, and phase state of multipolymer is of particular interest in conjunction with the control of the functional properties of food protein systems, even though it remains as yet little studied.

D. Denaturation

Studies on the denaturation of proteins have attracted researchers of different backgrounds for decades, for it provides information on the structure, properties, and functions of proteins. Consequently, denaturation of proteins has meant different things to different people. Because proteins in living organisms are believed to be at a unique state and conformation to carry out their biological functions, native proteins can exist only in living organisms. With respect to the overall structure of a native protein, any conformational changes could mean a certain degree of denaturation. Therefore, the term *unfolding* is often used in place of *denaturation*. As biologically active macromolecules, proteins respond to environmental changes through their conformational adjustment. However, during the processes of extraction and purification, the changes from physiological environment to buffer conditions have caused some irreversible conformational changes that can hardly be measured. Therefore, all in vitro proteins are theoretically denatured by this definition. In addition, since a minute fraction of the protein conformation may undergo change without affecting the functional property of the protein, it is difficult to decide to what extent the conformational change of the protein molecule can be included in the concept of denaturation. A more "restrictive" definition of denaturation, therefore, specifies the loss of the most characteristic properties of the protein (e.g., enzyme activity, solubility), which depends on the objectives of a research.

Theoretically, because proteins have to maintain their certain conformation to perform their specific functions, any factor that can change the conformation of proteins is a potential denaturation factor. Therefore, the denaturation of proteins can be brought about in many ways, including thermal denaturation, denaturation by changing pH, by high concentration of urea, by guanidinium chloride and other guanidinium salts, by inorganic salts, by organic solvents and solutes, and by detergents. Denaturation of proteins significantly changes their functional properties, including surface activity. It is well known that, when absorbed at an interface, proteins undergo conformational changes. Therefore, conformational changes caused by the denaturation of proteins usually are a necessary step for proteins to be absorbed and distributed at the interface.

Heat is the most common factor that causes denaturation of proteins. Heat treatment of proteins increases thermal motion, leading to the rupture of various intermolecular and intramolecular bonds stabilizing the native protein structure. Protein solubility decreases when proteins (especially globular proteins) are denatured. Therefore, precipitation of proteins has long been used as one of the guidelines for protein denaturation. Although denaturation of proteins usually results in precipitation, precipitation of proteins is not necessarily caused by denaturation. For instance, most proteins are insoluble at their isoelectric point, at which their net charge is zero, but they may not be denatured, although their conformation may undergo certain changes during the process of precipitation.

As a physicochemical process, protein denaturation can be reversible or irreversible, depending on the process of denaturation and the conditions after denaturation. Denatured proteins may refold back to their original structure and resume their biological functions. The process of protein folding, unfolding, and refolding is still an attractive research area, although it has been extensively studied for decades.

II. NATURALLY OCCURRING SURFACE-ACTIVE PROTEINS

Proteins are highly complex polymers with sophisticated three-dimensional structures. During the formation of a protein-stabilized emulsion, the protein molecules must first reach the water/lipid interface and then unfold so that their hydrophobic groups can contact the lipid phase. The primary reaction during homogenization is a rapid binding of protein molecules to the newly created fat surface. The initial binding of proteins is a diffusion-controlled process. Once adsorbed, the protein molecules adopt the most favorable conformation through rearranging their structure. This phenomenon occurs between the hydrophobic sidechains of the protein molecules and the surface of the nonaqueous material. Because all proteins contain both hydrophobic and hydrophilic amino acid residues, all proteins should theoretically be surface active. Therefore, proteins are natural polymeric surfactants. However, because the diffusion/adsorption of proteins on the interface is a complicated physicochemical process affected by several factors, most native proteins demonstrate very week surface activity due to their slow diffusion onto the interface and their inability to adsorb there. Therefore, among thousands of proteins that can be recovered from raw materials and agricultural by-products, only a few are naturally surface active. Most naturally occurring surface-active proteins contain nonprotein components covalently linked with protein molecules. Although proteins can be synthesized only from the combination of the 20 different amino acids, some of the amino acid residues are designed to be modified by the cytoplastic enzymes. The process of posttransitional modification may produce protein complexes such as glycoproteins and lipoproteins, which may possess excellent surface activity.

A. Native Proteins with Flexible Structure

Although a rapid adsorption of proteins is necessary to facilitate the reduction in surface tension, it is not a rate-limiting step under dynamic flow conditions. The rate of conformational rearrangement/reorientation of proteins at the interface is a rate-limiting step in reducing the interfacial tension. Results from the adsorption behavior of denatured and reduced proteins indicate that the effectiveness of a protein film in reducing the interfacial energy is dependent on its conformational flexibility at the interface [12]. Therefore, the flexibility of proteins has profound effects on their surface activity. If a protein molecule possesses sufficient flexibility, then there is a greater chance that a maximum number of its hydrophobic residues will project from the surface as loops or tails. Therefore, flexible proteins will favor maximum hydrophobic interactions between sidechains and the interface and have better surface activity.

Among thousands of native proteins, caseins are well-known examples that possess excellent surface activity due to their molecular properties. The casein molecules comprise a group of acidic, proline-rich phosphoproteins with an intrinsic tendency to bind calcium ions. In solution, caseins have relatively little regular secondary structure and are characterized by a large fraction of random coil conformation [13]. For instance, the primary structure of β-casein (consisting of about 35% of the total caseins in milk) exhibits the typical characteristics of amphiphile. Among the 209 amino acids, the N-terminal portion of β-casein is rich in polar and charged residues, and the C-terminal consists of dominant nonpolar hydrophobic amino acids (Fig. 1). At pH 7.0 its net charge is about -12, and this net charge, originating from the ionization of 6 glutamic acid and 4 phosphoserine residues, is completely confined within the first 21 N-terminal residues; the rest of the molecule has zero net charge [14]. On the other hand, the average hydrophobicity increases toward the C-terminal end, from about 3.5 kJ to the extremely high figure of 8.3 kJ per residue. This distribution suggests that the unfolded peptide chain should exhibit soaplike characteristics. Therefore, β-casein is able to unfold and rearrange/reorient as soon as it arrives at the interface. Neutron reflectance studies of adsorbed β-casein revealed that the adsorbed protein anchors to the interface with a relatively long tail of protein protruding into the solution [15].

B. Glycosylated Proteins

Glycoproteins contain carbohydrate units covalently attached to their amino acid residues, mostly lysine or asparagine through N-glycosidic bonds (Fig. 2a). Less frequently is the attachment of sugars to serine and threonine side-

Signal					
MKVLILACLV	ALALARELEE	LNVPGEIVES	LSSSEESITR	INKKIEKFQS	EEQQQTEDEL
QDKIHPFAQT	QSLVYPFPGP	IPNSLPQNIP	PLTQTPVVVP	PFLQPEVMGV	SKVKEAMAPK
HKEMPFPKYP	VEPFTESQSL	TLTDVENLHL	PLPLLQSWMH	QPHQPLPPTV	MFPPQSVLSL
SQSKVLPVPQ	KAVPYPQRDM	PIQAFLLYQE	PVLGPVRGPF	PIIV	

FIG. 1 Amino acid sequence of β-casein. 244 amino acids, including signal peptide (15 residues). The bold letters are polar residues at the N-terminal. (Protein Data Base: Swissport.)

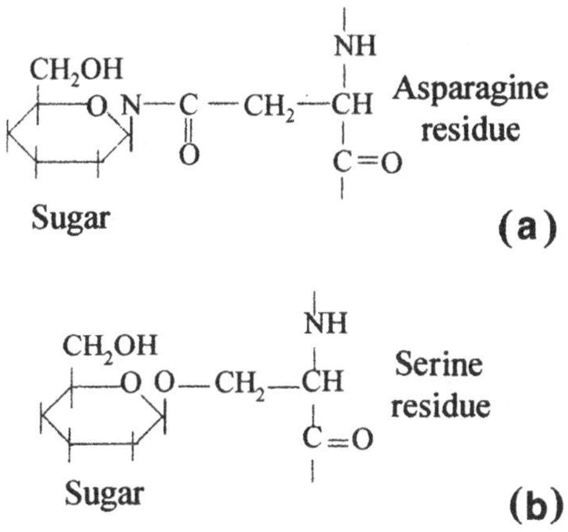

FIG. 2 Glycosidic linkage between sugar and asparagine residue in glycoproteins. (a) N-glycosidic linkage. (b) O-glycosidic linkage.

chains through O-glycosidic bonds (Fig. 2b). Glycoproteins are particularly widespread in nature and make up a large part of the membrane coating around living cells. Collagen, a family of fibrous proteins that exists in all multicellular organisms and that is also the most abundant protein in mammals, contains carbohydrate units covalently attached to its hydroxylysine residues.

Glycophorin A, for example, is a well-characterized transmembrane protein in red-cell membrane, which consists of 16 oligosaccharide units (60% of the mass of this glycoprotein) attached to a single polypeptide chain. Human glycophorin A is an abundant erythrocyte membrane protein that is the first membrane protein to be sequenced. The molecular structure of glycophorin A consists of three domains: (1) an amino-terminal region containing all of the carbohydrate units, which is located on the outer surface of the membrane; (2) a hydrophobic middle region, which is buried within the hydrocarbon core of the membrane; and (3) a carboxyl-terminal region rich in polar and ionized sidechains, which are exposed to the inner surface of the red-cell membrane [16]. Because sugars are highly hydrophilic, glycoproteins usually possess excellent surface activity. The carbohydrate parts of glycoproteins are always located on the external surface of plasma membrane. On cell surfaces, the carbohydrates are also important in intercellular recognition.

C. Lipoproteins

Lipoproteins, in general, refer to those that contain a fatty acid moiety. Because lipoproteins contain highly hydrophobic lipid units, they are thermodynamically unfavorable in a single phase, and thus are usually distributed at interfaces. Most lipoproteins, such as the well-known plasma lipoproteins, however, are complexes of a hydrophobic lipid core surrounded by polar lipids and then by a shell of apoproteins. The apoproteins, which are lipid-free proteins, serve primarily as detergents to solubilize lipids. These complexes transport cholesterol, cholesteryl esters, triglycerides, and phospholipids for bodily functions [17].

A typical example of covalently linked lipoprotein is a small protein in the outer membrane of gram-negative bacterial envelopes that contributes to the mechanical stability of the cell envelope. This lipoprotein contains only 58 amino acid residues containing three covalently attached fatty acids, all joined to the amino-terminal cysteine [18]. In the cell envelope, the amino group of the carboxyl-terminal lysine is linked to a carboxyl group of peptidoglycan. Thus, the lipoprotein ties the outer membrane to the peptidoglycan layer and thereby contributes to the mechanical stability of the cell membrane.

In summary, although there are several native proteins that have excellent surface activity, most proteins from agricultural by-products are not good surfactants. Proteins are designed and synthesized for their specific biological functions, other than forming emulsions. Therefore, many proteins with good foaming and emulsifying capacity often do not possess the ability of stabilizing foams and emulsions. On the other hand, proteins with poor foaming and emulsifying capacity often display the ability to stabilize the dispersed systems. Therefore, most proteins from agricultural by-products need to be modified in order to improve their functionality. The term *modification* here refers to *in vitro* treatments of proteins through physical, chemical, physicochemical, and biochemical approaches.

III. MODIFICATION OF PROTEINS BY ENZYMATIC HYDROLYSIS

Proteins can be modified by a group of peptide hydrolyses (peptidases) commonly called *proteases* (or *proteinases*). Based on their ability to hydrolyze specific proteins, proteases are classified as collagenase, keratinase, elastase, etc. On the basis of the pH range over which they are active, they are classified as either acidic, neutral, or alkaline. However, according to their mechanism of action, the Enzyme Commission classifies proteases into the four distinct classes of serine, cysteine, aspartyl, and metalloproteases. Serine proteases, for example, always contain serine residue at their catalytic center, which is essential for the action of proteolysis.

Proteolytic reactions have been studied scientifically for more than 200 years, beginning with observations on peptic digestion [19]. Therefore, proteolysis has been one of the most extensively investigated approaches to obtain the improved functionality of proteins in recent decades. Proteolytic reaction produces protein hydrolysate with smaller-molecular-weight polypeptides that are generally very flexible in their structure. In addition, proteolysis with some specific proteases may produce charged hydrolysates. Many investigators have reported that well-controlled proteolysis to a certain degree can improve the surface properties of some proteins. However, extensive hydrolysis of proteins may produce protein hydrolysate with poor surface activity. A well-controlled process of proteolysis with a specific enzyme is critical to a desired product.

A. Chemistry of Peptide Bond Cleavage

By definition, all proteinases catalyze the hydrolytic degradation of peptide bonds of proteins (Fig. 3). In solution, the equilibrium lies so far to the right that the degradation and not the synthesis to large molecules is thermodynamically favored [20]. The shift toward hydrolysis can be attributed to the free-energy contribution from the ionization of the carboxyl and amino groups.

The hydrolysis of peptide produces free carboxyl and free amino groups, which can be ionized, depending on the pH of the hydrolysis system. At 25°C the pK values of —COOH and ^+H_3N- in peptides are in the range of 3.1–3.6 and 7.5–7.8, respectively [21]. Therefore, the carboxyl group will be fully dissociated above pH 5.0, undissociated below pH 2.0, and partially dissociated at pH 2–5. On the other hand, the amino group will be fully protonated below pH 6, unprotonated above pH 9.5, and partially protonated at pH 6–9.5. This means that outside the region of about pH 5–6, in which the uptake and release of protons cancel each other, the release or uptake of H^+ during proteolysis will change pH. The change of pH during proteolysis, therefore, can be used for monitoring the process of proteolysis under certain conditions.

If pH is kept constant during the process of proteolysis, the proteolytic reaction will proceed under the consumption of a considerable amount of acid or base groups. The relation between equivalent peptide bonds cleaved and

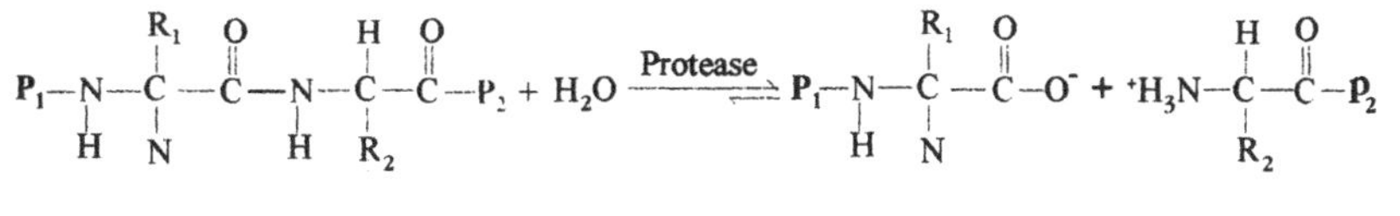

FIG. 3 Proteolytic reaction catalyzed by protease. P_1 and P_2 refer to polypeptides 1 and 2, respectively. R_1 and R_2 refer to the amino acid residues of the protein.

equivalent base consumed is proportional. The proportionality constant is simply the degree of dissociation, α, which is given as follows:

$$\alpha = \frac{10^{(pH-pK)}}{1 + 10^{(pH-pK)}} \tag{2}$$

where pK is the average of the α-NH_2 liberated during the hydrolysis, which can be determined from a direct assay of the liberated amino groups. The pK values change with temperature and peptide chain length.

B. Degree of Hydrolysis (DH)

Experimental evidence has shown that if the extent of proteolysis is limited, it is possible to achieve an improvement in functionality. Therefore, a very important factor that has to be well controlled during the production of protein hydrolysate is the extent of the proteolytic degradation (the number of peptide bonds cleaved during the proteolysis). Although several methods have been developed for determining the extent of proteolysis, practical control of the reaction is still a problem. The commonly used trichloroacetic acid (TCA) solubility index, for instance, determines the percentage of nitrogen that is soluble in TCA under certain conditions. This method has been used for characterizing the molecular weight distribution of protein hydrolysates, for it precipitates all large protein molecules and some fractions of the peptides [22]. However, the popularity of using the TCA solubility index for following the proteolytic reaction is undeserved. The solubility of protein hydrolysates does not depend on their molecular weight only. The amino acid sequence of polypeptides also affects the solubility of protein hydrolysates in TCA solution. Therefore, using the TCA index for characterizing the molecular weight distribution of protein hydrolysates and comparing specific proteolytic activity of different proteases is unreliable.

The degree of hydrolysis (DH) is defined as the percentage of peptide bonds cleaved [23], which is the most commonly used approach for determining the extent of proteolysis under certain conditions. The number of peptide bonds cleaved during proteolysis is called the *hydrolysis equivalents* (*h*) and is expressed as equivalents per kilogram of protein (meqv/kg protein). By determining the increase in free amino (or carboxyl) groups, the hydrolysis equivalents can be assayed. The most widely accepted method for determining the free amino acid groups is the reaction with trinitrobenzene-sulphonic acid (TNBS) developed by Adler-Nissen [24]. Therefore, according to the definition, the degree of hydrolysis is:

$$DH = \frac{h}{h_{tot}} \times 100\% \qquad (3)$$

where h_{tot} is the total number of peptide bonds in a given protein substrate, which can be determined from the amino acid composition of the protein by summing up of the moles of the individual amino acids per gram of protein. Therefore, the determination of h_{tot} relies on the reliability of amino acid analysis. Because the average molecular weight of amino acids in many proteins is about 125, the h_{tot} per kilogram of protein ($6.25 \times N$) can often be approximately 8.0.

Theoretically, the DH value can be used for calculating the average peptide chain length (PCL) of protein hydrolysate. The PCL is the average number of amino acid residues in the polypeptides, which is also commonly used for describing the size of protein hydrolysate. If a protein has a polypeptide chain length of PCL_0, then the total number of peptide bonds in the protein is $PCL_0 - 1$. When this protein is hydrolyzed into "n" peptides by the cleavage of $(n - 1)$ peptide bonds, the definition of DH as the percentage of peptide bonds cleaved gives:

$$DH - \frac{n - 1}{PCL_0 - 1} \times 100\% \qquad (4)$$

The value of "n," thus, can be obtained from the DH determined. The average polypeptide chain length becomes:

$$PCL = \frac{PCL_0}{n} \qquad (5)$$

Because the size of polypeptides produced from proteolysis relate directly to their functional properties, the application of DH is the most common approach to monitor the process of proteolysis. However, it should be pointed out that the calculation of PCL from the DH is a theoretical concept only. Although it has been widely used and documented in almost all of the literature, it cannot be used for comparing protein hydrolysates from different proteolytic reaction systems. When a protein substrate is hydrolyzed by different proteases, the differences in the specificity of enzymes may produce protein hydrolysate with significantly different-size distributions, although they have the same DH values.

C. Kinetics of Proteolysis

Proteolysis is the most important biochemical reaction *in vivo* and has been extensively studied during recent decades. Historically, the reaction of proteolysis has been used as a model for studying the mechanism of enzyme reaction. However, the kinetics of proteolytic reactions has not been fully understood. In the most general case, the theoretical foundation for describing the hydrolysis reaction still does not match practical situations. The hydrolysis of many proteins, such as soy protein, occurs as a simultaneous hydrolysis of both soluble and insoluble substrates in the native as well as the denatured form [23]. The theoretical description of the kinetics of protein hydrolysis is still fairly rudimentary. While almost all proteolytic reactions have been treated using steady-state kinetics, the conditions of proteolysis commonly applied in studies do not satisfy the basic assumptions of steady-state kinetics [25].

1. Basic Assumptions of Steady-State Kinetics

According to the model proposed by Michaelis and Menten to accommodate the kinetics they observed:

$$E + S \underset{k_{-1}}{\overset{k_{+1}}{\rightleftharpoons}} ES \overset{k_2}{\rightarrow} E + P \qquad (6)$$

where $k_2 = k_{cat}$ is a turnover number and P refers to product, none of the product reverts back to the substrate, and $K_M = k_{-1}/k_{+1}$ is an equilibrium dissociation constant for the first step [26]. The classical Briggs–Haldane treatment of enzyme reaction systems removed the restrictive assumption that the enzyme-substrate complex is in equilibrium with free enzyme and substrate, and supposed that the enzyme system is approximately in a steady state for most of the time that it is working, though the substrate is in fact being progressively used up [27]. Therefore, steady-state kinetics assumes the following.

1. There is not a significant fraction of the substrate bound to the enzyme during the assay (it is not that the enzyme must be saturated with substrate):

    $$[S_0] \gg [E_0] \qquad (7)$$

    where $[E_0]$ is the total enzyme added and $[S_0]$ is the total substrate before the reaction.
2. The total activity of the enzyme does not change during a valid assay, which is also termed the *enzyme conservation* expression:

    $$[E_0] = [E] + [ES] \qquad (8)$$

3. The enzyme system is approximately in a *steady state*. Because $[S_0] \gg [E_0]$, the change of [ES] during the enzyme reaction should be small compared to the rate of substrate utilization. That is,

$$\frac{d[E]}{dt} \approx \frac{d[ES]}{dt} \approx 0 \qquad (9)$$

4. The product [P] is insignificant at the beginning of enzymatic reaction:

$$[P] = 0 \qquad (10)$$

Under this assumption, the [ES] depends solely on the [E_0] and [S].

2. Proteolytic Reaction System Contains Mixed Chain Reactions

Proteolytic reaction on a protein substrate is a mixed enzymatic reaction with several pathways of chain reactions ended differently, depending on the nature of an enzyme and the structure of the substrates (Fig. 4). At the moment that enzyme and substrates are mixed together, the exact concentration of either

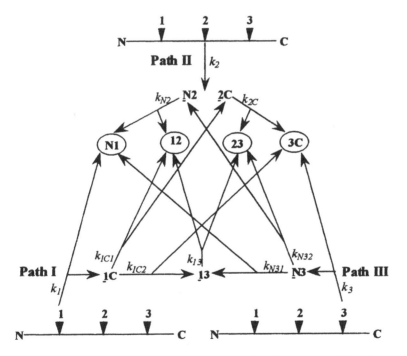

FIG. 4 Possible proteolytic reactions on a protein substrate containing three potentially sensitive bonds. The underlined polypeptide pieces are processing intermediates. The circled polypeptide is final products. The k_i (i = 1, 2, 3, N2, 2C, . . .) are k_{cat} values for all potential substrates in the reaction system. (From Ref. 25.)

substrate or product becomes variable. The breaking of peptide bonds results either in a continuous degradation or in the formation of large fragments. When such a multiple fragmentation occurs, the kinetic data show the build-up of the fragment(s) of the first generation, followed by their decay, with the simultaneous appearance of smaller fragments. Because of the characteristics of chain reaction, the kinetics of proteolysis cannot fit the steady-state kinetic model. Proteolytic reaction is not a single enzyme–substrate (S + E) reaction, but reactions with mixed pathways (E + Σ S$_i$) of chain reactions (E + Σ P$_i$). The same enzyme, in terms of its quantity, undergoes reactions with varying substrates (Σ S$_i$) at different reaction rates, depending on their molecular size and structure.

3. Steady State Does Not Exist in Proteolytic Reaction Systems

As a consequence of the propagated chain reaction during proteolysis, the enzyme molecules are distributed among all available substrates that are a collection of unhydrolyzed substrate molecules and intermediate polypeptides produced from proteolytic degradation:

$$[E_0] = [E] + [ES] + \Sigma\ [EP_i] \tag{11}$$

where the $[EP_i]$ are complexes of the enzyme and polypeptides produced from the process of proteolysis, and are a function of time (t), dissociation constant (K_d), and k_{cat}. That is,

$$\Sigma[EP_i] = f(t, K_d, k_{cat}) \tag{12}$$

As long as those potential substrates produced during the reaction are present, they will immediately build equilibrium with free enzymes. Because all newly released potential substrates compete with free intact substrate $[S_0]$ for binding with free enzyme $[E]$ to form enzyme–substrate complexes $[EP_i]$, the $[ES]$ may decrease dramatically with the ongoing proteolytic reaction. The competitive binding limits the rebinding of released free enzymes to their original substrate. In addition, the relative quality and quantity of $[ES]$ and $[EP_i]$ are dependent on the time (t), the dissociation constant $(K_d)_i$, and the $(k_{cat})_i$ of each polypeptide with respect to the enzyme. Therefore, both the quality and the quantity of substrates change dramatically in the proteolytic reaction system. The concentrations of $[S_0]$, $[P_i]$, $[E]$, $[EP_i]$, and $[ES]$ in the proteolytic reaction system are all time dependent during the entire process of reaction:

$$\frac{d[E]}{dt} = \frac{d\{[E_t] - [ES] - \Sigma\ [EP_i]\}}{dt} \neq 0 \tag{13}$$

and

$$\frac{d[ES]}{dt} = \frac{d\{[E_t] - [E] - \Sigma [EP_i]\}}{dt} \neq 0 \tag{14}$$

Therefore, a proteolytic reaction that is a mixed chain reaction system cannot enter a steady-state phase. Therefore, steady state does not exist in proteolytic reaction systems.

4. Initial Rate is a Sum of Proteolytic Reactions on Different Peptide Bonds

Even at the very beginning of proteolytic reactions, there is more than one reaction occurring in the system. Assuming, for instance, that a protein substrate contains only three cutting sites for a specific protease, there will be three possible pathways to begin the chain reactions, which contain 10 possible reactions (Fig. 4). The kinetic data, therefore, should show the build-up of the fragment(s) of the first generation, followed by their decay, with the simultaneous appearance of smaller fragments. The rate of each reaction, according to the figure, will be:

$$-\frac{d(S)}{dt} = (k_1 + k_2 + k_3)(S) \tag{15}$$

$$\frac{d(N1)}{dt} = k_1(S) + k_{N2}(N2) + k_{N31}(N3) \tag{16}$$

$$\frac{d(N2)}{dt} = k_2(S) + k_{N32}(N3) - k_{N2}(N2) \tag{17}$$

The rate of proteolysis in the reaction system thus is:

$$v_t = -\frac{d(S)}{dt} = \Sigma \left| \frac{d(P_i)}{dt} \right| \tag{18}$$

5. Total Proteolytic Activity May Change During Proteolysis

For an intact protein substrate, proteolysis extends its compact structure, so more potential sensitive bonds become available. The breakdown of intact proteins results in a more flexible structure of polypeptide intermediates. Because the secondary enzyme–substrate interactions involving substrate residues not involved in the sensitive bond significantly affect the rate of proteolysis [28], the rate of proteolysis on the hydrolyzed intermediates may be significantly different from each other and from the intact protein substrate. In addition, the newly released polypeptides cannot have the same reactivity as either the intact protein substrate or their counterparts. Therefore, all k_i values in Fig. 4 are theoretically different, due to the different nature of their sensitive peptide bond and the secondary binding during proteolysis ($k_1 \neq k_2 \neq k_3 \neq k_{1C1} \neq k_{1C2} \neq k_{N2} \neq k_{13} \neq k_{N31} \neq k_{N32} \neq k_{2C}$). The total proteolytic activity is dependent on the distribution of enzyme substrates ($\Sigma[ES_i] + \Sigma[EP_i]$). Because the distribution of enzyme molecules in potential substrates is a function of reaction time, dissociation constant, and turnover number, the total proteolytic activity changes during the entire process of reaction:

$$[E_0] \neq [E_t]^t = [E] + f(t, k_d, k_{cat}) \tag{19}$$

where $[E_t]^t$ is the total proteolytic activity at time t. Therefore, during proteolytic reaction, the sum of actual proteolytic activity may change. That is: the assumption of *enzyme conservation* cannot be satisfied during the process of proteolysis.

D. Preparation of Protein Hydrolysates

According to the definition, proteolytic reaction requires the presence of aqueous phase. Because substrate proteins always contain peptide bonds sensitive to a protease, theoretically the proteolytic reaction will occur in any system containing proteins and protease(s) in solution. However, because proteins have a three-dimensional structure, those sensitive peptide bonds buried inside a compact tertiary structure cannot be hydrolyzed. Even some peptide bonds on the surface of a tertiary-structured protein may become buried if the protein associates with other proteins to form quaternary structures (either mono- or heteroassociates). Therefore, some proteins are highly resistant to certain proteases, although they contain peptide bonds sensitive to the enzyme. The resistance of native proteins to certain kinds of proteolysis is essential for them to carry out biological functions.

Proteolytic reaction can be carried out isothermally in a well-stirred reactor that is scale independent (Fig. 5). The reaction proceeds spontaneously, with negligible changes in the density or enthalpy. The concentration of substrate can be as high as it can be prepared. The ratio of enzyme to substrate is usually in the range 0.1–1%, depending on the specific activity of the enzyme. Although most proteolytic reactions are carried out in systems containing soluble substrate proteins and enzyme, this does not mean that insoluble protein substrates cannot be hydrolyzed. Azocoll, for instance, is an insoluble protein substrate that is commonly used for determining the activity of proteases [29].

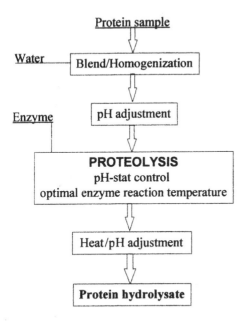

FIG. 5 Preparation of protein hydrolysate. Protein powder can be blended directly with water, while textured protein samples need to be homogenized before the adjustment of pH.

Proteolytic reaction usually can solubilize insoluble substrate proteins. On the other hand, soluble substrate proteins can also be hydrolyzed by insoluble proteases. By passing soluble protein substrates through an insoluble matrix containing immobilized proteases, the substrate can be hydrolyzed to a certain DH [30]. In general, the viscosity of the reaction system decreases significantly with the process of proteolysis.

The reaction temperature is usually set at the range of optimal reaction temperature for the enzyme. Because the optimal reaction temperature for proteases can be in the range of 25°C up to 65°C, depending on the source of the enzyme, the property of protein substrate should also be considered. Most proteolytic reactions are terminated by a rapid heat treatment to inactivate enzymes. During the heat treatment process, proteolytic reactions proceed at a very high rate due to the increasing reaction temperature. Therefore, proteolysis during the heat treatment process needs to be considered, especially for those reaction systems involving heat-stable proteases.

The reaction pH is usually adjusted according to the optimal reaction pH of the enzyme. However, other factors, such as the method of monitoring the reaction, the solubility of the substrate, and the requirement of final product, need to be considered. The adjustment of pH is also a common approach to terminate proteolytic reactions. Most acidic proteases, for instance, are inactive at alkaline pH. On the other hand, many alkaline proteases lose their proteolytic activity at acidic pH. Termination of proteolytic reactions by pH adjustment is mild and has little effect on final DH, although it brings additional salts in final products.

For the proteolysis of using alkaline protease, the pH-stat is a common approach to monitor the reaction process and control the degree of hydrolysis. The principle behind the pH-stat technique, that is, continuous monitoring of the DH during proteolysis, is as follows: At pH values outside the range given by the pK values of the amino and carboxyl groups, the proteolysis at constant pH will liberate carboxyl or amino groups. If the reaction is unbuffered, the drop or increase in pH can be several tenths of a pH unit in a few minutes. The pH-stat overcomes this problem by intermittently adding small amounts of base (or acid) to the reaction mixture to keep the pH constant. The signal is the milliliters of titrant added, and this is usually continuously recorded to give a curve of base (viz. acid) consumption versus time. The automatic pH-stat was originally developed at the Carlsberg Laboratory [31].

In a pH-stat-controlled reaction system, the uptake of CO_2 from the atmosphere is one of the error sources need to be considered. The degree of proteolysis can be calculated from the assumption of base (B) during the reaction. The number of peptide bonds cleaved is equal to the mols of base consumed. That is:

$$h = \frac{B \times N_b}{MP \times \alpha} \tag{20}$$

where N_b is the normality of protein and α is the average degree of dissociation of the α-NH_2 groups at a certain pH. Under the designed batch reaction, α, MP, h_{total}, and N_b are all known. The degree of hydrolysis (DH), thus, can be calculated from the assumption of the base (B). At very low value of α, the determination of DH becomes much more uncertain. In addition, the ionization of amino acid sidechains and the buffering capacity of polypeptides play a role in the functioning of the pH-stat. Therefore, at extreme pH values (e.g., pH > 11 or pH < 3), the pH-stat is inoperable due to the relatively high buffering capacity of proteins.

In the early 1970s, Puski carried out a systematic investigation on functional soy protein hydrolysates. Puski hydrolyzed soy protein isolate with *Aspergillus oryzae* protease at pH 7 for 3 hours with different levels of enzyme/

substrate (E/S) at 50°C [32]. The enzyme was then inactivated by heat, followed by lyophilization of the hydrolysate mixture. It was found that the solubility of protein hydrolysates increased with an increasing extent of proteolysis, whereas the effect on the other functionalities varied. Emulsifying capacity increased, but the emulsion stability decreased. The work of Puski is an early example of applying a limited hydrolysis for improving the functionalities of proteins.

Adler-Nissen and Olsen hydrolyzed soy protein isolate with Alcalase at pH 8 and with Neutrase at pH 7. They found that by limited hydrolysis the emulsifying capacity and foaming capacity were all increased. However, if the reaction was continued to DH values higher than 5%, the emulsifying and foaming capacities decreased again [24].

E. Surface Activity of Protein Hydrolysates

Proteolysis generates hydrolysates with flexible structure that can easily be distributed and absorbed at the interface. Therefore, proteolysis under a certain control should theoretically facilitate surface activity of proteins by increasing their emulsifying capacity. Proteolysis produces flexible hydrolysates that have more chance to be adsorbed at the interface. Assuming a proteolytic reaction produces "n" number of polypeptides, the interface may be occupied by all the polypeptides instead of only one native protein molecule (Fig. 6). Because hydrolyzed polypeptides are small in size and unfolded in their structure, they have many ways of inserting a hydrophobic group into a nonpolar environment. The interaction/arrangement of hydrolysate at the interface cannot be as tight as for an intact protein. The surface area (A) that can be covered by protein hydrolysates (h) is:

$$A = A_{h1} + A_{h2} + A_{h3} + \cdots + A_{hn} = \Sigma A_{hi} \tag{21}$$

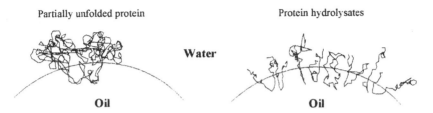

Partially unfolded protein   Protein hydrolysates

Water

Oil   Oil

FIG. 6 Distribution of partially unfolded protein molecule and protein hydrolysates at oil–water interface.

where *hi* is the individual hydrolysate segment produced from the proteolysis. Compared with the area that can be covered by an intact protein molecule (A_p),

$$\Sigma \, A_{hi} > A_p \tag{22}$$

Therefore, proteolysis usually increases emulsifying capacity.

The surface activity (ψ) of each hydrolysate segment (h_i) released from the proteolysis may differ significantly, depending on their size and amino acid sequence/structure. That is:

$$\psi^{h1} \neq \psi^{h2} \neq \psi^{h3} \neq \cdots \neq \psi^{hn} \tag{23}$$

The surface activity of protein hydrolysates, therefore, is a sum of all surface activities from every individual segment in the system ($\Sigma\psi^{hi}$). Although the quality and quantity of amino acids in the proteolytic reaction system do not change, their surface activity changes significantly. That is:

$$\Sigma \, \psi^{hi} \neq \psi^p \tag{24}$$

where ψ^p is the surface activity of the substrate proteins.

On the other hand, the stability of colloids may decrease due to the reduced molecular size of proteins/polypeptides. There are two main mechanisms that are widely used in considering the stability of colloids: the interactions between charged molecules, and steric stabilization. Extensive proteolysis usually gives peptides that are poor emulsifiers, presumably because there is a limiting length below which the emulsifying property of the peptide is poor [33]. The small size of short peptides cannot sufficiently project into the solution from the droplet surface, thus precluding steric stabilization of the emulsion system [34]. An increase in both emulsifying capacity and stability can be obtained only through a well-controlled proteolysis, which should include the selection of substrates and specific proteases, the degree of proteolysis, and the condition of proteolytic reaction.

IV. ENZYMATIC CROSSLINKING OF PROTEINS

Another possibility for improving the surface activity of proteins is to form covalently bonded protein conjugates through enzymatic reactions. Polymerization of proteins may influence their surface properties. It is well recognized that proteins with good foaming and emulsifying capacity often do not possess the ability to stabilize foams and emulsions. On the other hand, proteins with poor foaming and emulsifying capacity often display the ability to stabilize the dispersed systems. Therefore, if the preceding two proteins can be covalently linked together to form heterogeneous conjugates, they should logically combine both of the properties just mentioned. A well-known enzyme that can catalyze the formation of protein conjugates is transglutaminase.

Transglutaminase can catalyze several reactions both in vivo and in vitro [35], and therefore it has been extensively studied by biochemists. One of the catalytic properties of transglutaminase is the acyl transfer reaction between protein-bound glutaminyl residues and a variety of primary amines. It can also catalyze several reactions in addition to the crosslinkage of proteins, which include: aminolysis of the carboxamide group of peptide-bound glutamine residues; hydrolysis at the carboxamide group of peptide-bound glutamine when the level of amine substrate is low or when amine substrate is absent; hydrolysis and aminolysis of certain aliphatic amides of active esters [35]. Extensive screening for transglutaminase-producing microorganisms during the 1980s led to commercial production of the enzyme from microorganisms [36].

A. Effects of Thermodynamic Compatibility of Protein Substrates on Crosslinkage

A fundamental requirement for transglutaminase-catalyzed protein crosslinkage is that glutamine and lysine residues have to be available in substrate proteins. The reactivity of substrate proteins, thus, is dependent on their amino acid sequence. Theoretically, any proteins with glutamine/lysine residues can be crosslinked to form homogeneous and/or heterogeneous conjugates. In practice, however, it is not so straightforward. Not all proteins with glutaminyl and lysyl residues can be crosslinked to form protein conjugates.

In the early 1960s, it was believed that serum albumin as well as ovalbumin did not serve as substrates for transglutaminase [37]. In the 1980s, protein substrates for transglutaminase were classified into the following four groups: (1) Gln type, in which only Gln residues are available; (2) Lys type, in which only Lys residues are available; 3) Gln-Lys type, in which both Gln and Lys residues are available; and (4) nonreactive type, in which both Gln and Lys residues are unavailable for crosslinking [38]. This classification was based on the accessibility of lysine and/or glutamine residues on the protein surface by the enzyme. According to this classification, it can be logically surmised that a mixture of two Gln-Lys type of proteins should be able to form heterogeneous conjugates in a transglutaminase-catalyzed reaction. However, this does not occur in experiments with a majority of binary protein systems. For instance, although both β-lactoglobulin and β-casein are the Gln-Lys type of substrate for transglutaminase, they cannot form heterogeneous conjugates in transglutaminase-catalyzed reactions [39]. Ikura et al. reported that native bovine serum albumin (BSA) and ovalbumin did not form transglutaminase-catalyzed protein–protein conjugates with maleylated a_{s1}-casein [38]. Similar behavior can also be observed in many other binary system containing thermodynamically incompatible proteins. Although the reaction of transglutaminase-catalyzed crosslinking requires the presence of glutamine and lysine residues, the presence of these residues does not necessarily guarantee the occurrence of crosslinkage. The protein substrates must be oriented and in close proximity to form a covalent bond at the active site of transglutaminase; that is, they must be at least microenvironmentally and thermodynamically compatible [39].

The thermodynamic compatibility of proteins is related to the nature and intensity of the interaction between the two protein macromolecules as they approach each other when dispersed in a continue phase. In most cases, when two proteins (protein i and protein j) are mixed together, the interaction energy, ΔG_{ij}, is usually positive. Therefore, proteins of different classes exhibit limited thermodynamic compatibility in aqueous solution. Thermodynamically incompatible native proteins, such as β-lactoglobulin and α-casein, cannot form heterogeneous conjugates in transglutaminase-catalyzed reactions. The thermodynamic incompatibility of proteins arises as a result of the repulsive interaction between polar and apolar surfaces. Therefore, two dissimilar proteins will approach each other only up to a distance at which the free energy of the interaction is zero. The inability to come close together at the enzyme's active center thus precludes the formation of heterogeneous conjugates in transglutaminase-catalyzed reactions.

B. Preparation of Transglutaminase Crosslinked Protein Conjugates

Simply incubating a protein solution with the enzyme at a certain temperature can carry out the crosslinking of proteins by transglutaminase. The outcome, however, depends on several factors, including the ratio of enzyme to substrate (E/S), the reaction conditions, and the reactivity of protein substrates. As it has already been mentioned, some proteins can hardly be crosslinked by transglutaminase, although they contain both glutamine and lysine residues. Therefore, the reactivity of the protein substrate is the most important factor that determines what reaction condition should be applied and the outcomes of the reaction. This is especially significant in reaction systems containing different substrate proteins.

Transglutaminase crosslinking requires a relatively high ratio of enzyme to substrate (E/S). When the formation of covalent crosslinkage between two macromolecules occurs, the viscosity of the reaction system increases. The increasing viscosity in the reaction system decreases the rate of diffusion of all molecules in the system, resulting in a decreased rate of reaction. Consequently, the concentration of enzyme has to be relatively high in order to compensate for the limited dynamic movement of reactants and the catalyst. Generally, E/S ratio must be in the range of 2–10 units per gram of substrate proteins.

The most common method used for determining transglutaminase activity is the calorimetric hydroxamate assay method described by Folk and Cole [40]. The reaction mixture contains 0.1 M Tris-acetate buffer, pH 6.0, 100 mM CBZ-L-glutaminylglycine, 100 mM hydroxylamine. The enzyme solution is added to the reaction mixture and incubated at 37°C for a certain period (5–30 min). The reaction is stopped by the addition of ferric chloride–trichloroacetic acid reagent. After removing the precipitate formed by centrifugation, the resulting red color is measured at 525 nm. L-Glutamic acid-c-monohydroxamic acid can be used as a standard for the calibration. One unit of TG activity is defined as the amount of enzyme that catalyzes the formation of 1 micromole of hydroxamic acid per minute in the test.

Transglutaminase-catalyzed protein crosslinkage can occur over a wide range of pH and temperature, although both of them significantly affect the reaction rate and products formed. Transglutaminase is active at an acidic pH range and also at alkaline conditions, as long as the status of glutamine and lysine residue dose not change. However, pH affects the reactivity of the substrate proteins and the physicochemical properties of the protein conjugates formed. Therefore, the selection of the reaction pH depends on the reactivity of the substrate proteins.

As with all other enzymatic reactions, increasing reaction temperature accelerates the rate of polymerization. More importantly, because the polymerized products are macromolecules, the temperature applied throughout the entire reaction will change the structure of the protein conjugates. An increased reaction temperature at a certain level can cause a partial unfolding (or conformational changes) of substrate proteins, resulting in a more flexible substrate structure and more exposed glutamine and lysine residues. Because the availability of substrate glutaminyl and lysinyl residues are mostly dependent on their conformation under certain conditions, changes in the reaction temperature may change the reactivity of the substrate proteins.

The most common approach to terminate the reaction of crosslinking is heat treatment. The inactivation of transglutaminase requires an increase in the temperature of the reaction system to 75–80°C for several minutes. However, while heat treatment denatures the enzyme, it also denatures other proteins in the system. Heat treatment significantly affects the functionality of protein conjugates.

Figure 7 shows the SDS-PAGE profile of protein conjugates produced by transglutaminase-catalyzed crosslinking. Lanes 1 and 2 indicate that caseins (both α-casein and β-casein) are good substrates for transglutaminase. Almost

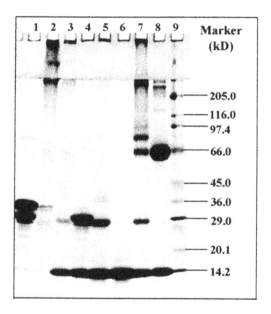

FIG. 7 SDS-PAGE profile of transglutaminase-catalyzed polymerization. Gradient gel (5–20% T). The conditions for crosslinking: 1% (w/v) of protein substrate in 0.1 M Tris-HCl buffer containing 5 mM CaCl₂ at pH 7.5; 10 units of transglutaminase per gram of substrate proteins; incubation at 37°C for 1 hour. The reactions were terminated by adding SDS-PAGE buffer directly to the reaction mixture. Lane 1: mixture of α-casein and β-casein; Lane 2: α-casein and β-casein treated with the enzyme; Lane 3: α-casein and α-lactoalbumin treated with TG; Lane 4: α-casein and α-lactoalbumin control; Lane 5: α-lactoalbumin treated with TG; Lane 6: α-lactoalbumin control; Lane 7: α-lactoalbumin and BSA treated with TG; Lane 8: α-lactoalbumin and BSA control; Lane 9: protein markers.

all protein substrates were crosslinked to form conjugates/polymers (Lane 2). When β-casein and α-lactalbumin were treated with transglutaminase, β-casein was much more reactive compared with the α-lactalbumin substrate (Lanes 3 and 4). While most of the β-casein substrate was consumed, very little α-lactalbumin was crosslinked by the enzyme (Lane 3). However, in the system that contained only α-lactalbumin as a substrate, it formed dimmer and other conjugates (Lanes 5 and 6). Lane 7 shows that very nice heterogeneous conjugates formed between α-lactalbumin and bovine serum albumin (BSA). The protein profiles presented on the SDS-PAGE gel suggest that the characteristics of protein substrates significantly affect the outcome of crosslinking reactions.

C. Surface Activity of Transglutaminase Crosslinked Protein Conjugates

A known fact of transglutaminase crosslinking is that it produces protein conjugates with large molecular size. This suggests that the steric effects of transglutaminase crosslinked protein conjugates on emulsion systems are likely to be predominant. Figure 8 shows an increased stability of β-casein polymer-stabilized emulsions. The stability of emulsion systems increased with the increasing degree of polymerization by transglutaminase. The emulsion system made with 20-min crosslinked β-casein conjugates was less stable. However, those samples stabilized by β-casein conjugates crosslinking in reaction periods longer than 40 min were much more stable than the β-casein control [41].

The stability of an emulsion depends on the physical and chemical properties of the adsorbed surfactant layer and its ability to prevent the flocculation and coalescence of oil droplets. The molecular factors, such as molecular weight, chain length, hydrophilicity/hydrophobicity, solubility, and conformation of proteins, will greatly affect its emulsifying properties. Although β-casein is highly surface active due to its flexible molecular structure, it gives a poorly cohesive film and relatively unstable emulsions. The surface viscosity of casein at the oil–water interface was found to be very low [42]. Because β-casein lacks the ability to form a strong surface film, emulsions stabilized by it do not possess the additional stability that can be conferred by the formation of rigid films. The increased emulsifying stability of β-casein polymers produced by transglutaminase, therefore, may possibly be due to increasing the mechanical/viscoelasitic properties of the interfacial film. The real mechanisms involved in the changes of surface activity of transglutaminase-

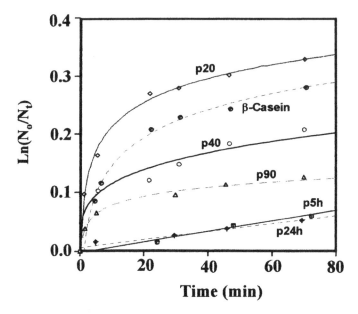

FIG. 8 Variation of ln (N_0/N_t) as a function of crosslinking time for emulsions stabilized by 0.05% β-casein polymers crosslinked by transglutaminase. N_t is the concentration of droplets at time t, N_0 is the initial concentration of droplets at time 0. The ratio of N_0/N_t was calculated from the turbidity measurement. The smaller the ratio change, the more stable the emulsion is. A typical crosslinking reaction system contained 1% β-casein in 0.1 M tris-HCl buffer, pH 7.5, at the ratio of enzyme to substrate of 4.25 units/g protein. The reaction mixtures were incubated at 37°C for the period shown in the figure, and the reaction was terminated at various time intervals by heating up to 85°C for 5 min. (From Ref. 41.)

catalyzed crosslinking of proteins are unclear, because the mechanism of transglutaminase-mediated crosslinking of proteins is still unclear.

A Surface area
α Degree of dissociation
C_i Concentration of an ion
C_s Molar concentration of salt
$[E_0]$ Total enzyme added
ΔG_{ij} Interaction energy between substrates i and j
H Hydration level, defined as grams of H_2O per gram of protein

h	Hydrolysis equivalents (meqv/kg)
h_{tot}	The total number of peptide bonds in a given protein
k_{cat}	Turnover number
K_d	Dissociation constant of enzyme-substrate intermediate [EP_i]
K_M	Michaelis–Menten constant
K_S	Salting-out constant
N_b	Normality of protein
Ψ	Surface activity
[S_0]	Substrate concentration before the reaction
μ	Ionic strength
v_t	Rate of proteolysis at time t
Z_i	Ion valance

REFERENCES

1. L. Pauling, R. B. Corey, and H. R. Branson. Proc. Natl. Acad. Sci. U.S.A. 37: 205 (1951).
2. P. N. Lewis, F. A. Monany, and H. A. Scheraga. Proc. Natl. Acad. Sci. USA. 68:2293 (1971).
3. C. M. Venkatachalam. Biopolymers 6:63 (1968).
4. J. S. Richardson. Protein Chem. 34:167 (1981).
5. J. F. Leszczynski and G. D. Rose. Science 234:849 (1986).
6. J. A. Rupley and G. Careri. Adv. Prot. Chem. 41:37 (1991).
7. H. B. Bull and K. Breese. Arch. Biochem. Biophys. 128:488 (1968).
8. R. B. Gregory. In: Protein-Solvent Interactions (R. B. Gregory ed.), Marcel Dekker, New York, 1993, pp. 191–264.
9. G. F. Cerofolini and M. J. Cerofolini. J. Colloid Interface Sci. 78:65 (1980).
10. P. Zaks and A. Klibanov. Science 224:1249 (1984).
11. R. Gregory and R. Lumry. Biopolymers 24:301 (1985).
12. S. Xu and S. Damodaran. J. Colloid Interface Sci. 159:124 (1993).
13. L. Sawyer and C. Holt. J. Dairy Sci. 76:3062 (1993).
14. T. A. J. Payens and H. J. Vreeman. In: Solution Behavior of Surfactants: Theoretical and Applied Aspects (K. L. Mittal and E. J. Fendler eds), vol. 1, Plenum Press, New York, 1982, pp. 543–571.
15. E. Dickinson, D. S. Horne, J. S. Phipps, and R. M. Richardson. Langmuir 9:242 (1993).
16. R. S. Weinstein, T. K. Khodadad, and T. L. Steck. J. Supramol. Struct. 8: 325 (1978).
17. J. L. Goldstein and M. S. Brown. Annu. Rev. Biochem. 41:897 (1977).
18. A. G. Steve, B. Heggeler, R. Muller, J. Kistler, and J. P. Rosenbusch. J. Cell Biol. 72:292 (1977).

19. R. N. Perham. FEBS Lett. 62 (suppl.): E20 (1976).
20. J. S. Fruton. In Hydrolytic Enzymes (A. Neuberger and K. Brocklehurst, eds.), Elsevier Science, New York, 1987, pp. 1–37.
21. J. Steinhart and S. Beychok. In: The Proteins (H. Neurath, ed.), vol. II, Academic Press, New York, 1964, pp. 139–304.
22. N. A. Greenberg and W. F. Ship. J. Food Sci. 44:735 (1979).
23. J. Adler-Nissen. J. Agric. Food Chem. 24:1090 (1976).
24. J. Adler-Nissen. J. Agric. Food Chem. 27:1256 (1979).
25. X. Q. Han. Ph.D. dissertation, University of Wisconsin-Madison, Madison, WI. 1997.
26. L. Michaelis and M. L. Menten. Biochem. Z. 49:333 (1913).
27. G. E. Briggs and J. B. S. Haldane. Biochem. J. 19:338 (1925).
28. J. S. Fruton. Adv. Enzymol. 44:1 (1976).
29. R. Chavire, T. Burnett, and J. H. Hagman. Anal. Biochem. 136:446 (1984).
30. X.-Q. Han and F. Shahidi. Food Chem. 52:285 (1995).
31. C. F. Jocobsen, J. Leonis, K. Linderstrom, and M. Ottesen. Meth. Biochem. Anal. 4:171 (1957).
32. G. Puski. Cereal Chem. 53:655 (1975).
33. K. Ochiai, Y. Kamata, and K. Shibasaki. Agric. Biol. Chem. 46:91 (1982).
34. Y. Kamata, K. Ochiai, and F. Yamauchi. Agric. Biol. Chem. 48:1147 (1984).
35. J. E. Folk. In: Advances in Enzymology, vol. 54 (A. Meister, ed.), Wiley, New York (1983), pp. 1–56.
36. H. Ando, M. Adachi, K. Umeda, A. Matsuura, M. Nonaka, R. Uchio, H. Tanaka, and M. Motoki. Agric. Biol. Chem. 53:2613 (1989).
37. H. Waelsch. In: Monoamines et Système Nerveux Central, Masson, Paris, 1962, pp. 93–104.
38. K. Ikura, M. Goto, and M. Yoshikawa. Agri. Biol. Chem. 48:2347 (1984).
39. X.-Q. Han and S. Damodaran. J. Agri. Food Chem. 44:1211 (1996).
40. J. E. Folk and P. W. Cole. J. Biol. Chem. 240:2951 (1965).
41. M. Liu. Masters Thesis, University of Wisconsin-Madison, 1997.
42. J. Castle, E. Dickinson, B. S. Murray, and G. Stainsby. ACS Symp. Ser. 343: 118 (1987).

3
Protein Interaction at Interfaces

YASUKI MATSUMURA   Kyoto University, Kyoto, Japan

I. INTRODUCTION

There are several types of surfaces or interfaces that are of great practical importance and attracting fundamental interest. These general classifications include liquid–gas, liquid–liquid, liquid–solid, solid–gas, and solid–solid interfaces. Many examples of interfaces of natural and technological importance are available, for instance, oil–water interfaces in emulsions, air–water interfaces in foams, solid–gas interfaces in gas chromatography, and solid–water interfaces in cleaning systems [1].

Proteins, naturally occurring macromolecular surfactants with amphiphilic nature, are adsorbed onto interfaces, thereby affecting the physical states of interfaces. Many enzymes are involved in catalytic reaction at interfaces. For enzymatic reaction at interfaces, different from the reaction in homogeneous systems, interfacial contact and subsequent conformational change of enzymes are important events determining their catalytic activity. In this chapter, I will describe the conformation of proteins and their interaction (protein–protein and protein–surfactant) at interfaces (mainly liquid–liquid interfaces). The characteristics of enzymatic reaction at liquid–liquid and solid–liquid interfaces, especially lipase reaction, will also be described.

II. CONFORMATION OF ADSORBED PROTEINS

Proteins are highly complex polymers made up of combinations of 20 different amino acids with varying hydrophilic (or hydrophobic) properties. During adsorption, the exposure of hydrophobic residues from the interior to the molecular surface occurs that leads to the conformational rearrangement of protein molecules at the interfaces. The conformation of adsorbed proteins plays an important role in determining the nature of adsorbed layers at the interfaces, thereby influencing the physical properties of interfaces. Therefore, special attention has been given to understanding the conformation of adsorbed proteins at interfaces.

Caseins are typical surface-active proteins with a flexible structure [2], and their adsorption behavior is different from that of globular proteins with a compact rigid structure. The structure of adsorbed layers of β-casein (the main component of casein fraction) at interfaces has been investigated theoretically using self-consistent-fields theory [3] and experimentally by the techniques of ellipsometry [4], small-angle x-ray scattering [5], dynamic light scattering [6], neutron reflectivity [7], and proteolysis [8]. A schematic representation of the possible adsorbed conformation of β-casein is shown in Fig. 1a [9]. The conformation of β-casein can be explained by the "train-loop-tail" model, which was devised for the synthetic flexible homopolymers and copolymers.

About three-quarters of the β-casein molecule is close to the interface in trains and small loops, whereby the rest, consisting of N-terminal 40–50 residues, is in a tail dangling well into the bulk aqueous phase (Fig. 1a). The tail length, corresponding to the thickness of the adsorbed layer of β-casein, has been shown to be in the range of 10–15 nm, varying according to the pH in the aqueous phase [7]. The region of trains and small loops is rich in hydrophobic amino acid residues, whereas the N-terminal tail region is highly charged. The high charge of the N-terminal tail is due to the high content of phosphorylserine residue. Dephosphorylation caused a decrease of about 25% in film thickness of β-casein at the air–water and oil–water interfaces [10]. This means that the N-terminal tail containing serine residues attached by phospho-

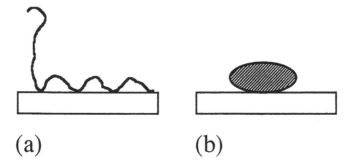

(a)   (b)

FIG. 1 Sketches of representative adsorbed configurations of (a) β-casein and (b) β-lactoglobulin. (From Ref. 9.)

ryl residues changed to a more compact structure when dephosphorylated. Furthermore, stability toward the coalescence of oil droplets in emulsions was substantially reduced when dephosphorylated β-casein was used as an emulsifier instead of native β-casein [10]. This is either as a direct result of charge repulsion between β-casein N-terminal regions or more probably as an indirect result of the reduced N-terminal charge, permitting dephosphorylated β-casein to adopt a different, more compact conformation and resulting in a loss or reduction of a steric barrier.

Conformations of adsorbed globular proteins are quite different from those of caseins. Experimental evidence by neutron reflectivity indicates that the β-lactoglobulin (the main component of milk whey proteins) monolayer thickness is 2–3 nm and substantially less than the β-casein layer thickness (Fig. 1b) [7]. Similar results have been obtained for other globular proteins, such as bovine serum albumin [11]. These results with other available data suggest that rigid compact conformations of globular proteins are not unfolded extensively; i.e., much of the secondary structure is retained after adsorption. Therefore, it is not appropriate to describe the adsorbed state of globular protein in terms of trains, loops, and tails. It is more realistic to regard an adsorbed monolayer of globular protein molecules as a two-dimensional system of interacting deformable particles that are not very different in structure from the native protein molecules in solution (Fig. 1b).

However, partial unfolding, at least at the level of tertiary structure, should be caused by adsorption of globular proteins. Dickinson and Matsumura [12] demonstrated the exposure of a reactive SH group originally embedded in the interior of the β-lactoglobulin molecule to the molecular surface by the adsorption of the protein at the oil–water interface. The increases in susceptibility to the enzyme-catalyzed reaction of adsorbed proteins also suggest that the native rigid conformation is partially lost and that reactive sites become available to enzymes during adsorption [13,14]. Moreover, structural analysis of adsorbed β-lactoglobulin using monoclonal antibodies clearly indicates the unfolding of local structure corresponding epitope region [15]. These examples are consistent with the speculation that the partially unfolded conformation of adsorbed globular proteins is similar to the *molten globule state*, which is defined as a protein configurational state with a nativelike secondary structure and a disordered tertiary structure [16]. In addition to biochemical approaches, direct measurements of secondary and tertiary structures using various spectroscopic techniques would confirm the partial unfolding of adsorbed proteins, although these techniques are applicable mainly for analysis at solid–liquid interfaces.

III. PROTEIN INTERACTION

Adsorbed protein molecules interact at the interfaces to form viscoelastic films. The viscoelastic properties of protein films adsorbed at fluid interfaces in food emulsions and foams are important in relation to the stability of such systems with respect to film rupture and coalescence. Interfacial rheology techniques are very sensitive methods to measure the viscoelastic properties of proteins, thereby evaluating the protein–protein or protein–surfactant interactions at the interfaces. There was an excellent review about the principal and methods of interfacial rheology [17].

The adsorption process is normally monitored by the decrease of interfacial tension (the increase of interfacial pressure). Steady-state values of interfacial tension are reached in several hours, even in the case of proteins with low surface activity, such as lysozyme and ovalbumin [18]. However, the situation is different for parameters of interfacial rheology, such as interfacial viscosity. Measurements with a variety of proteins have shown that a steady-state interfacial viscosity is never reached over the normal experimental time scale (several days) [17]. The reason for the larger time scale in interfacial rheology is that this method reflects intermolecular interactions as well as intramolecular re-arrangements of adsorbed proteins.

The value of interfacial viscosity for adsorbed β-casein film at the oil–water interface was very low, i.e., less than 1 mNsm^{-1}, even after 24 h [17]. However, interfacial viscosity for adsorbed β-lactoglobulin reached 1.2×10^3 mNsm^{-1} under the same condition [17]. Similar results were obtained for other globular proteins. The high values of interfacial viscosity for globular proteins can be attributed to the high packing density of globular protein molecules in the adsorbed layer, as compared with the rather loose packing for disordered β-caseins. As shown in Fig. 1b, a conformation such as a deformable particle of globular protein may be advantageous for strong interaction among adsorbed molecules and dense packing within two-dimensional film. In contrast, a highly charged tail dangling into the aqueous phase (Fig. 1a) may prevent dense packing among adsorbed β-caseins. However, a transglutaminase-catalyzed cross-linking reaction was found to increase the viscoelasticity of adsorbed β-casein film substantially [19]. This means that the incorporation of covalent bonding can modify the mechanical properties of adsorbed protein films.

The interactions between proteins and low-molecular-weight surfactants at interfaces have crucial effects on physical states of interfaces, such as interfacial energy, interfacial rheological properties, ζ potential, and thickness of adsorbed layer. The competitive displacement of globular proteins by surfactants at liquid interfaces (normally, oil–water interfaces) has been extensively studied by several groups, particularly the group at Leeds University. A systematic approach by the Leeds group has shown the more efficient displacement of proteins by water-soluble surfactants (i.e., surfactants with high HLB) rather than by oil-soluble surfactants, such as monoglycerides, phospholipids, and synthetic surfactants with low HLB [20–22]. Figure 2 shows the result for the case of a simple combination of a protein and a water-soluble surfactant, i.e., the competitive adsorption of β-lactoglobulin and Tween 20 (polyoxyethylene (20) sorbitan monolaurate) at the n-tetradecane–water interface [20,23]. β-Lactoglobulin was almost completely displaced from the emulsion droplet surface by Tween 20 at R (Tween 20/β-lactoglobulin molar ratio) > 8. It is generally accepted that the displacement of proteins by the excess amounts of surfactants is caused by the decrease in interfacial tension to a much lower level than that attained by the adsorption of proteins. Despite no or negligible displacement of β-lactoglobulin at $R = 1$ (and a little decrease in interfacial tension), the interfacial viscosity of the protein at the planar interface dramatically decreased in the range $0 \leq R \leq 1$ [20,23]. Fluorescence

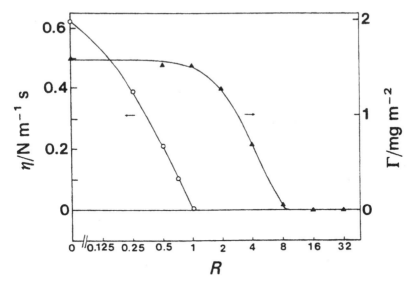

FIG. 2 Competitive adsorption of β-lactoglobulin + Tween 20 at the oil–water interface. R is the Tween 20/β-lactoglobulin molar ratio. Γ (triangles), concentration of β-lactoglobulin on the surface of oil droplets in the emulsion (0.45 wt% protein, 10 wt% n-tetradecane, pH 7); η (circles), apparent surface shear viscosity after 5 h adsorption from a 10^{-3} wt% aqueous protein solution (pH 7, 25°C). (From Ref. 23.)

recovery after a photobleaching (FRAP) experiment showed that the diffusion coefficient of β-lactoglobulin at the interfaces increased sharply at $R = 1.0$, indicating an increase in the molecular mobility of the protein in the adsorbed layer [24]. These results suggest that Tween 20 breaks the interaction among adsorbed β-lactoglobulin molecules, thereby decreasing the cohesive nature of adsorbed film. Dickinson et al. also pointed out that the coalescence of droplets in β-lactoglobulin-stabilized emulsions was remarkably enhanced at $R = 1.0$ when the emulsions were sheared continuously after the addition of various amounts of Tween 20 [23]. This means that the orthokinetic instability, i.e., enhanced coalescence of β-lactoglobulin-coated emulsion droplets by low concentrations of Tween 20, should be due to a loosening of the protein film structure by the breakage of protein–protein interaction. Similar results were obtained for other proteins with respect to the relation of the protein–nonionic surfactant interaction and the instability of emulsions or foams [25]. In the case of ionic surfactants capable of binding to proteins, the competitive adsorption is more complicated than that of the nonionic surfactants [26].

Competitive adsorption among proteins is discussed elsewhere [27]. In fact, the protein already adsorbed can be displaced by another protein with more surface activity at the oil–water interface as well as the solid–water interface. It should be emphasized, however, that the displacement among proteins is possible only at relatively high temperatures, that is, above 40°C [27,28].

IV. ENZYMATIC REACTION AT INTERFACES

A. Lipase Reactions at Lipid–Water Interfaces

The most important and typical enzymes that function at lipid–water interfaces in micelles, liposomes, emulsions, etc., are lipases. Lipases are carboxylic ester hydrolases and have been termed glycerol ester hydrolases (EC3.1.1.3) in the international system of classification. They differ greatly as regards both their origins (bacterial, fungal, plant, mammalian, etc.) and their properties, and they can catalyze the synthesis as well as the hydrolysis of a wide range of different carboxylic esters. Numerous reports have appeared about the structure and function of pancreatic lipases, because they are ubiquitous in mammalian species and play important roles in dietary fat absorption [29,30]. In this part, I will describe a structural feature and its relation to catalytic mechanism at the interfaces of lipases, particularly pancreatic lipases.

The most unique characteristic of lipases is the "interfacial activation" phenomenon. This means that the activity of lipases is enhanced on insoluble substrates (aggregates such as emulsions) as compared to the same substrate in monomeric solutions [31]. The mechanism of interfacial activation had been an enigma for researchers involved in studies of lipases. The resolution of the three-dimensional structure of human pancreatic lipase (HPL) in the absence or presence of colipase and lipid micelles (corresponding to interfaces in emulsion systems) is a breakthrough in solving such an enigma [32–34].

Figure 3 shows the structure of the HPL-colipase complex [30,34]. Human pancreatic lipase consists of 449 amino acid residues and has 85% homology with that of porcine pancreatic lipase. Two distinct domain exists in HPL, that is, a larger N-terminal domain (N-domain) comprising 1–335 residues and a smaller C-terminal domain (C-domain). Of these domains, the N-domain con-

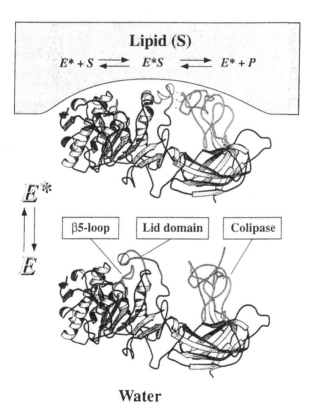

FIG. 3 Structure of human pancreatic lipase-colipase complex in the closed conformation (E), and structure of the complex in the open conformation (E*S). These two figures show the conformational changes in the lid, the β5-loop, and the colipase during "interfacial activation." (From Ref. 30.)

tains the active site with a catalytic triad formed by Ser152, Asp176, and His263. The catalytic triad of these three amino acid residues is conserved widely, not only in lipases with various origins and but also in serine proteases. Ser152 is the nucleophilic residue essential for catalysis, which is also supported by the results of chemical modification [35]. The C-domain has the binding site for colipase, which is a small protein with the "three-finger" structure common to snake toxins topologically. Colipase functions as the co-factor of lipase; that is, it binds to the lipid surface and enhances the affinity of the lipase for the lipid surface [36].

The most interesting observation in Fig. 3 is the presence of a "lid" covering the active site of the HPL. The lid corresponds to the region between Cys237 and Cys261. In the closed conformation "E" (in solutions), Trp252, which is located in the short α-helix (residues 248–255) of the lid domain, is embedded directly over Ser152, active site. Furthermore, the β5-loop makes van der Walls contact exclusively with the lid. Therefore, the active site is completely inaccessible to solvent or substrate in the closed conformation.

However, the approach of HPL to the lipid surface gives rise to the substantial changes in the conformation of the lid domain and the β5-loop (in the open conformation, "E*") [30,34]. The α-helix (residues 248–255) partially unwinds, and the new helices are formed (residues 241–246 and 251–259). The conformational change in the lid results in a maximal main-chain movement of 2.9 nm for Ile248. According to such a conformational change of lid domain, the interaction of the β5-loop to the lid is also broken and the β5-loop is lifted back onto the core of the molecule. The substrate now becomes accessible to the active site. In the open conformation, colipase also binds to the lid domain. As the results of all these conformational changes, the open lid and the colipase form a continuous hydrophobic plateau, extending over 5.0 nm. This surface might be able to interact strongly with a lipid/water interface. The overall process of conformational change of HPL described here, which results in the opening of the lid domain and subsequent exposure of the active site to substrate lipids, is the machinery of "interfacial activation."

The importance of the lid domain in interfacial activation was confirmed by the experiments using deletion mutants of HPL [37]. Two mutants were created in the region of the lid domain. One, the 240–260 deletion mutant, removed virtually all of the lid domain to test the importance of the domain in lipase function. The second mutation, the 247–258 deletion mutant, was designed to remove the α-helix covering the active site and to minimize disruption of the remaining lid domain residues. Interfacial activation of wild and mutant HPLs was determined by measuring the activity over a range of tributyrin concentrations in the presence of colipase (Fig. 4). The activity of

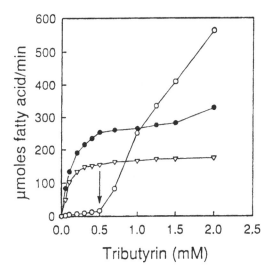

FIG. 4 Absent interfacial activation in the lid deletion mutants. Activity was determined against tributyrin by the pH-STAT technique. Colipase was included to prevent interfacial inactivation of the lipases. The solid arrow shows the tributyrin saturation point determined by turbidity measurements. Open circles, wild human pancreatic lipase; closed circles, 248–257 deletion mutant; triangles, 240–260 deletion mutant. (From Ref. 37.)

wild HPL increased sharply when tributyrin reached its saturation point (i.e., the point changing from monomeric form to aggregate or emulsion form), demonstrating interfacial activation. However, both mutants reached maximal activity at tributyrin concentrations below the saturation point, indicating that they hydrolyzed monomeric substrate and did not require an interface for activation. These findings are consistent with the participation of the lid domain in interfacial activation.

The common three-dimensional structure of Fig. 3 has been found for lipoprotein lipase [38] and hepatic lipase [39]. Interestingly, a similar lid domain is found in the structure of a lipase from fungi, *Rhyzomucor miehei* [40], and interfacial activation is also demonstrated in this enzyme. Recently, new types of lipases have been found, such as pancreatic lipase from coypu and lipases from *Pseudomonas glumae*, *Candida antarctica* B, etc., which have a lid covering the active site but show no interfacial activation [30].

Since interfacial activation is essential for the catalytic activity of pancreatic lipases or *Rhizopus* lipases, the activity of lipases at interfaces can be inhibited by a number of other surface-active molecules via a competitive adsorption mechanism. Verger et al. showed that proteins such as bovine serum albumin and β-lactoglobulin inhibited the action of lipases from the pancreas and from *Rhizopus dolman* [41]. Using monolayer systems, they clearly demonstrated that the inhibition of lipases by proteins was due to the protein coverage over the lipid surface and was not caused by direct protein–enzyme interaction in the aqueous phase. The proteins were also able to cause dissociation of the lipases from the interface when added afterwards. Bile salts can also inhibit lipase reaction by displacing the enzyme at the interface in the absence of colipase [29,36].

We have recently shown that lipoxygenases from plants can act on substrate lipids emulsified by proteins or bile salts [42]. This means that the interfacial activation is not essential for lipoxygenase reaction in emulsion systems. The data of the three-dimensional structure of lipoxygenases have shown the absence of a "lid" but the presence of the "cavity" functioning as a path for a substrate lipid from the exterior of the enzyme molecule to the catalytic site [43]. It is thought that lipoxygenase uses this cavity for the uptake of substrate lipid from the oil droplet surface into the interior catalytic site without substantial conformational change, i.e., interfacial activation.

B. Surfactant-Coated Enzymes in Organic Solvents or Supercritical Fluids

Proteins including enzymes normally lose their native conformations and activity in organic solvent media. Therefore, in principle it is thought that an enzymatic reaction in organic media is impossible or difficult. Nevertheless, considerable efforts have been made to mimic the catalytic action of enzymes in organic solvents, especially for lipase reaction in which both substrates and products are lipophilic substances soluble in organic solvents. Several methods have been shown to be useful for enzymatic reaction in organic media: (1) the method of direct dispersion of powdered enzymes in organic solvents [44], (2) the use of immobilized enzymes [45], (3) the water-in-oil microemulsion or the reversed micellar system in which enzymes are solubilized in the inner aqueous phase [46], (4) the use of enzymes modified with polyethylene glycol (PEG) [47]. However, there are disadvantages for these methods, as discussed later.

The Okahata group has recently developed a new method to use surfactant-coated enzymes, which are freely soluble in organic solvents and show high catalytic activity in organic media. The surfactant-coated enzymes are prepared by the following very simple procedure [48]. Typically, the enzyme aqueous solution is mixed with an aqueous dispersion of surfactants at 4°C and then kept, under stirring, for 20 h at 4°C. The resultant precipitates are collected by centrifugation at 4°C and lyophilized. The obtained powder is a surfactant-coated enzyme.

Various types of surfactants have been tested with respect to yield and enzymatic activity. Although cationic and anionic amphiphiles produced the complexes of surfactants and enzymes in a fair yield, their enzyme activities were very low. Probably, the strong electrostatic interaction between the ionic head groups of amphiphiles and the charged amino acid residues on the surface of enzymes caused the denaturation of enzymes. On the other hand, the nonionic glycolipid, didodecyl N-D-glucono-L-glutamate (Fig. 5) gave the complex in a good yield, and the obtained complex showed a high enzymatic activity in organic solvents [48]. Since N-D-glucono-L-glutamate was shown to be suitable for the formation of surfactant-coated enzymes, the mode of interaction between the surfactant and enzymes was analyzed. The example of N-D-glucono-L-glutamate-coated lipase D (from *Rhizopus delemar*) will be shown. Based on the results of gel permeation chromatography in organic solvent (dichloromethane), the molecular weight of the surfactant-coated lipase was ca. 13×10^4. Since the molecular weight of a native lipase D and the glycolipid are ca. 3.3×10^3 and 661, respectively, it can be calculated that the surfactant-coated lipase contain ca. 150 surfactant (glycolipid) molecules per one lipase. Estimating roughly from the molecular area of the surfactant (0.45 nm^2) and lipase (diameter: ca. 3 nm), 150 surfactant molecules probably covers the surface of a lipase as a monolayer. ^{1}H-NMR spectra of the surfactant-coated lipase D were compared with those of the free surfactant. The chemical shift of only hydrophilic OH groups of the surfactant in the complex moves to the down field (from 3.7 to 4.1 ppm) relative to that of the

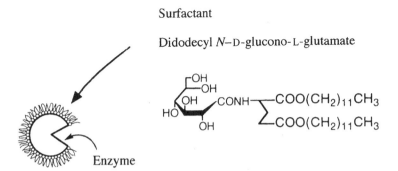

FIG. 5 Schematic illustration of a surfactant-coated enzyme. (From Ref. 48.)

free surfactant. This indicates that the hydroxyl head groups of N-D-glucono-L-glutamate interact with the hydrophilic surface of lipase through hydrogen bonds and lipophilic tails of lipids are thought to solubilize the complex in hydrophobic organic solvents, as shown in Fig. 5.

The superiority of N-D-glucono-L-glutamate-coated enzyme system over other systems is shown in Fig. 6 [48]. In this figure, lipase B (from *Pseudomonas fragi* 22-39B) was used as an enzyme. Di- and trilaurin syntheses from 1-monolaurin and lauric acid were carried out in the dry benzene with molecular sieves in the presence of the same amount (1 mg of protein) of the surfactant-coated lipase B, PEG-grafted lipase B, lipase B in water-in-oil emulsion, and powdered lipase B. When the surfactant-coated lipase B was used, the conversion from 1-monolaurin to di- and trilaurin was completed within 7 h [initial rate: $v_0 = 12.2$ mM min^{-1} (mg of protein)$^{-1}$]. However, powdered lipase B showed a very slow reaction rate (one hundredth) compared with that of the surfactant-coated lipase. In the dispersion method, the specific surface area accessible to substrate is low and a high amount of enzymes may be required to get a high reaction rate. In other words, the homogeneously soluble surfactant-coated lipase has much higher activity than the dispersed lipase when the same amount of enzyme is used. In the water-in-oil system, the rate of ester syntheses was very slow [$v_0 < 0.1$ mM min^{-1} (mg of protein)$^{-1}$] and decreased with

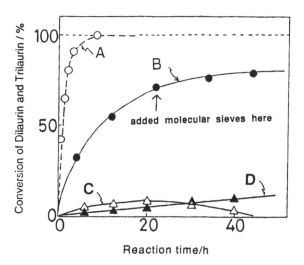

FIG. 6 Comparison of catalytic activities for di- and trilaurin syntheses of (A) the surfactant-coated lipase B, (B) PEG-grafted lipase B, (C) Lipase B in W/O emulsion, and (D) dispersion of lipase B powder, in benzene solution at 40°C. (From Ref. 48.)

increasing reaction time. This may be due to the existence of a small amount of water in the w/o emulsion system, causing the reverse hydrolysis reaction.

When PEG-grafted lipase B was employed, the esterification proceeded at a fair rate [v_0 = 6.0 mM min^{-1} (mg of protein)$^{-1}$]. However, the conversion reached a plateau at 70% after 40–80 h. The further addition of molecular sieves could not improve the conversion. This indicates that the PEG lipase can keep the produced water near the amphiphilic PEG-grafted chains so that the water molecules may not be removed by the addition of molecular sieves. The water solubilized near the lipase surface inhibits the ester synthesis in organic solvents. In contrast, the hydrophobic surfactant-coated lipase cannot solubilize the produced water, and the glyceride synthesis can proceed completely when molecular sieves from the solution remove the produced water.

Therefore, it was proved that the surfactant-coated lipase could catalyze effectively and completely the glyceride synthesis in dry organic solvents, as compared with other enzyme systems, such as water-in-oil emulsion, enzyme dispersion, and PEG-grafted enzyme.

In addition to the synthesis of triglycerides, the surfactant-coated lipase can be employed as an enantioselective catalyst for the esterification of racemic alcohols with aliphatic acids in homogeneous dry isooctane [49]. Moreover, Okahata's group reported that phospholipases [50] and glycosidases [51] coated by the same glycolipids (*N*-D-glucono-L-glutamate) are soluble in organic solvents and can catalyze reverse hydrolysis reactions such as esterification and transglycosylation in homogeneous organic media. In the case of the surfactant-coated β-D-galactosidase, first transgalactosylation to alcohols from *p*-nitorophenyl β-D-galactopyranoside (Gal-*p*NP) was performed in dry isopropyl ether [51]. However, this system could not be applied for a large-scale synthesis because of the low solubility of the galactosyl donor (Gal-*p*NP) in hydrophobic organic media. To overcome this problem and explore the more practical method, Mori et al. [52] used a new transgalactosylation system with the same surfactant-coated β-D-galactosidase in a two-phase aqueous-organic system in which both the surfactant-coated enzyme and hydrophobic acceptor alcohols existed in the organic phase and an excess amount of lactose as galactose donor was present in the aqueous phase. A schematic illustration of the transgalactosylation in two phases are shown in Fig. 7 [52].

Figure 8 shows a typical time course of transgalactosylation from the 10-fold excess of lactose as a galactosyl donor to 5-phenyl-1-pentanol (PhC$_5$OH) as a galactosyl acceptor in two phases of isopropyl ether and phosphate buffer (10 mM, pH 5.1) at 30°C [52]. In the case of the surfactant-coated β-D-galactosidase, the transglycosylated product (5-phenyl-1-pentyl β-D-galactopyranoside, Gal-O-C$_5$Ph) was obtained in 66% yield after 8 days, and hydrolyzed

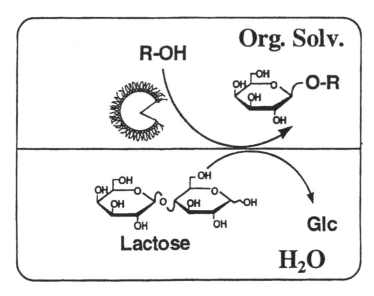

FIG. 7 Schematic illustration of transgalactosylation catalyzed by a surfactant-coated β-D-galactosidase in aqueous-organic two phases. (From Ref. 52.)

product (galactose) was not detected during 8 days. In contrast, the use of unmodified β-D-galactosidase resulted in no conversion, indicating the denaturation of the enzyme at the interface between aqueous phase and isopropyl ether. Therefore, it was demonstrated that the surfactant-β-D-galactosidase is applicable to the transglycosylation between water-soluble sugars and organic solvent-soluble alcohols in the aqueous-organic two phases. They also discussed the effects of chemical structures of acceptor alcohols on the transgalactosylation [52].

Recently, it has been reported that surfactant (N-D-glucono-L-glutamate)-coated enzymes (lipase D and β-D-galactosidase) are soluble and act as more efficient catalysts in supercritical carbon dioxide (scCO$_2$) than in conventional organic solvents [53,54]. Figure 9 shows typical time courses of the transgalactosylation from Gal-pNP to PhC$_5$OH catalyzed by the surfactant-coated β-D-galactosidase at 40°C both in scCO$_2$ at 150 atm and in isopropyl ether at atmospheric pressure [53]. In scCO$_2$, the transglycosylated product (Gal-O-C$_5$Ph) was obtained in 72% yield after 3 h. The conversion reaction reached a plateau at 6 h. The surfactant-coated β-D-galactosidase could catalyze the transgalactosylation in isopropyl ether, but the conversion rate was very low compared with the reaction in scCO$_2$. That is, the transgalactosylation rate

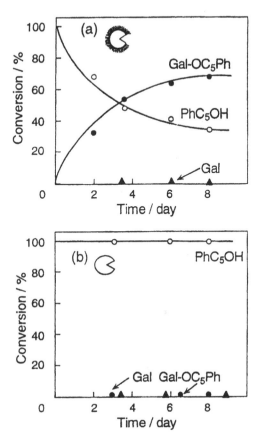

FIG. 8 Typical time courses of transgalactosylation from lactose (10 mM) to 5-phe-nyl-1-pentanol (PhC$_5$OH, 1.0 mM) catalyzed by (a) the surfactant-coated β-D-galactosi-dase and (b) a native β-D-galactosidase from *E. Coli*, in an isopropyl ether-aqueous buffer two-phase system: 30°C, 10 ml of isopropyl ether and 10 ml of buffer solution (0.01 M phosphate, pH 5.1), and [Enzyme] = 0.1 mg of protein. (From Ref. 52.)

in scCO$_2$ was 15-fold higher than that in isopropyl ether. Unmodified β-D-galactosidase could not catalyze the transglycosylation in scCO$_2$. This may be due to the insolubility and instability of unmodified enzyme in scCO$_2$. Mori et al. also found that the surfactant-coated lipase D catalyzed triglyceride syn-thesis 10-fold faster in scCO$_2$ than in atmospheric isooctane [54].

These results indicate that scCO$_2$ is a good media for the catalytic reaction by the surfactant-coated enzymes, although several studies have reported no

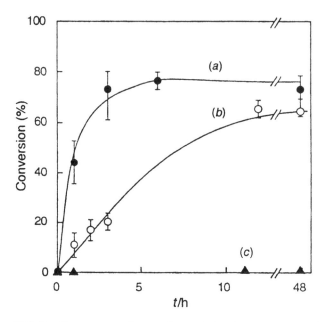

FIG. 9   Time courses of transgalactosylation from 1-O-p-nitrophenyl-β-D-galactopyr-anoside (0.1 mM) to 5-phenylpentan-1-pentanol (1 mM) at 40°C catalyzed by β-D-galactosidase (1 mg of protein) in 10 ml: (a) the surfactant-coated enzyme in scCO$_2$ with 150 atm, (b) the surfactant-coated enzyme in isopropyl ether, and (c) a native enzyme in scCO$_2$ with 150 atm. (From Ref. 53.)

improvement in the reactivity of enzymes in scCO$_2$ when unmodified and im-mobilized enzymes are used. Therefore, the combination of the surfactant-coated enzymes and scCO$_2$ reaction media should be a promising system for production of useful materials. The advantages of this system are summarized as follows: (1) the activity of the surfactant-coated enzyme was 5–10 times higher than that in conventional organic media, (2) the reaction can be switched on and off very simply by adjusting the pressure or temperature of CO$_2$, (3) scCO$_2$ media is nontoxic and easily removed by decompression, compared with organic solvents, and suitable for the biotransformation of food products and pharmaceuticals.

C.   Activity of Immobilized Enzymes (Reaction at Solid–Liquid Interfaces)

Enzymes immobilized microporous matrices, such as organic polymer beads, silica particles, and ceramic carriers, have been widely used for the production of various useful substances [55]. Immobilization of an enzyme facilitates its recovery for possible reuse and continuous application, and improves its stability as well. In addition, the interactions between the support and the enzyme may favorably alter the chemical or physical properties of the enzyme, and often a new property is generated that can be exploited for industrial purposes. The use of immobilized enzymes is advantageous, particularly in organic media, in which the stability of the enzymes is dramatically reduced. Of various physicochemical parameters of solvents, water activity is thought to be the most important one affecting the activity of immobilized enzyme [56]. Supports (carriers) and solvents modify the hydration states of immobilized enzymes, thereby affecting enzymatic activity and stability. There have been only a few reports dealing with the stability of immobilized enzymes in organic solvents in combination with water activity, although stability is the most important factor for the construction of a bioreactor [57].

Using the reverse of the hydrolytic reaction, proteases are capable of catalyzing the formation of peptide bonds. Thermolysin is a powerful enzyme in catalyzing the synthesis of various useful oligopeptides, either in a free form or in an immobilized form in organic solvent media [58]. Recently, the systematic study about the stability of immobilized thermolysin has been reported [59]. In this study, the authors described the mechanism and kinetics for the inactivation of immobilized thermolysin with respect to the effects of organic solvents, kinds of support, water content, and so on. The strategy and results of their approach are discussed next.

The immobilized thermolysin was prepared by adsorption of the enzyme onto the supports, polyacrylic ester resins (Amberlite series) and pore glass particles (CPG series), followed by crosslinking with glutaraldehyde in aqueous solutions. The activity of the immobilized enzyme was assayed by measuring the initial synthetic rate of N-(benzyloxycarbonyl)-L-phenylalanyl-L-phynylalanine methyl ester in the biphasic reaction system.

First, the effects of the kinds of organic solvents on the stability of thermolysin immobilized onto Amberlite XAD-7 were investigated. Table 1 shows the relative remaining activity of thermolysin immobilized onto XAD-7 after 5 h of incubation at 70°C in water-immiscible and water-micsible organic solvents [59]. The water content in the bulk organic solvent phase was usually adjusted to 4% using 2-(N-morpholino) ethanesulfonic acid (MES) buffer. For tert-amyl alcohol, the results for the water content of 12% is also shown. In methyl alcohol, the stability of the immobilized enzymes was extremely low. The remaining activity increased with the increase of the carbon chain number up to 5, but decreased again in n-hexyl and n-heptyl alcohols. In alcohols having the same number of the carbon chain, the immobilized enzyme was more stable in tertiary alcohols than in normal alcohols. In particular, the im-

TABLE 1   Effect of Organic Solvents on Relative Remaining Activity of Thermolysin Immobilized onto XAD-7 and Solubility of Water

No.	Organic solvent	Relative remaining activity (%)	Solubility of water[a] (g water/100 mL sol$'_n$)
1	Methyl alcohol	7	Miscible[b]
2	Ethyl alcohol	63	Miscible[b]
3	n-Propyl alcohol	84	Miscible[b]
4	n-Butyl alcohol	88	17.2
5	tert-Butyl alcohol	93	Miscible[b]
6	n-Amyl alcohol	83	10.0
7	tert-Amyl alcohol	98	12.8
8	tert-Amyl alcohol	42	—
9	n-Hexyl alcohol	70	6.1
10	n-Heptyl alcohol	56	4.7
11	Ethyl acetate	38	3.9
12	Acetonitrile	72	Miscible[b]

The immobilized enzyme was incubated for 5 h at 70°C in various organic solvents containing 4% of water (50 mM MES-NaOH buffer, pH 6.0, containing 5 mM CaCl$_2$), except tert-amyl alcohol, for which the organic solvent contained 12% water.
[a] The solubility of MES buffer at 70°C.
[b] Water-miscible organic solvent.
Source: Ref. 59.

mobilized enzyme was quite stable in tert-amyl alcohol including 4% of water, but the elevating water content up to 12% dramatically reduced the stability. In ethyl acetate, which is often used for enzymatic synthesis of oligopeptide precursors because of its ability to show high activity, the stability was low.

The relationship between the remaining activity and the amount of enzyme immobilized onto the support (Amberlite XAD-7) was investigated to learn the inactivation mechanism of immobilized enzyme in organic solvents (Fig. 10) [59]. In water-miscible organic solvents, the amount of enzyme immobilized was only slightly decreased by incubation, despite the changes in the relative remaining activity of enzyme according to solvents. During incubation, any appreciable protein and peptide fragments were not liberated. These results suggest that the immobilized enzyme was inactivated, mainly suffering from irreversible thermal denaturation, without the leakage of the enzyme from the support, caused probably either by physical desorption or by autolysis.

On the other hand, in water-immiscible organic solvents, there was a positive correlation between the relative remaining activity and remaining amount

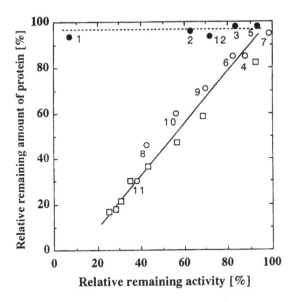

FIG. 10 Relationship between relative remaining activity and relative remaining amount of enzyme after 5 h of incubation in various organic solvents at 70°C. Closed circles, in water-miscible organic solvents; open circles, in water-immiscible organic solvents; open squares, data from time course during incubation in ethyl acetate containing 4% water. The numbers in the figure correspond to the kinds of organic solvents shown in Table 1. (From Ref. 59.)

of enzyme (Fig. 10), indicating that inactivation of the immobilized enzyme arose from the loss of the amount of enzyme during incubation. In these solvents, various peptide fragments were liberated during incubation. Therefore, it is thought that autolysis of thermolysin inside the support is responsible for the inactivation in the case of reaction in water-immiscible organic solvents. The involvement of autolysis was confirmed by the addition of phosporamidon, a specific inhibitor for thermolysin, into the reaction mixture, which could prevent the leakage of the enzyme from the support. It is very interesting that the inactivation process of immobilized thermolysin differs according to the miscibility of solvent with water.

Kinetic analyses were performed for the inactivation of the immobilized enzyme onto Amberlite XAD-7 in ethyl acetate and *tert*-amyl alcohol at 70°C, varying the water content in the bulk organic solvent phase. The results are shown in Table 2 [59]. The kinetic parameter k_2^* reflects the effect of the conformational change of enzyme, that is, a change from the active to inactive

TABLE 2 Kinetic Parameters for Inactivation of Enzymes Immobilized onto Amberlite XAD-7 at Different Water Contents at 70°C

Solvent	Water content[a][%]	$k_2^*[E_0]$ [min^{-1}]
Ethyl acetate	1.0	0.8×10^{-3}
	2.0	1.0×10^{-3}
	3.5	2.0×10^{-3}
	4.0	8.2×10^{-3}
	8.0[c]	17.0×10^{-3}
tert-Amyl alcohol	2.0	[b]
	4.0	0.2×10^{-3}
	8.0	1.2×10^{-3}
	12.0	3.8×10^{-3}

[a] Water content in the bulk organic solvents.
[b] No decrease in activity was detected after 5-h incubation at 70°C.
[c] The bulk organic solvent phase was separated into organic and water phases.
Source: Ref. 59.

form as well as that of a proteolytic reaction. [E$_0$] means the initial concentration of total enzyme. The $k_2^*[E_0]$ values determined for the immobilized enzyme in tert-amyl alcohol were much lower than those in ethyl acetate at each water content. For instance, at 4% water content, the $k_2^*[E_0]$ value in ethyl acetate was 40-fold higher than that in tert-amyl alcohol. Figure 11 [61] shows that, at 4% water content in the bulk organic solvent, the water content inside the enzyme immobilized onto Amberlite XAD-7 was around 7% in ethyl acetate and 4.2% in tert-amyl alcohol, respectively. The solubility of water in ethyl acetate was only 3.9% (Table 1). Therefore, in the case of ethyl acetate, the phase inside the support might be separated into aqueous and organic phases. Probably, the enzymes are concentrated in this aqueous phase, and as a result autolysis was accelerated.

With an increase in the water content in the bulk organic solvent, the $k_2^*[E_0]$ value appreciably increased, as shown in Table 2. The increased $k_2^*[E_0]$ values with an increase in the water content in the bulk organic solvent might be ascribed to the increased fraction of the aqueous phase in which the enzyme aggregates exist, which promoted autolysis. To test this speculation, the relative remaining activity after 5-h incubation at 70°C at 4% water content in the bulk organic solvent was plotted against the water content inside the support, for enzyme immobilized onto various supports in addition to Amberlite XAD-7 in ethyl acetate and in tert-amyl alcohol (Fig. 12) [59]. The values of

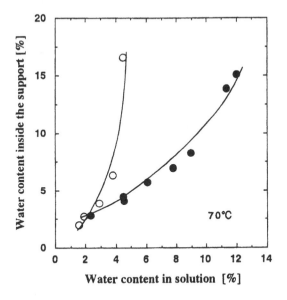

FIG. 11 Adsorption isotherms of water for thermolysin immobilized onto Amberlite XAD-7 in ethyl acetate (open circles) and in *tert*-amyl alcohol (closed circles) at 70°C. (From Ref. 59.)

the relative remaining activity tended to decrease with an increase in the water content inside the support, irrespective of the supports, although the tendency was different in ethyl acetate and in *tert*-amyl alcohol. Therefore, for immiscible organic solvents, it can be concluded that the stability of the thermolysin immobilized in the different supports is closely related to the water content inside the immobilized enzyme. This suggests that the excess water in the supports causes the separation of the solvent phase and the water phase, in which the concentrated enzymes become susceptible to autolysis.

For water-miscible solvents, the correlation of the stability of immobilized enzyme with physicochemical parameters of organic solvent was investigated. Two parameters, log P (the partition coefficient between the n-octanol-water phases, i.e., reflecting the hydrophobicity of organic solvents) and E_T (30) (the polarity of organic solvents), were chosen. As shown in Fig. 13 [59], either the log P or E_T (30) value was found to be applicable to correlate the stability of the immobilized enzyme in the water-miscible organic solvents. Therefore, the inactivation of the immobilized enzyme in water-miscible organic solvents was affected by the extent to which they remove water from the surface of the enzyme molecule.

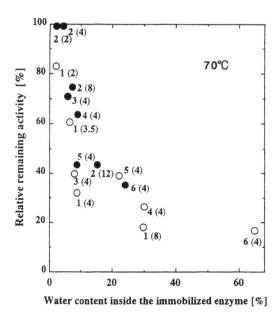

FIG. 12 Relationship between the relative remaining activity of thermolysins immobilized onto Amberlite XAD-7 [1,2], Amberlite XAD-8 [3], Amberlite XAD-5003 [4], and CPG-7.5 [5], and CPG-300 [6] and water content inside the immobilized enzyme after 5-h incubation at 70°C in ethyl acetate (open circles) and in *tert*-amyl alcohol (closed circles). The figures in parentheses indicate the water content in the bulk organic solvent phase. (From Ref. 59.)

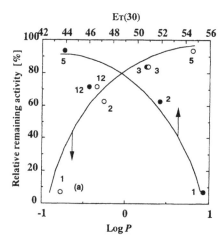

FIG. 13 Correlation of the relative remaining activity of the enzyme immobilized onto Amberlite XAD-7 with log P and E_T [30] values for water-miscible organic solvents. The numbers in the figure correspond to the kinds of organic solvents shown in Table 1. All the organic solvents contained 4% water. (From Ref. 59.)

V. EXAMPLES OF BIOLOGICALLY IMPORTANT NONENZYME PROTEINS WITH INTERFACIAL FUNCTIONS

There are a lot of nonenzymatic proteins that exist and function at interfaces in biological systems. Before closing this chapter, I will refer to several examples of such naturally occurring protein surfactants of biological and/or industrial importance that exist in mammalian tissues, microorganism, and plants.

Plasma lipoproteins (LPs) are soluble aggregates of lipids and proteins that deliver hydrophobic, water-insoluble lipids (triglycerides and cholesteryl esters) from the liver and intestine to other tissues in the body for storage or utilization as an energy source [60]. All LP particles have a common structure of a neutral lipid core surrounded by a surface monolayer of amphipathic lipids (phospholipids and unesterified cholesterol) and some specific apoproteins (Fig. 14). The LPs are usually classified according to density, from very low-density lipoprotein (VLDL) to high-density lipoprotein (HDL). The size of LPs varies from 5–12 nm for HDL to 30–80 nm for VLDL.

A common structural feature of most apolipoproteins is the presence of amphipathic helical structures, which are thought to be responsible for binding apoproteins to lipids of the surface monolayer [60]. Each α-helix contains a hydrophilic and a hydrophobic face; the hydrophilic face is exposed to water, whereas the hydrophobic face associates with lipids of the monolayer. Such amphiphilic structure enables apolipoproteins to function as natural surfactants

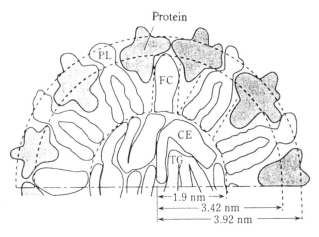

FIG. 14 Structure of a lipoprotein particle. PL: phospholipid, FC: unesterified cholesterol, CE: cholesteryl ester, TG: triglyceride.

stabilizing lipid particles in water. Additionally, apolipoproteins demonstrate specialized functions at the surface LP particles, such as ligand for cell surface receptors, cofactors for plasma lipases, and competitive inhibitors of endocytosis or metabolism [61].

Another important natural protein surfactant in mammalian are lung surfactants, which are secreted by type II pneumocytes and reduce surface tension at the air–water interface of distal airways and the alveoli of lungs, thereby decreasing the work of breathing and the tendency for alveoli to collapse at low lung volumes. The components of lung surfactants as well as the structure and functions of surfactant proteins have been precisely reviewed recently [62].

A wide variety of microorganisms also produce many kinds of surface-active lipoproteins or lipopeptides [63]. Mostly they exhibit the typical amphiphilic character and are generally extracellular. Representative of such surface-active lipopeptides is surfactin produced by *Bacillus subtilis*. It is composed of a heptapeptide cycle closed by a C_{14-15} β-hydroxy fatty acid that forms a lactone ring system (Fig. 15) [64]. This structure resembles those of iturins, another class of lipopeptides also produced by *Bacillus subtilis* [65].

Surfactin is a biosurfactant with a high industrial and commercial potential because of its superb surface and interfacial activity and because it has a diversity of bioactive properties. Surfactin forms large micelles at a very low concentration and the cmc is of the order of 10^{-5} M [66]. The bioactivities of surfactin include: inhibition of blood clotting [67], hemolytic activity [68], channel formation in membranes [69], synergistic antifungal activity when combined with iturins, antibiotic activity [70], antitumor [71], and anti-HIV effects [72].

Many other surface-active lipopeptides, such as rhamnolipids and viscosin from *Pseudomonas* species, have been found [73]. More surface-active lipopeptide, arthrofactin, the cmc of which is extremely less than that of surfactin, has been purified from *Arthrobacter* species [74]. These lipopeptides, including surfactin, have great industrial potential for pharmatheuticals, cleaning detergents, cosmetics, foods, and so on, because of their superior surface activity and antimicrobial effects.

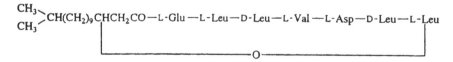

FIG. 15 Structure of surfactin from *Bacillus subtilis*. (From Ref. 64.)

Finally, I introduce the naturally occurring protein surfactant originated from plants, called oleosin [75,76]. In plant seeds, oils are stored in discrete organelles called *oil bodies*. Oil bodies are small spherical particles approximately 1 μm in diameter. Notice in Fig. 14 the difference in size between plant oil bodies and mammalian lipoproteins (from approximately 10 nm to 80 nm in diameter) despite the similarity in structure. Each oil body has a core of triglycerides surrounded by a layer of phospholipids (Fig. 16) [76]. Oil bodies inside the cells of mature seeds or in isolated preparations are remarkably stable and do not aggregate or coalesce. This stability cannot be attained by only a layer of phospholipids. Seed oil bodies are stable because, in addition to the phospholipid layer, a layer of unique proteins, termed *oleosin* shields their surface.

Oleosins from many plant species have molecular weights ranging from 15 to 26 KD [75–76]. The structure of maize oleosin is also shown in Fig. 16 [76]. Each oleosin molecule from various species has a highly conserved central hydrophobic domain of 72 amino acid residues. This domain is likely organized into an antiparallel β-structure and embedded in the hydrophobic acyl moieties of phospholipids and triglyceride matrix. The other two domains, the N and C regions, are much less conserved in their sequence homologies. The N-terminal domain is usually about 50–70 residues and may form an amphipathic α-helix. The C-terminal region may have a similar structure but is more variable in length (55–98 residues). It has been suggested that these terminal domains lie at or on the surface of an oil body, with the positively charged residues orientated toward the lumen of the organelle and the negatively charged residues facing the cytosol. Such charged surfaces and steric hindrance effects may prevent the coalescence of the oil body. In addition to playing this structural role, oleosins (particularly, the C-terminal domain) may also serve as recognition signals on the surface of oil bodies for the binding of newly synthesized lipase during germination [77].

Oleosins are expressed at high levels in seeds and are specifically targeted to oil bodies. Based on these properties, van Rooijen and Moloney attempted to use oleosins as carrier proteins for high-value peptides and proteins [78]. It was demonstrated that an oleosin-β-glucuronidase fusion protein under the control of an *Arabidopsis* oleosin promoter was correctly targeted to the oil body membrane. The same group succeeded in the expression and targeting to oil bodies of other recombinant protein [79]. Since oil bodies are isolated by the simple procedure, the recovery of desired proteins is very easy using an endoprotease and further purified. Alternatively, a fusion protein, which is enzymatically active and resides on the oil bodies, can be used directly in heterogeneous catalysis. Therefore, this oleosin carrier system should be a

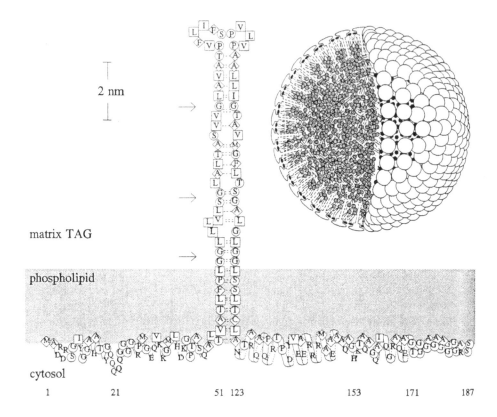

FIG. 16 Model of a maize oleosin on the surface of an oil body. The shaded area represents the phospholipid layer, with head groups facing the cytosol. Amino acid residues are highlighted according to their decreasing hydrophobicity: square, diamond, circle, and no enclosure. The molecular can be divided into four portions: the amphipathic N-terminus (residues 1–50), the central antiparallel β-strands (residues 51–122), the carboxylic α-helix (residues 123–170), and the C-terminal extension (residues 171–187). Arrows indicate where folding of the antiparallel strands could occur. The inset shows a model of an oil body, with a quarter-cut open. Mushrooms, dark spheres attached to two lines, and shaded spheres attached to three lines represent oleosins, phospholipids, and triglycerides, respectively. (From Ref. 76.)

promising method of production and purification of recombinant proteins or peptides that are biologically and pharmaceutically active, such as enzymes, and opioid peptides in plants. The low cost of seed production and the compatibility with existing agricultural processing procedures make it an attractive alternative to conventional bacterial and yeast fermentation systems.

ACKNOWLEDGMENTS

The author is very grateful to Prof. Nakanishi of Okayama University for his permission to use his original data in this manuscript before publication. His research "On the Stability of Immobilized Thermomlysin in Organic Solvents" was partially supported by the Program for Promotion of Basic Research Activities for Innovative Biosciences (PROBRAIN), Japan. The author wishes to thank Prof. Okahata of Tokyo Institute of Technology for the discussion relating to the text.

REFERENCES

1. D. Myers, in Surfaces, Interfaces, and Colloids, VCH, New York, 1991, pp. 7–24.
2. P. F. Fox, in Developments in Dairy Chemistry—4 (P. F. Fox, ed.), Elsevier, London, 1989, pp. 1–53.
3. E. Dickinson, D. S. Horne, V. J. Pinfield, and F. A. M. Leermakers, J. Chem. Soc. Faraday Trans. 93:425 (1997).
4. T. Kull, T. Nylander, F. Tiberg, and N. M. Wahlgren, Langmuir 13:5141 (1997).
5. A. R. Mackie, J. Mingins, and A. N. North, J. Chem. Soc. Faraday Trans. 87:3043 (1991).
6. D. G. Dalgleish, Colloids Surf. 46:141 (1990).
7. P. J. Atkinson, E. Dickinson, D. S. Horne, and R. M. Richardson, J. Chem. Soc. Faraday Trans. 91:2847 (1995).
8. D. G. Dalgleish and J. Leaver, J. Colloid Interface Sci. 141:288 (1991).
9. E. Dickinson, J. Chem. Soc. Faraday Trans. 94:1657 (1998).
10. F. A. Husband, P. J. Wilde, A. R. Mackie, and M. J. Garrood, J. Colloid Interface Sci. 195:77 (1997).
11. A. Eaglesham, T. M. Herrington, and J. Penfold, Colloids Surf. 65:9 (1992).
12. E. Dickinson and Y. Matsumura, Int. J. Biol. Macromol. 13:26 (1991).
13. M. Shimizu, T. Kamiya, and K. Yamauchi, Agric. Biol. Chem. 45:2491 (1981).
14. Y. Chanyongvorakul, Y. Matsumura. A. Sawa, N. Nio, and T. Mori, Food Hydrocoll. 11:449 (1997).
15. M. Shimizu, in Food Macromolecules and Colloids (E. Dickinson and D. Lorient, eds.), The Royal Society of Chemistry, Cambridge, 1995, pp. 34–42.

16. O. P. Ptitsyn, in Protein Folding (T. E. Creighton, ed.), Freeman, New York, 1992, pp. 243–300.
17. B. S. Murray and E. Dickinson, Food Sci. Technol. Int. 2:131 (1996).
18. T. Koseki, N. Kitabatake, and E. Doi, J. Biochem. 103:425 (1988).
19. M. Faergemand and B. S. Murray, J. Agric. Food Chem. 46:884 (1998).
20. J. L. Courthaudon, E. Dickinson, Y. Matsumura, and D. C. Clark, Colloids Surf. 56:293 (1991).
21. E, Dickinson and S. T. Hong, J. Agric. Food Chem. 42:1602 (1994).
22. E. Dickinson and S. Tanai, J. Agric. Food Chem. 40:179 (1992).
23. E. Dickinson, R. K. Owusu, and A. Williams, J. Chem. Soc. Faraday Trans. 89: 865 (1993).
24. M. Coke, P. J. Wilde, E. J. Russell, and D. C. Clark, J. Colloid Interface. Sci. 138:489 (1990).
25. D. C. Clark, P. J. Wilde, and D. R. Wilson, Collids. Surf. 59:209 (1991).
26. J. Chen and E. Dickinson, Colloids Surf. A 101:77 (1995).
27. E. Dickinson and Y. Matsumura, Colloids Surf. B 3:1 (1994).
28. Y. Matsumura, S. Mitsui, E. Dickinson, and T. Mori, Food Hydrocoll. 8:555 (1994).
29. F. K. Winkler and K. Gubernator, in Lipases (P. Woolley and S. B. Peterson, eds.), Cambridge University Press, Cambridge, 1994, pp. 139–158.
30. S. Ransac, F. Carriere, E. Rogalska, and R. Verger, in Molecular Dynamics of Biomembranes (J. A. F. Op den Kamp, ed.), Nato ASI Series, Springer-Verlag, Heidelberg, 1996, pp. 265–304.
31. L. Sarda and P. Desnuelle, Biochim. Biophys. Acta 30:513 (1958).
32. F. K. Winkler, A. D'arcy, and W. Hunziker, Nature 343:771 (1990).
33. H. van Tillbeurgh, L. Sarda, R. Verger, and C. Cambillau, Nature 359:159 (1992).
34. H. van Tillbeurgh, M. P. Egloff, N. Rugani, R. Verger, and C. Cambillau, Nature 362:814 (1993).
35. A. Guidoni, F. Benkouka, J. deCaro, and M. Rovery, Biochim. Biophys. Acta 666:148 (1981).
36. B. Borgstrom and C. Erlanson-Albertsson, in Lipases (B. Borgstrom and H. L. Brockman, eds.), Elsevier, Amsterdam, 1984, pp. 151–184.
37. M. L. Jennens and M. E. Lowe, J. Biol. Chem. 269:25470 (1994).
38. H. van Tilburgh, A. Roussel, J. M. Lalouel, and C, Cambillau, J. Biol. Chem. 269:4626 (1994).
39. W. A. Hide, L. Chan, and W. H. Li, J. Lipid Res. 33:167 (1992).
40. L. Brady, M. Brzozowski, Z. S. Derewenda, E. Dodson, G. Dodoson, S. Tolley, J. P. Turkenburg, L. Chirstiansen, B. Huge-Jensen, L. Norskov, L. Thim, and U. Menge, Nature 343:767 (1990).
41. Y. Gargouri, G. Pieroni, C. Riviere, L. Sarda, and R. Verger, Biochemistry 25: 1733 (1986).
42. Y. Matsumura, N. Matsuo, J. Kimata, K. Matsui, and T. Mori, in Hydrocolloids 2: Fundamentals and Applications of Dispersion Systems in Food, Biology and Medicine (K. Nishinari, ed.), Elsevier, Amsterdam, 1999, pp. 57–62.

43. J. C. Boyington, B. J. Gaffney, and L. M. Amzel, Science 260:1482 (1993).
44. A. M. Kilbanov, Trends in Biochem. Sci. 14:141 (1989).
45. M. Norin, J. Boutelje, E. Homberg, and K. Hult, Apple. Microbiol. Biotechnol. 28:527 (1988).
46. D. G. Hayes and E. Gulari, Biotechnol. Bioeng. 35:793 (1990).
47. A. Matsushita, Y. Kodera, K. Takahashi, Y. Saito, and Y. Inada, Biotechnol. Lett. 8:73 (1986).
48. Y. Okahata and I. Kuniharu, Bull. Chem. Soc. Jpn. 65:2411 (1992).
49. Y. Okahata, Y. Fujimoto, and K. Ijiro, J. Org. Chem. 60:2244 (1995).
50. Y. Okahata, K. Niikura, and K. Ijiro, J. Chem. Soc. Perkin Trans. 1:919 (1995).
51. Y. Okahata and T. Mori, J. Chem. Soc. Perkin Trans. 1:2861 (1996).
52. T. Mori, S. Fujita, and Y. Okahata, Carbohydr. Res. 298:65 (1997).
53. T. Mori and Y. Okahata, Chem. Commun. :2215 (1998).
54. T. Mori, A. Kobayashi, and Y. Okahata, Chem. Lett. : 921 (1998).
55. E. Ruckenstein and X. Wang, Biotechnol. Bioengin. 42:821 (1993).
56. M. Reslow, P. Adlercreutz, and B. Mattiasson, Eur. J. Biochem. 172:573 (1988).
57. M. Miyanaga, T. Tanaka, T. Sakiyama, and K. Nakanishi, Biotechnol Bioeng. 46:631 (1995).
58. K. Nakanishi, M. Kondo, and R. Matsuno, Appl. Microbiol. Biotechnol. 28:229 (1988).
59. M. Miyanaga, M. Ohmori, K. Imamura, T. Sakiyama, and K. Nakanishi, J. Biosci. Bioeng. 87:463 (1999).
60. R. A. Davis and J. E. Vance, in Biochemistry of Lipids, Lipoproteins and Membranes (D. E. Vance and J. E. Vance, eds.), Elsevier, Amsterdam, 1996, pp. 473–494.
61. P. Fielding and C. J. Fielding, in Biochemistry of Lipids, Lipoproteins and Membranes (D. E. Vance and J. E. Vance, eds.), Elsevier, Amsterdam, 1996, pp. 495–516.
62. J. Johansson and T. Curstedt, Eur. J. Biochem. 244:675 (1997).
63. S. Lang and F. Wagner, in Biosurfactants (N. Kosaric, ed.), Marcel Dekker, New York, 1993, pp. 251–268.
64. A. Kakinuma, A. Ouchida, T. Shima, H. Sugino, M. Isono, G. Tamura, and K. Arima, Agric. Biol. Chem. 33:1669 (1969).
65. F. Peypoux, G. Guinand, G. Michel, L. Delcambe, B. C. Das, and E. Ledrer, Biochemistry 17:3992 (1978).
66. K. Ishigami, M. Osman, H. Nakahara, Y. Sano, R. Ishiguro, and M. Matsumoto, Colloids Surf. B 4:341 (1995).
67. K. Arima, A. Kakinuma, and G. Tamura, Biochem. Biophys. Res. Commun. 31: 488 (1968).
68. L. Thimon, F. Pyepoux, and G. Michel, Biotech. Lett. 14:713 (1992).
69. J. D. Sheppard, C. Jumarie, D. G. Cooper, and R. Laprade, Biochim. Biophys. Acta 1064:13 (1991).
70. J. Vater, Progr. Colloid Polym. Sci. 72:12 (1986).

71. Y. Kameda, S. Ouhira, K. Matsui, S. Kanamoto, T. Hase, and T. Atsusaka, Chem. Pharm. Bull. 20:938 (1974).
72. H. Itokawa, T. Miyashita, H. Morita, K. Takeya, T. Hirano, M. Homma, and K. Oka, Chem. Parm. Bull. 42:604 (1994).
73. S. Itoh, H. Honda, F. Tomita, and T. Suzuki, J. Antibiot. 24:855 (1971).
74. M. Morikawa, H. Daido, T. Takao, S. Murata, Y. Simonishi, and T. Imanaka, J. Bacteriol. 175:6459 (1993).
75. J. A. Napier, A. K. Stobart, and P. R. Shewry, Plant Mol. Biol. 31:945 (1996).
76. A. H. C. Huang, Plant Physiol. 110:1055 (1996).
77. A. H. C. Huang, Annu. Rev. Plant Physiol. Mol. Biol. 43:177 (1992).
78. G. J. H. van Rooijen and M. M. Moloney, Biotechnology 13:72 (1995).
79. G. J. H. van Rooijen and M. M. Moloney, Plant Physiol. 109:1353 (1995).

4

Amino Acid Surfactants: Chemistry, Synthesis, and Properties

JIDING XIA Wuxi University of Light Industry, Wuxi, China

IFENDU A. NNANNA Proliant, Inc., Ames, Iowa

KAZUTAMI SAKAMOTO Ajinomoto Company, Inc., Kanagawa, Japan

I. INTRODUCTION

Every living system on earth has universal structures organized as molecular assemblies in aqueous medium, and amphiphilic molecules play a crucial role in sustaining such organized structures. Many of the various types of naturally occurring amphiphilic molecules, that is, natural surfactants, have amino acid–based structures. Most of these natural surfactants have *N*-acylamino acids structures, with fatty acid residue as a hydrophobic moiety connected to the amino group of the amino acid.

Amino acid surfactants (AAS), both natural and synthetic types, have been the subject of many studies, due mostly to their huge potential application in pharmaceutical, cosmetic, household, and food products. The AAS are derived from acidic, basic, or neutral amino acids. Amino acids such as glutamic acid, glycine, alanine, arginine, aspartic acid, leucine, serine, proline, and protein hydrolysates have been used as starting materials to synthesize AAS commercially and experimentally. Methods of preparation include chemical, enzymatic, and chemoenzymic processes, although chemical processes have been prevalent due to their relatively low cost of production. In recent years, more research papers have focused on the use of enzymatic methods to synthesize AAS. It is our opinion that the enzymatic approach would be more attractive to manufacturers in the near future.

This chapter will discuss the state of knowledge of AAS, also known as lipoamino acids, with an emphasis on the types with *N*-acylamino acid structures.

II. BRIEF HISTORY OF AMINO ACID SURFACTANTS

Macfarelane [1] defined lipoamino acid as a single amino acid combined with a lipid, excluding lipopeptide and lipoprotein. It has long been known that there are some lipids that contain amino acid in their structures that are inseparable by physical methods [2–4]; such substances are called lipoamino acids or lipopeptides. These natural lipoamino acids are found to exist widely in animals, plants, and microorganisms. The general structures and origins of these lipoamino acids are shown in Table 1, but the exact structures and functions of these lipoamino acids are yet to be investigated. Various acylamino acids have been synthesized and their properties studied as model molecules of lipoamino acid.

Bondi reported the first research on N-acylamino acid in 1909 [5], in it he synthesized lipoamino acid from glycine and alanine as a form of N-lauroylamino acid. In that research, Bondi speculated that the necrosis of fat in the cell should be attributed to the decomposition of acylamino acid to amino acid and fatty acid. To elucidate this theory, he studied the biodegradation of synthesized acylamino acids. Since then, research into acylamino acids has appeared periodically in the journals, with two aspects of interest: one is the investigation of natural existence, and the other is to explore the functionality of such molecules.

Kameda and Toyoura [6] and Utino et al. [7] studied the natural existence of acylamino acids and found them to be degraded by some sorts of microorganisms. Later, Nagai [8] and Hazama et al. [9,10] conducted a biodegradation study of acylamino acids with enzymes obtained from microorganisms and kidney of animals. These research efforts provided the evidence that acylamino acids are widely existing in living systems, from microorganisms to animals. Fukui and Axelrod [11] indicated the possible natural existence of N-acylamino acids by converting phenylalanine and oleic acid or palmitic acid to corresponding N-acylamino acid with an enzyme from rat liver. Cartwright identified N-(D-3-hydroxydecanoyl)-L-serine, extracted and purified it from a microorganism strain of *Serratia* in 1955, and named it saratamic acid [12,13]. This is the first confirmation of the natural existence of a lipoamino acid.

Since then, many lipoamino acids and lipopeptides have been found in various microorganisms [14–20], plants [21], digestive juices of invertebrates [22,23], chicken oviduct [24], and human skin [25,26]. The lipoamino acids found in microorganisms often contain 3-hydroxy fatty acid, as shown in Table 1. Furthermore, such lipoamino acids (except seratamic acid) are derivatives of lysine and ornithine with acylamino acid ester structure. These lipoamino acids, with the two hydrophobic chains similar to phospholipids, are found localized in the cell membrane, which indicates their role to regulate membrane function.

Lipoamino acids found in the digestive juices of invertebrates have taurine or cysteine, amino acids with a sulfur atom, favorable to be surface active at low pH (4–5). Thus these lipoamino acids are considered to be counterparts of colic acids in vertebrates that emulsify and disperse fatty materials to be digested and absorbed as nutrients. Research on the structure and role of lipoamino acids in vertebrates has yet to be put forward, though some have speculated that lipoamino acids found in human skin might have a role in controlling the barrier function of skin.

As already indicated, research on naturally existing lipoamino acids has not extended to their properties as surface-active molecules or roles and potential applications. On the other hand, many acylamino acids have been synthesized and their properties studied. Hentrich et al. [27] initiated series of investigations and first patented an application of acylsarcosinate and acylaspartate as surfactants for detergent, shampoo, and toothpaste in 1930. Since then, Staudinger and Becer [28] and many other researchers [29–31] have studied the synthesis and physicochemistry of acylamino acids. These studies have confirmed that acylamino acids were potent surfactants with desirable surface activities and safety and mildness features. Takehara et al., led by Yoshida, established an economical process for synthesizing acylglutamate from monosodium glutamate and fatty acid chloride [32], both of which are abundant sources for the production of commercial surfactants.

Takehara et al. also investigated the physicochemical properties of acylamino acid and found them to have superior surface activities at the weakly acidic pH around 5–6, similar to skin pH, even its anionic character [33,34]. Further, they developed various application formulas, such as solid cleansers, shampoo, and creamy or paste-type facial cleansers for acylglutamate [35]. These various research efforts led to the commercialization of acylglutamate in 1972; less than a decade later, a new market was established in Japan for mild cleansers [36]. This is the first commercial application of acylamino acid to have a significant impact on the personal care industry. Since then, many amino acid derivatives, such as acylsarcosinate and acylmethytaurate, have been utilized for personal care products. The applications of acylglutamate have been further extended to almost all types of personal care applications due to the mildness features that the market demands.

TABLE 1 Types of Lipoamino Acids

Class	Structure	Lipid moiety	Amino acid moiety	Location
N-Acylamino acid	CH₂OH | R'CONHCH₂COOH	—3—Hydroxydecanoic acid	Ser	Microorganism (Serratia)
	R'CO(NHCHCO)₂OH R etc.	5-Hydroxydodecanoic acid	Taurine Sarcosyltaurine Hydroxyprolyltaurine Hydroxyprolylcysteic acid	Digestive fluid of invertebrate
	(CH₂)₃NH + R' CONHCH00—	RCHCH₂COOH | O | R″CHCO | OH R: C₁₄H₂₇— R″: C₆H₁₃CH—CH—CH(CH₂)₈— CH₂	Orn Ornityltaurine	Microorganism

N-Acylamino acid ester	$(CH_2)_nNH_2$ $R^1CONHCHCOOR^2$	R^1CO: linear and branched R^2: $—CH_2CH_2OCOR^3$ CH_3 $—CHCH_2OCOR^3$ R^3: branched	$n = 3$: Orn $n = 4$: Lys	Microorganism
Phospholipid with amino acid	CH_2OCOR $CHOCOR$ $=O=$ $CH_2OPOCH_2CHCOOH$ OH NH_2	Phosphatidic acid	Ser	Animal tissue
	CH_2OCOR $CHOCOR$ $=O=$ $CH_2OPO—CH_2$ OH CHOH CH_2OAA	Phosphatidylglycerin	AA: Ala Orn Lys etc.	Anaerobic micro-organism

Source: K. Sakamoto and M. Takehara. *Fragrance. J.* 31:60 (1978).

III. STRUCTURAL SCHEMES AND METHODS FOR THE PREPARATION OF AMINO ACID SURFACTANTS

Amino acids have two functional groups, the carboxylic group and the amino group, which can be converted to surfactant with a reactive molecule bearing a hydrophobic chain. Amino acids with reactive sidechains, such as lysine, arginine, aspartic acid, and glutamic acid, offer opportunities for the molecular design of AAS.

Structurally, amino acid–based surfactants may be depicted as shown nearby. The fundamental structures may be considered as (1) N-acylated amino acid, which is essentially an anionic surfactant, or modified as in (b), where the carboxylic group is converted to ester or amide, as seen in cationic surfactants. The structures in (c) and (d) represent those of amphoteric surfactants, in which both amino and carboxyl groups represent hydrophilic moieties.

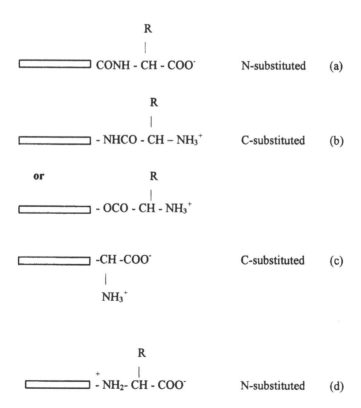

$$
\begin{array}{l}
\overset{\displaystyle R}{\underset{\displaystyle |}{}} \\
\boxed{}\; CONH - CH - COO^{-} \qquad \text{N-substituted} \qquad (a)
\end{array}
$$

$$
\begin{array}{l}
\overset{\displaystyle R}{\underset{\displaystyle |}{}} \\
\boxed{}\; - NHCO - CH - NH_3^{+} \qquad \text{C-substituted} \qquad (b)
\end{array}
$$

$$
\textbf{or} \qquad \begin{array}{l}
\overset{\displaystyle R}{\underset{\displaystyle |}{}} \\
\boxed{}\; - OCO - CH - NH_3^{+}
\end{array}
$$

$$
\boxed{}\; -CH - COO^{-} \qquad \text{C-substituted} \qquad (c)
$$
$$
\phantom{\boxed{xxxx}\;}\;\; |
$$
$$
\phantom{\boxed{xxxx}\;}\; NH_3^{+}
$$

$$
\begin{array}{l}
\phantom{\boxed{xxxx}\; - NH_2-}\overset{\displaystyle R}{\underset{\displaystyle |}{}} \\
\boxed{}\; - \overset{+}{N}H_2- CH - COO^{-} \qquad \text{N-substituted} \qquad (d)
\end{array}
$$

There are various synthetic routes to introduce hydrocarbon long chains into amino acid–based surfactants. For example, a long-chain fatty acyl group is introduced on the amino part of amino acids by using an acid chloride. To obtain amino acid esters or amides, the carbonyl parts of amino acids are reacted with fatty alcohol or amines, respectively. For example, C-alkylation of an amino acid is obtained by the reaction of α-bromo fatty acid with ammonia or by a transmission reaction of the amino part of the amino ester with a stable Schiff base followed by deprotonation with a strong base. This is followed by alkylation with an alkyl halide. N-Alkylation of an amino acid is generally obtained by the reaction of fatty amines with monochloroacetic acid, methyl acrylate, or maleic acid or by the addition of 1,2-epoxy alkane to amino acids.

A schematic of routes for the synthesis of amino acid surfactants is shown in Fig. 1a; representative structures are illustrated in Fig. 1b. The Schotten–Baumann method is a typical and useful way to synthesize *N*-acylamino acid.

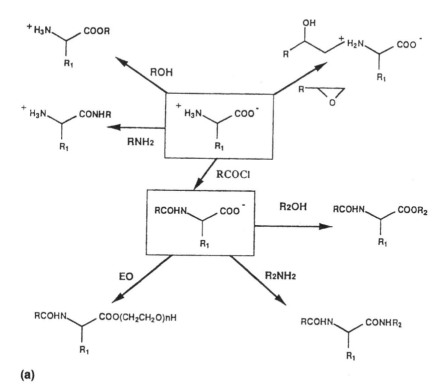

(a)

FIG. 1   (a) Schematic of routes for the synthesis of amino acid surfactants.

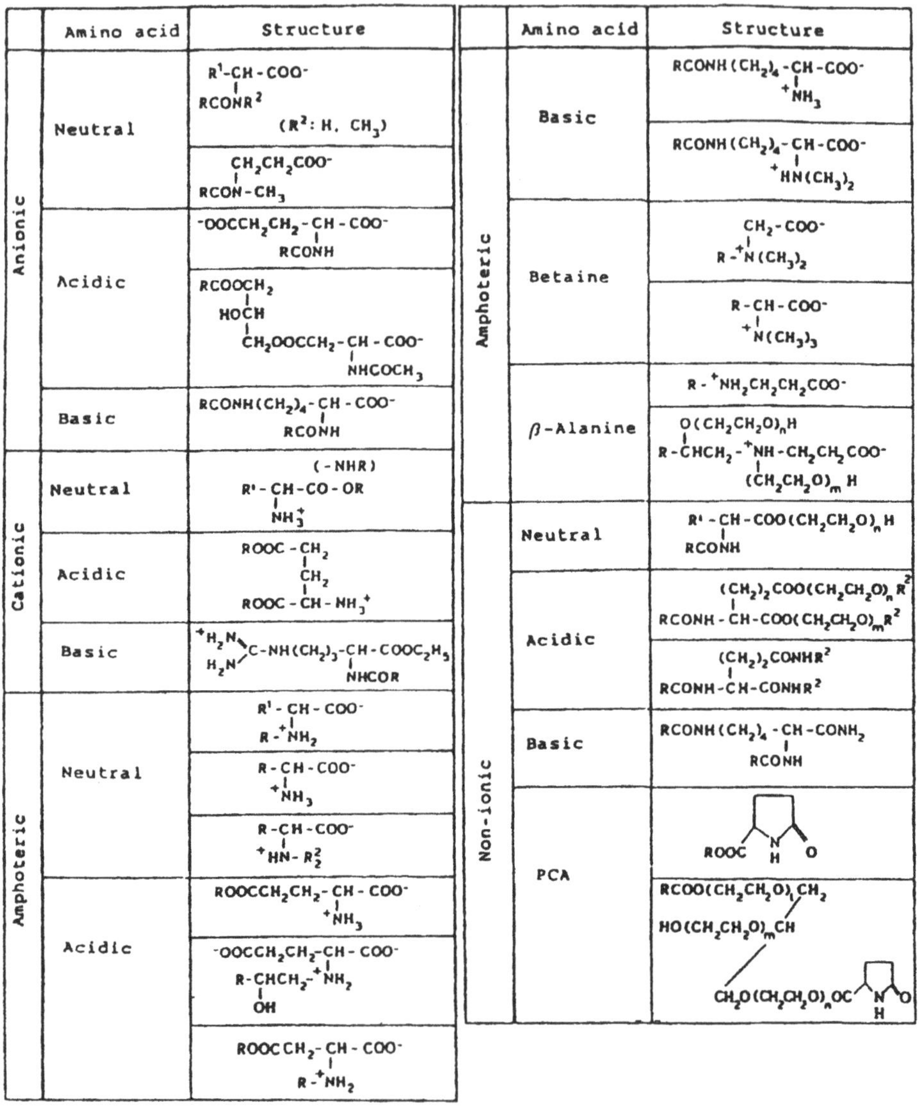

	Amino acid	Structure		Amino acid	Structure
Anionic	Neutral	$R^1-CH-COO^-$ $\mid$ $RCONR^2$ (R^2: H, CH_3) $CH_2CH_2COO^-$ $\mid$ $RCON-CH_3$	**Amphoteric**	Basic	$RCONH(CH_2)_4-CH-COO^-$ $\mid$ $^+NH_3$ $RCONH(CH_2)_4-CH-COO^-$ $\mid$ $^+HN(CH_3)_2$
	Acidic	$^-OOCCH_2CH_2-CH-COO^-$ $\mid$ $RCONH$ $RCOOCH_2$ $\mid$ $HOCH$ $\mid$ $CH_2OOCCH_2-CH-COO^-$ $\mid$ $NHCOCH_3$		Betaine	CH_2-COO^- $\mid$ $R-^+N(CH_3)_2$ $R-CH-COO^-$ $\mid$ $^+N(CH_3)_3$
	Basic	$RCONH(CH_2)_4-CH-COO^-$ $\mid$ $RCONH$		β-Alanine	$R-^+NH_2CH_2CH_2COO^-$ $O(CH_2CH_2O)_nH$ $\mid$ $R-CHCH_2-^+NH-CH_2CH_2COO^-$ $\mid$ $(CH_2CH_2O)_mH$
Cationic	Neutral	$(-NHR)$ $R'-CH-CO-OR$ $\mid$ NH_3^+	**Non-ionic**	Neutral	$R'-CH-COO(CH_2CH_2O)_nH$ $\mid$ $RCONH$
	Acidic	$ROOC-CH_2$ $\mid$ CH_2 $\mid$ $ROOC-CH-NH_3^+$		Acidic	$(CH_2)_2COO(CH_2CH_2O)_nR^2$ $\mid$ $RCONH-CH-COO(CH_2CH_2O)_mR^2$ $(CH_2)_2CONHR^2$ $\mid$ $RCONH-CH-CONHR^2$
	Basic	^+H_2N $\quad\searrow$ $\quad\quad C-NH(CH_2)_3-CH-COOC_2H_5$ $\quad\nearrow\quad\quad\quad\quad\quad\mid$ $H_2N\quad\quad\quad\quad\quad NHCOR$		Basic	$RCONH(CH_2)_4-CH-CONH_2$ $\mid$ $RCONH$
Amphoteric	Neutral	$R^1-CH-COO^-$ $\mid$ $R-^+NH_2$ $R-CH-COO^-$ $\mid$ $^+NH_3$ $R-CH-COO^-$ $\mid$ $^+HN-R^2$		PCA	$ROOC$—(pyrrolidinone ring, N–H, =O) $RCOO(CH_2CH_2O)_xCH_2$ $HO(CH_2CH_2O)_mCH$ $CH_2O(CH_2CH_2O)_nOC$—(pyrrolidinone ring, N–H, =O)
	Acidic	$ROOCCH_2CH_2-CH-COO^-$ $\mid$ $^+NH_3$ $^-OOCCH_2CH_2-CH-COO^-$ $\mid$ $R-CHCH_2-^+NH_2$ $\mid$ OH $ROOCCH_2-CH-COO^-$ $\mid$ $R-^+NH_2$			

(b)

FIG. 1 Continued. (b) Representative structures of amino acid surfactants. R: Long chain alkyl radical; R^1: Amino acid's sidechain; R^2: Alkyl radical, etc.

This method consists of condensing amino acid with fatty acid chloride under higher pH, around 11–12. Neutral amino acids are relatively easy to convert to *N*-acylamino acid in good yield by this method. Additional considerations are required for the amino acids with a reactive hydrophilic group, such as —OH, —SH, —COOH, or —NH$_2$ in sidechain [37]. Acylation of glutamic acid was first reported by Kester [38], and Jungerman explained that the low yield of the Schotten–Baumann method should be attributed to the production of acid anhydride as a by-product [39]. Since then, several procedures have been reported to synthesize *N*-acylglutamic acid efficiently (Fig. 2). Ueda et al. [40] and Weiss [41] used glutamic acid esters for acylation in the organic solvent and then converted to acylglutamate by hydrolysis. Yoshida applied fatty acid and sulfur trioxide to make acylglutamic acid through fatty acid–sulfonic acid mixed anhydride [42]. None of these methods can satisfactorily produce acylglutamic acid commercially. Takehara et al. [34] found an innovative method utilizing a mixture of water-soluble organic solvents such as acetone, ethyl alcohol, and dioxane with water for the Schotten–Baumann acylation to achieve stoichiometric yield. The optimum reaction condition differs, depending on the length of the fatty acid, as shown in Fig. 3. This modified

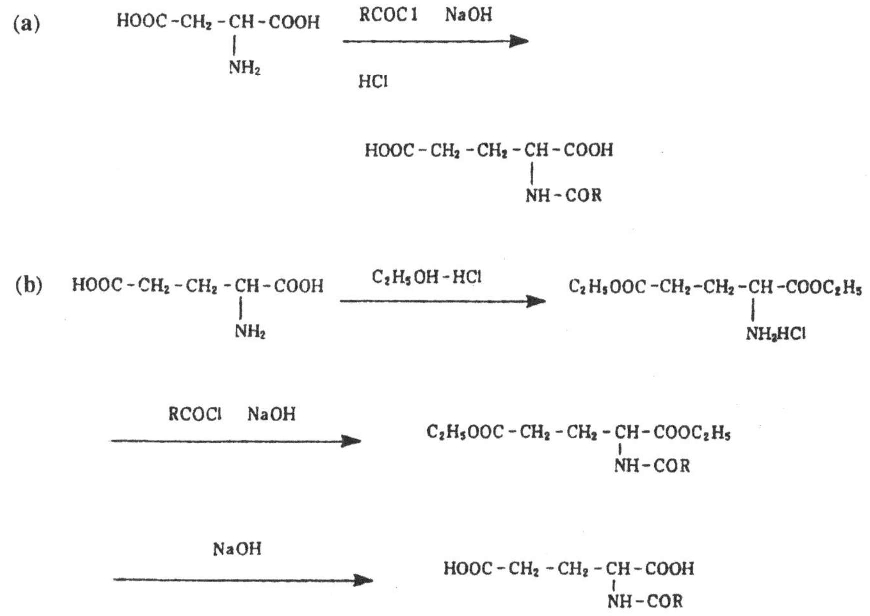

(a) HOOC -CH$_2$ - CH - COOH $\xrightarrow[\text{HCl}]{\text{RCOCl NaOH}}$
  |
  NH$_2$

HOOC - CH$_2$ - CH$_2$ - CH - COOH
  |
  NH - COR

(b) HOOC - CH$_2$ - CH$_2$ - CH - COOH $\xrightarrow{\text{C}_2\text{H}_5\text{OH - HCl}}$ C$_2$H$_5$OOC -CH$_2$-CH$_2$-CH -COOC$_2$H$_5$
  | |
  NH$_2$ NH$_2$HCl

$\xrightarrow{\text{RCOCl NaOH}}$ C$_2$H$_5$OOC - CH$_2$ - CH$_2$ - CH - COOC$_2$H$_5$
                                                          |
                                                          NH - COR

$\xrightarrow{\text{NaOH}}$ HOOC - CH$_2$ - CH$_2$ - CH - COOH
                                                     |
                                                     NH - COR

FIG. 2 Preparation of *N*-acylamino acid.

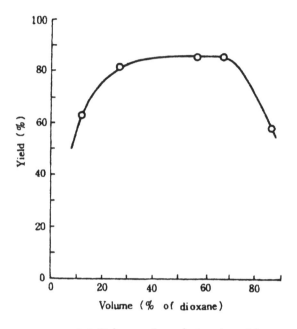

(a) N-lauroyl-L-glutamic acid

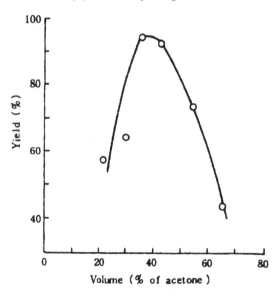

(b) N-palmitoyl-L-glutamic acid

FIG. 3 Preparation of N-acyglutamic acid by an improved Schotten–Baumann method, between the yield of N-acyl-L-glutamic acid and solvent composition. (From Ref. 32.)

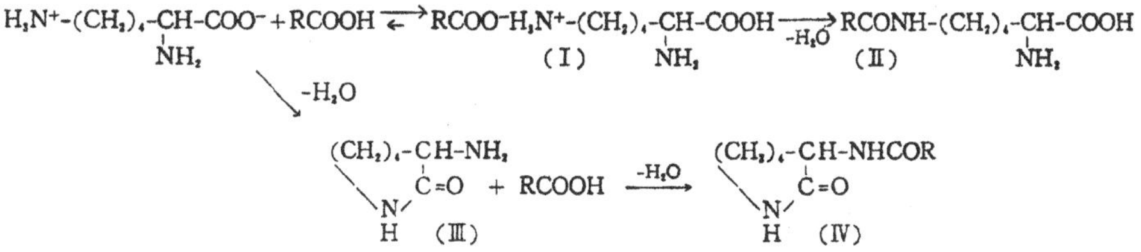

$$H_2N^+-(CH_2)_4-CH-COO^- + RCOOH \rightleftharpoons RCOO^-H_3N^+-(CH_2)_4-CH-COOH \xrightarrow[-H_2O]{} RCONH-(CH_2)_4-CH-COOH$$

NH₂   (I)   NH₂   (II)   NH₂

\\ $-H_2O$

$$(CH_2)_4-CH-NH_2$$
C=O + RCOOH $\xrightarrow[-H_2O]{}$ $(CH_2)_4-CH-NHCOR$
C=O

N / H (III)   N / H (IV)

FIG. 4 Preparation of *N*-acyllysine through fatty acid salt of lysine. Yield for lysine and lauric acid: *N*ᵉ-lauryllysine (II), 63%; *N*-lauroyl-α-amino-ε-caprolactum (IV), 10%. (From Ref. 45.)

Schotten–Baumann method can generally be applied for the acylation of neutral amino acid to improve the reaction yield.

Paquet reported the acylation of methionine, lysine, threonine, and tryptophan as difficult amino acids to get high yield and selectivity via the Schotten–Baumann method, by the utilization of succinimidyl esters of fatty acid [43]. Although he achieved improved results, this method is still not satisfactory enough for commercial-scale production. Acylation of lysine with fatty acid chloride gives mixtures of diacyl, unselective monoacyl for its α and ε amino groups because of the similar reactivity of each amino group to fatty acid chloride [39,44]. To have selective acylation for either the α or the ε amino group in lysine, one of the amino groups must be protected. Sakamoto et al. found a novel method for synthesizing the ε amino group by the thermal dehydration of the fatty acid salt of lysine [45] (Fig. 4). Because of the lower reactivity of fatty acid compared to acid hydrochloride, neutralization occurs selectively with the ε amino group, whose basicity is higher than its α counterpart; i.e., pKa is 10.3 for ε and 8.9 for α (Fig. 5).

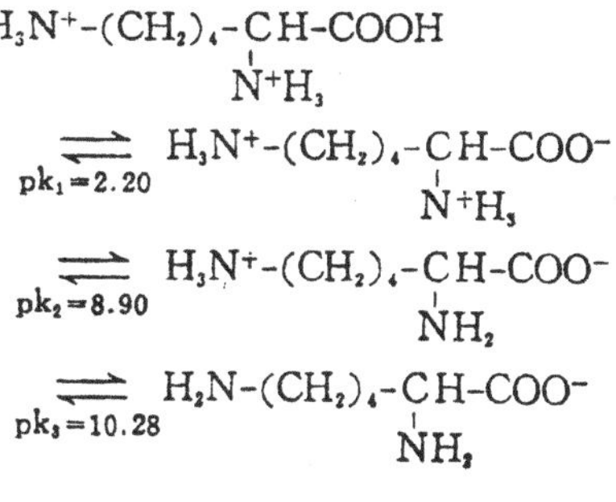

$$H_2N^+-(CH_2)_4-CH-COOH$$
$$N^+H_3$$

$\underset{pk_1=2.20}{\rightleftharpoons}$ $H_3N^+-(CH_2)_4-CH-COO^-$
$$N^+H_3$$

$\underset{pk_2=8.90}{\rightleftharpoons}$ $H_3N^+-(CH_2)_4-CH-COO^-$
$$NH_2$$

$\underset{pk_3=10.28}{\rightleftharpoons}$ $H_2N-(CH_2)_4-CH-COO^-$
$$NH_2$$

FIG. 5 Ionic species of lysine. (From Ref. 45.)

IV. PHYSICOCHEMICAL PROPERTIES OF SELECTED GROUPS OF SURFACTANTS

A. N-Acylamino Acids: General Overview

The history of the physicochemical properties of N-acylamino acid started from the patent by Hentrich et al. [27]. Since then, Staudinger and Becer reported the solubility and viscosity of N-acylsarcosinate [28]; Naudet measured the surface tension and interfacial tension of aqueous solutions of N-acylserinate and N-acylleusinate [46]. Tsubone measured the surface tension and critical micelle concentration (cmc) of various N-acylamino acids and investigated the effect of structural differences of amino acids and length of fatty acid residue on these surface activities [29]. Heitmann [30] further investigated the cmc of N-acyl cysteine, serine, and glycine and indicated that the–SH in cysteine stabilized the micelle structure with its hydrophobic nature. Ooki reported the surface activities of N-acylsarcosinate [47].

Research on the physicochemical properties of N-acylamino acids has been periodic, except for the effort at Ajinimoto Company, lead primarily by Yoshida, Takehara, and Sakamoto. As a result, there are few systematic studies on the structure and property relationship of N-acylamino acids. Sakamoto introduced the hydrophobicity of amino acid as a measure of such structure and activity relationship [48,49]. The hydrophobicity of each amino acid is proposed by Tanford as a free-energy change from ethyl alcohol to water and normalized to Gly, as shown by the Eq. (1) [50] (Table 2).

$$\Delta_{gt} \text{ Ala} = \Delta G_t \text{ Ala} - \Delta G_t \text{ Gly}$$

$$= -3900 - (-4630) = 730 \text{ (cal/mol)} \tag{1}$$

where

ΔG_t Ala = Free-energy change for Ala from ethyl alcohol to water

ΔG_t Gly = Free-energy change for Gly from ethyl alcohol to water

Δg_t Ala = Normalized hydrophobicity of Ala

Figure 6 shows the linear relationship between log cmc of sodium salt of N-lauroylamino acids and their Δg_t values. The following equation, obtained from Fig. 6, shows that the hydrophobicity of the sidechain of amino acid contributes partly to the micelle formation:

$$\log \text{cmc} = 1.1 - 2.0 \times 10^{-4} \, \Delta g_t \tag{2}$$

The surface tension of the sodium salt of N-lauroylamino acids increased with the increase of hydrophobicity of amino acid, as shown in Fig. 7. There

TABLE 2 Free-Energy Change of Each Amino Acid from Ethyl
Alcohol (cal/mol, 25°C)

	ΔG_t, Whole molecule	Δg_t, Sidechain contribution
Nonpolar sidechains		
Glycine	−4630	0
Alanine	−3900	+730
Valine	−2940	+1690
Leucine	−2210	+2420
Isoleucine	−1690	+2970 (20°)
Phenylalanine	−1980	+2650
Proline	−2060	+2600 (19°)
Other sidechains		
Methionine	−3330	+1300
Tyrosine	−930	+2870 (95% EtOH)
Threonine	−4190	+440
Serine	−4590	+40
Asparagine	−4540	−10
Glutamine	−4730	−100
Aspartic acid	−4090	+540
Glutamic acid	−4080	+550
Contribution of a CH$_2$ group		
Ethane	+3020	—
Methane	+2260	—
Ethane-methane	—	+760
Alanine-glycine	—	+730
Leucine-valine	—	+730

Source: Ref. 50.

is a reverse linear correlation between surface tension and cmc, as shown Fig. 8. Figure 8 reveals two fundamental surface phenomena. It shows that amino acids with a large sidechain and a high hydrophobicity contribute effectively to the formation of micelles. It also reveals that the larger sidechain interferes with close packing at the air/water interface and increases the unit surface area, as reflected in the higher surface tension. Takahashi et al. [51] reported supporting data for this speculation.

Other physicochemical properties of *N*-lauroylamino acid that correlate with the hydrophobicity of each amino acid are shown in Fig. 9–12 [48,49].

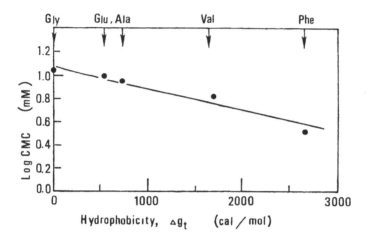

FIG. 6 Relationship between hydrophobicity and log cmc of sodium salt of *N*-lauroyl-L-amino acids: 40°C by electroconductivity method. (From Ref. 49.)

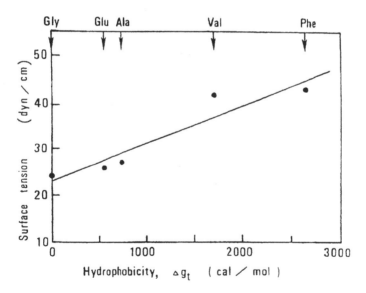

FIG. 7 Relationship between the hydrophobicity and the surface tension of the sodium salt of *N*-lauroyl-L-amino acids: 0.25%, 40°C by Du Nouy method. (From Ref. 49.)

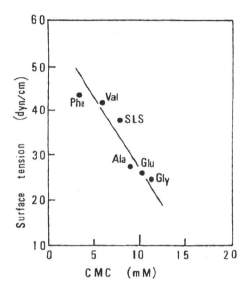

FIG. 8 Relationship between the cmc and the surface tension of the sodium salt of *N*-lauroyl-L-amino acids: 40°C from Fig. 6 and 7. (From Ref. 49.)

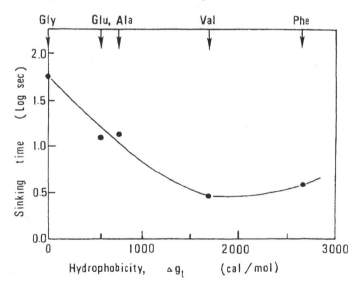

FIG. 9 Relationship between the hydrophobicity and the wetting ability (sinking time) of the sodium salt of *N*-lauroyl-L-amino acids: 0.25%, 40°C by semimicro-disk method. (From Ref. 49.)

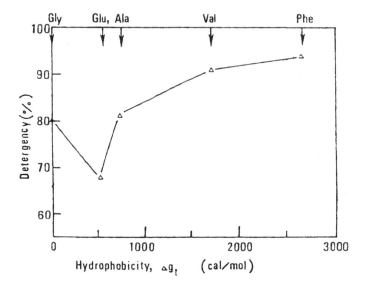

FIG. 10 Relationship between the hydrophobicity and the detergency of the sodium salt of N-lauroyl-L-amino acids: 0.25%, 40°C by Tergo-o-tometer. (From Ref. 49.)

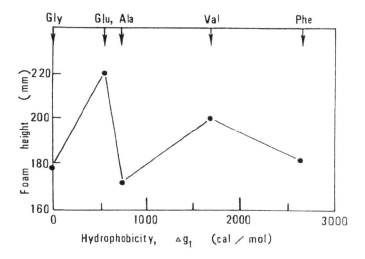

FIG. 11 Relationship between the hydrophobicity and the foaming power of the sodium salt of N-lauroyl-L-amino acids: 0.25%, 40°C by Ross and Miles method. (From Ref. 49.)

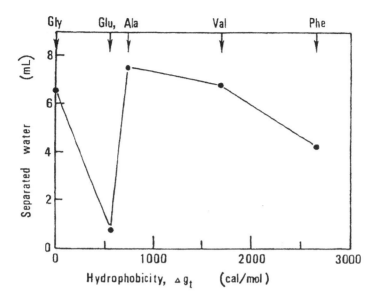

FIG. 12 Relationship between the hydrophobicity and the emulsifying power of the sodium salt of N-lauroyl-L-amino acids: 0.25%, 40°C by test tube method. (From Ref. 49.)

In terms of the neutral amino acid series, wetting ability and detergency increased with the increase in hydrophobicity (Figs. 9 and 10, respectively), whereas the relationship between foaming power and emulsifying power and hydrophobicity was rather erratic (Figs. 11 and 12, respectively). For acidic amino acid, as in N-acylglutamate, its ability to change surface tension depended on the degree of neutralization. Surface tension increased gradually over monoequivalent neutralization [52], which followed a steep increase at 1.6° of neutralization and then reached saturation at 2.0, as shown in Fig. 13. Foaming power reached a maximum at a degree of neutralization of about 1.6, whereas emulsification power decreased with the increase in neutralization degree, as shown in Figs. 14 and 15, respectively. Analysis by ^{13}C-NMR showed that the α-carboxylic group was preferentially neutralized then the γ-carboxylic group followed successively (Fig. 16). By the fluorescence probe method, one can see that the micropolarity and aggregation number increased rapidly when the degree of neutralization exceeded 1.6 (Fig. 17). These results suggest that the intermolecular hydrogen bond between amide hydrogen in the acyl group and carbonyl in the α-carboxylic group contribute to the forma-

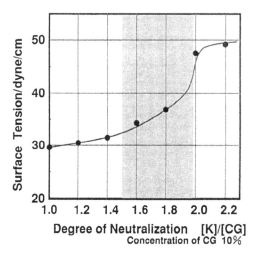

FIG. 13 Relationship between the surface tension and the degree of neutralization of sodium *N*-lauroyl-L-glutamate: 10%, 40°C.

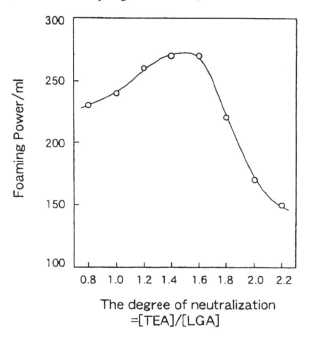

FIG. 14 Relationship between the foaming power and the degree of neutralization of triethanolammonium *N*-lauroyl-L-glutamate: 10%, 40°C.

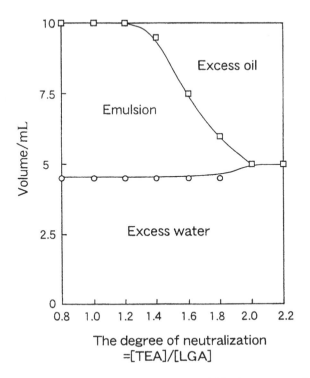

FIG. 15 Relationship between the emulsification power and the degree of neutralization of triethanolammonium *N*-lauroyl-L-glutamate: 10%, 40°C.

tion of closely packed small micelle when the neutralization degree is less than 1.6. Further neutralization beyond 1.6 would lead to the ionization of the α-carboxylic group to carboxylate, thereby changing the micelle structure to a loosely packed large aggregate [53].

N-Acylamino acids with basic amino acids such as lysine and arginine have an amphoteric structure. However, the solubility of the salt of these molecules is too poor to utilize them as a practical surfactant. Several modifications are applied to improve their solubility. Sakamoto et al. [45] reacted N^{ε}-acyllysine with ethylene oxide to introduce the polyoxyethylene group, whereas Sagawa et al. introduced additional the *N*-methyl group. Both methods led to lysine derivatives with surface activities that were pH dependent (Fig. 18 and 19) because of their amphoteric structures [54–57].

Sagawa et al. reported on a N^{ε}-lauroyllysine as a solid surface-active agent [58,59]. Inorganic powders treated with N^{ε}-lauroyllysine became hydrophobic

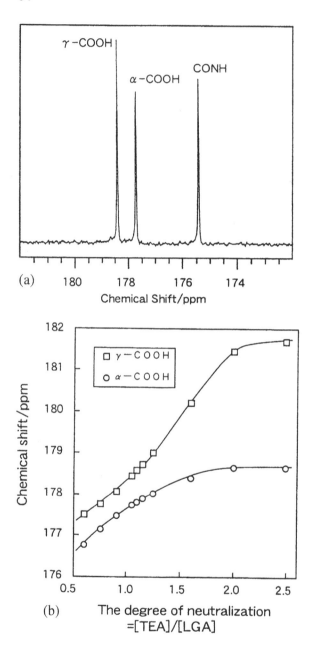

FIG. 16 (a) Chemical shift for each carbonyl group of triethanolammonium *N*-lauroyl-L-glutamate. (b) Relationship between the chemical shift of the carboxylic group and the degree of neutralization of triethanolammonium *N*-lauroyl-L-glutamate.

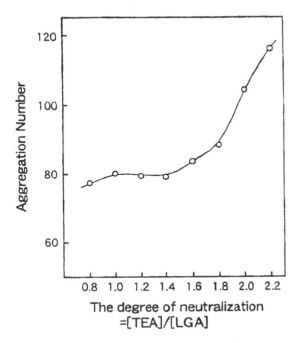

FIG. 17 Change of aggregation number of micelles of triethanolammonium *N*-lauroyl-L-glutamate by the fluorescence probe method.

because of the strong chelating effect of the amino acid residue to coat hydrophilic surfaces with the N^ϵ-acyl group as a hydrophobic chain. The friction coefficient and contact angle are good measures for this surface modification, as shown in Figs. 20 and 21, respectively. Conversion of *N*-acylarginine to esters leads to cationic surfactants, because the guanidyl group at the γ-position of arginine is a strong base. Yoshida et al. reported surface activities (Table 3) and antimicrobial properties (Table 4) of such cationic surfactant derivatives [60]. Infante et al. [61,62] have reported further studies on these types of cationic surfactants.

Xia et al. [63] developed new acyl AAS having the general structure α-amino-(*N*-acyl)-β-alkoxypropionate. The structure–function relationship of the anionic amino-alkoxypropionates was evaluated, with an emphasis on their antimicrobial properties. The anionic amino-alkoxypropionate was synthesized, starting from propionic acid methyl ester, through alkoxylation with fatty alcohol, bromination, saponification, amidation, acylation with R′OCl (acyl chloride), and saponification to obtain the final product. For the detailed procedure, see Xia et al. [63].

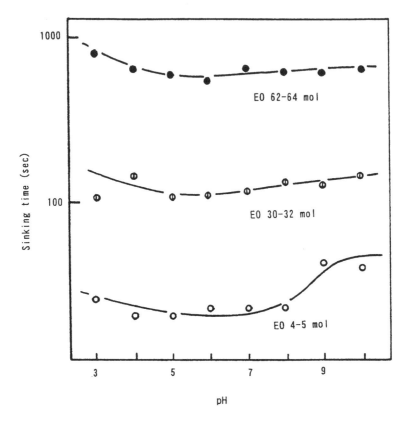

FIG. 18 Relationship between the pH and the surface tension of EO adducts of N^ε-lauroyllysine; 0.25%, 40°C. (From Ref. 48.)

Table 5 shows the surface tension, cmc, and Krafft point of the final product. As expected, the Krafft point of this amino surfactants solution increased with an increase in acyl chain length, and increases in acyl chain length (from C_{10} to C_{14}) resulted in a linear reduction in surface tension. Also, dramatic decreases in cmc occurred as the acyl chain length increased from C_{10} to C_{14}. A strong linear relationship existed between the chain length of the acyl group and log cmc. From the Table 5, it is seen that the addition of a CH_2 group to the MN14 structure to obtain EN14 caused a little change in the cmc, but the Krafft point was reduced by one-third. This implies that EN14 would be more soluble in water at a lower temperature than MN14. The study demonstrated that a simple case such as the removal of CH_2 from the MN14 structure or

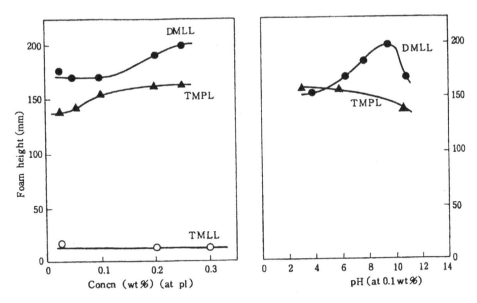

FIG. 19 Relationship between the pH and the foaming power of N^α-di- or trimethyl-N^ε-lauroyllysine; 0.1%, 40°C by Ross and Miles method. DMLL = N,N-dimethyl-N^ε-lauroyllysine, TMLL = N,N,N-trimethyl-N^ε-lauroyllysine, and TMPL = N,N-trimethyl-N^ε-palmitoyllysine. (From Ref. 54.)

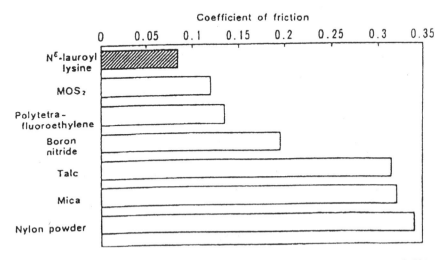

FIG. 20 Friction coefficient of powders treated by N^ε-lauroyllysine. (From Ref. 59.)

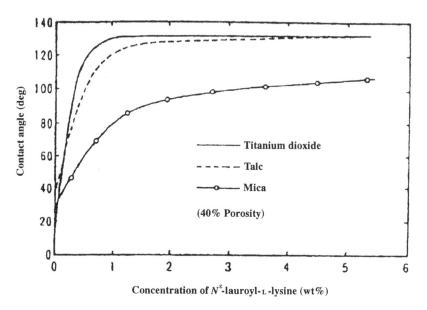

FIG. 21 Changes of contact angles of water on inorganic powders treated with N^ε-lauroyllysine. (From Ref. 59.)

the addition of CH_2 to it altered its surface activity remarkably. Equally important was the finding that changes in hydrophobic acyl chain length caused a greater effect on surface properties than changes in the β-alkoxyl group.

The anionic amino-alkoxypropionates exhibited broader antimicrobial effects than the reference standard, methyl *p*-hydroxybenzoate. The degree of effectiveness of these surfactants was dependent on their acyl chain length and on the microbial species used (Table 6). Mostly, MN14, EN14, and MN14 + $C_{12}E_8$ showed the highest antimicrobial activity against bacteria and fungi. To assess the effect of the β-alkoxy group on antimicrobial activity (i.e., H, CH_3, and C_2H_5; see Table 6), SN14, MN14, and EN14, respectively, were examined. SN14 was found to be less effective than the MN14 and EN124 counterparts. It was found that a substitution of CH_3 by a hydrogen atom could change the antimicrobial activity dramatically [63]. Such a simple substitution increased the MIC values greatly (MN14 vs SN14; see Table 6). Theoretical viewpoints have been provided elsewhere [63] for postulating the molecular basis to explain the differences in antimicrobial performance of MN14, EN14, and SN14.

The antimicrobial efficiency of the anionic amino-alkoxypropionates against the microorganisms was in the following order. *S. aureus* > *A. nigar* =

TABLE 3 Surface-Active Properties of Salts of N-Acyl-L-arginine Esters

Sample[a]	pH	KP (°C)	cmc (mmol/L)	Surface tention (dyne/cm)	Wetting power (s)
DAM[HCl]	4.1	5>	15.6	40.0	14.0
LAM[HCl]	4.1	17.6	6.3	32.1	4.4
MAM[HCl]	4.4	32.4	2.3	33.6	5.6
PAM[HCl]	4.0	43.2	1.0	35.0	12.3
LAE[HCl]	2.7	5>	6.2	30.2	5.0
MAE[HCl]	3.3	34.0	2.4	37.3	3.4
PAE[HCl]	3.2	44.5	1.0	36.4	9.4
LAE[PCA]	5.0	35.6	6.0	39.7	7.3
MAE[PCA]	5.1	49.6	2.4[b]	39.2	7.8
CAE[PCA]	5.0	26.0	—	37.5	8.2
BZC[c]	5.8	5>	4.1	32.3	2.5

[a] Measured in 10 mmol/L at 45°C.
[b] Measured at 50°C.
[c] BZC: Benzethonium chloride.
XAM: N-acyl (X)-L-arginine (A) ester (Y) salts (HCl or PCA).
X: D = Decyl (C10).
 L = Lauryl (C12).
 M = Myristyl (C14).
 P = Palmityl (C16).
 C = Cocoyl
Y: M = Methyl.
 E = Ethyl.
PCA: DL-Pyrrolidon carboxylic acid.
Source: Ref. 60.

S. cerevisiae > *E. coli* > *P. aeruginosa*. They have ~2.6–60-fold activity against the gram-negative and gram-positive bacteria and fungi. It is noteworthy that the incorporation of nonionic surfactant ($C_{12}E_8$; see Table 6) into MN14 not only facilitated more uniform dispersions of water-insoluble compounds, but endowed it with a positive synergistic effect against the most insensitive bacteria, *P. aeruginosa*.

More information on surface activities of N-acylamino acids is in reports by Yahagi and Tsujii [64] and Okabayashi et al. [65]. These authors discussed the self-aggregation of N-acylglycinate through hydrogen bonding. Recently, the structure and surface activities of other N-acylamino acids and N-acyl oligopeptide have been reported [66–73].

Many N-acylamino acids have been developed and utilized as mild and environmentally friendly surfactants. It is expected that further scientific in-

TABLE 4 Minimum Inhibition Concentrations (MICs)
of *N*-Acyl-L-arginine (γ/mL) Esters

Sample	*Ps. aerug.*[a]	*E. coli*[b]	*St. aure*[c]
DAM[HCl]	250	100	50
LAM[HCl]	500<	50	20
MAM[HCl]	500<	100	20
PAM[HCl]	500<	500<	100
LAE[HCl]	250	50	10
MAE[HCl]	500<	100	10
PAE[HCl]	500<	250	50
DAE[PCA]	250	100	20
LAE[PCA]	250	50	10
MAE[PCA]	500<	100	10
CAE[PCA]	250	50	10
LA[d]	1000<	1000<	1000<
Gly C$_{12}^{PCA}$[e]	500<	100	20
BZC	250	50	5>

[a] *Ps. aerug.*: *Ps. aeruginosa* ATCC 10145.
[b] *E. coli*: *E. coli* ATCC 10798.
[c] *St. aure.*: *St. aureus* ATCC 6538-P.
[d] LA: Lauroyl-L-arginine.
[e] Gly C$_{12}^{PCA}$: DL-Pyrrolidone carboxylic acid salt of lauryl glycinate.
Source: Ref. 60.

vestigations with advanced measurement instruments would lead to a better understanding of structure and activity relationships and to optimum development and applications of these surfactants.

The following discusses the properties of selected *N*- acylamino acids in greater detail.

1. Alanine-Based Surfactants

Alanine is a nonpolar neutral amino acid with one amino group and one carboxyl group, that is, α-amino propionic acid.

$$Ch_3-CH-COOH$$
$$|$$
$$NH_2$$

Alanine-type surfactants are well known as useful and mild anionic surfactants. It has been reported that N-lauroyl-β-alanines (LBA) have a low potential for inducing epidermal cell inflammation. They permeate skin much less than soap, sodium dodecyl sulfate, sodium cocoyl isothionate, acylmethyl tau-

TABLE 5 Surface Properties of Acyl Amino Acid Surfactants in Aqueous Solution

Compound	Surface tension, γ (mN/m) 0.1%, 25°C	cmc (mM)	Krafft point (°C)
MN10	43.0	17.90	6
MN12	40.3	5.19	22
MN14	36.6	1.49	31
MN16	36.0	0.43	42
SN14	47.2	4.17	17
EN14	36.0	1.20	21

MN10: $CH_3OCH_2CH(NHCOC_9H_{19})COONa$
MN12: $CH_3OCH_2CH(NHCOC_{11}H_{23})COONa$
MN14: $CH_3OCH_2CH(NHCOC_{13}H_{27})COONa$
MN16: $CH_3OCH_2CH(NHCOC_{15}H_{31})COONa$
SN14: $HOCH_2CH(NHCOC_{13}H_{27})COONa$
EN14: $C_2H_5OCH_2CH(NHCOC_{13}H_{27})COONa$
Source: Ref. 63.

TABLE 6 Minimum Inhibitory Concentration (MIC) of Amino Acid Surfactants for Different Microorganisms

	MIC (μg/ml) for microorganisms				
	Gram (+)	Gram (−)		Fungi	
	SA	EC	PA	AN	SC
Methyl Paraben	3200	>3200	>3200	800	800
MN10	800	200	3200	1600	1600
MN12	400	400	3200	800	800
MN14	50	400	1600	100	200
MN16	800	800	>800	200	200
SN14	>800	>800	>800	1600	1600
EN14	50	200	1600	200	200
MN14 + $C_{12}E_8$	50	400	800	50	400

Chemical formula of compounds are presented in footnote of Table 5. SA, *S. aureus*; EC, *E. coli*; PA, *P. aeruginosa*; AN, *A. Niger*; SC, *S. cerevisiae*; Methyl Paraben, Methyl *p*-hydroxybenzoate. *Source*: Ref. 63.

rine, or monoalkyl phosphates. For these advantages the surfactants of alanine type have been applied in shampoos, cosmetics, emulsion paints, and textiles and as a corrosion inhibitor and industrial cleaners, among many other applications. They are less toxic to higher animals and less irritating to human skin. Takai et al. [74] reported on a series of amphoteric surfactants, sodium salts of N-(2 hydroxyethyl)-N-(2-hydroxyalkyl)-β-alanines (Na-HAA), prepared by adding methylacrylate to N-(2-hydroxyalkyl)-ethanolamine followed by saponification and neutralization with HCl to obtain N-(2-hydroxyethyl)-N-(2-hydroxyalkyl)-β-alanines; the addition of ethylene oxide to Na-HAA leads to oxyethylated derivatives. These series of compounds were synthesized according to the following scheme:

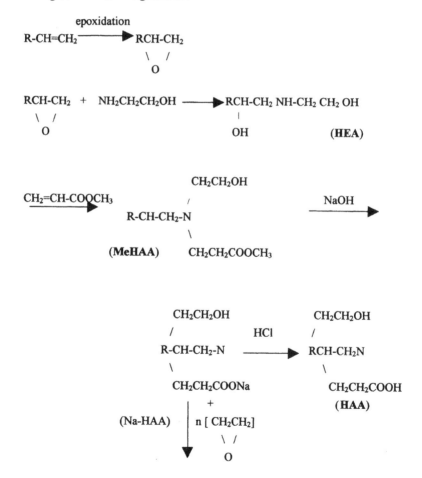

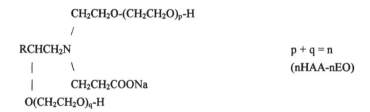

$$p + q = n$$

(nHAA-nEO)

The α-olefin was epoxidized with peracetic acid in benzene in the presence of sulfuric acid as a catalyst. A pure 1,2-epoxyalkane was obtained by repeated vacuum distillation. The monoethanolamine (9 moles) was reacted with 1,2 epoxyalkane (1 mole) dropwise for 1.5 hours. After removal of excess monoethanolamine under reduced pressure, the residue of reaction product was subjected to vacuum distillation in an atmosphere of nitrogen to give HEA in 70–80% yields based upon 1,2-epoxyalkane.

For preparation of N-(2-hydroxyethyl)-N-(2-hydroxyalkyl)-β-alanine (HAA) and its Na salts (Na-HAA), the HEA (1 mole), obtained as earlier, and methyl acrylate (1.2 moles) purified by the usual method were then subjected to the Michael addition at 70–80°C for 4 hours. After completion of the reaction and removal of excess methyl acrylate, the product was saponified with 20% aqueous NaOH solution for 2 hours at 85–95°C, and then water and methanol were removed by heating at reduced pressure. The crude Na-HAA thus obtained was dissolved in methanol containing a little water at 50–60°C, and to the solution HCl-methanol (conc. HCl/methanol = 1:1 v/v) was added gradually until the pH of the solution was brought to 6.0–6.5. The salt precipitate was filtered off and the filtrate concentrated and dried under reduced pressure; the dried material was dissolved in anhydrous ethanol at 30–40°C. The insoluble salt was again filtered off, and the filtrate was allowed to cool to room temperature for some time to precipitate white crystals of crude HAA. Triple recrystallization from anhydrous ethanol gave HAA of high purity.

The calculated amount of ethylene oxide (EO) was added directly to each Na-HAA in the absence of a catalyst at a temperature higher than the melting point of the substrate, 130–140°C, and at atmosphere pressure. Finally the ethylene oxide adduct of Na-HAA (Na-HAA-nEO) was obtained.

The calcium ion stability for Na-HAA is shown in Table 7. The order of calcium ion stability was: $C_{16} > C_{14} > C_{12} > C_{18}$. C_{12}-Na-HAA using N-dodecyl-β-alanine as a reference standard. There were no remarkable differences in the Krafft point of Na-HAA homologous series, as can be seen in Table 7.

TABLE 7 Calcium Ion Stability and Krafft Point of Na-HAA

Compound	Ca stability (ppm, CaCo$_3$)	Krafft point (°C)
C$_{12}$-Na-HAA	35	40.6
C$_{14}$-Na-HAA	60	41.1
C$_{16}$-Na-HAA	78	42.6
C$_{18}$-Na-HAA	15	44.8
C$_{12}$-β-alanine[a]	20	—

[a] Commercial grade of N-dodecyl-β-alanine.
Source: Ref. 74.

Enzymatic synthesis of N-acylalanine is possible and has been demonstrated recently. *Candida antarctica* lipase (CAL) has been widely used for the amidation of esters and amines. Izumi and coworkers [75] prepared a N-lauroyl-β-alanine homologous series in organic media, where the enzymatic amidation of 3-amino-propionitrile and β-alanine ethyl ester reacted with methyl caproate or methyl laurate by immobilized *C. antarctica* lipase as follows:

$$R_1R_2NH + R_3COOMe \xrightarrow{\textit{Candida Lipase}} R_1R_2NCOR_3$$
$$\text{(1a--d)} \qquad \text{(2a,b)} \qquad\qquad \text{(3a--d)}$$

1a: R_1 = H; R_2 = —CH$_2$CH$_2$CN
1b: R_1 = H; R_2 = —CH$_2$CH$_2$COOEt
1c: R_1 = H; R_2 = —CH(COOEt)CH$_2$CH$_2$COOEt
1d: R_1 = Me; R_2 = —CH$_2$COOEt
2a: R_3 = —(CH$_2$)$_4$CH$_3$
2b: R_3 = —(CH$_2$)$_{10}$CH$_3$
3a: R_1 = H; R_2 = —CH$_2$CH$_2$CN; R_3 = —(CH$_2$)$_4$ CH$_3$
3b: R_1 = H; R_2 = —CH$_2$CH$_2$CN; R_3 = —(CH$_2$)$_{10}$ CH$_3$
3c: R_1 = H; R_2 = —CH$_2$CH$_2$COOEt; R_3 = —(CH$_2$)$_{10}$ CH$_3$
3d: R_1 = H; R_2 = —CH$_2$CH$_2$COOH; R_3 = —(CH$_2$)$_{10}$ CH$_3$

Izumi et al. [75] found that the rate of amidation with the longer-chain ester (2b) was about two times faster than that of the shorter-chain ester (2a). The CAL-catalyzed reaction of equimolar amounts of 1b and 2b in various organic solvents resulted in the formation of N-lauroyl-β-alanine ethyl ester (3c). Generally, a CAL-catalyzed reaction of equimolar amounts of substrate and ester in di-isopropyl ether gave the best yield (99.3%) after a 24-hour incubation at 25°C. When 3-amino-propionitrile was used as substrate, diisopropyl ether was a suitable solvent. In contrast, when β-alanine ethyl ester hydrochloride was the substrate, the diisopropyl ether was unsuited as a solvent, due to the low solubility of the substrate in this solvent. In this case, the best yield (82.0%) was attained using dioxane as the solvent.

A process was recently patented [76] for the economical preparation of N-mixed saturated fatty acid -alanine, -glycine or -β-alanine in high yield. Briefly, the process involved the reaction of mixed saturated fatty acid (6- to 18-carbon) halide with an aqueous solution of alanine, glycine, or β-alanine in the presence of an alkali. A strong acid is then added to obtain an aqueous solution of the N-mixed saturated fatty acid acyl neutral amino acid salt. The solution is further separated into an organic and aqueous layers at ~60°C. The organic layer is mixed with aqueous triethanolamine to obtain an aqueous solution of mixed triethanolamine-N-acyl-alanine salt with a product purity of 92%.

2. Lysine-Type Surfactants

Lysine is a basic amino acid with two amino groups (positively charged at pH 6–7) and one carboxyl group with a chemical name α,ε-diamino caproic acid.

$$CH_2-(CH_2)_3-CH-COOH$$
$$\quad |\qquad\qquad\quad |$$
$$NH_2\qquad\quad NH_2$$

Industrial-scale synthesis of N^ε-acyllysine involves only a few chemical synthesis steps. N'-Lauroyl-L-lysine (Amihope LL) has been produced in white crystalline powder with specific gravity 1.2 and mean particle size 10–20 μm, having good antioxidant properties and high lubricity/low frictional coefficient (Ajimomoto Co. report, 1994–96). The chemical process takes advantage of the fact that the two substrates of the reaction are molecules with a common solvent. Lysine is soluble only in media containing high proportions of water. However, it is important to note that when a molecule contains two functional amino groups, such as lysine, acylation of one of them alone is almost impossible without previously protecting the other using a very meticulous protection/deprotection process.

It is possible to synthesize N^ε-acyllysine enzymatically. This has been demonstrated using a commercially available fixed lipase (Lipozyme, NOVO) bonded onto an anionic macroporous resin. Synthesis was performed by causing a triglyceride to react with the lysine, either with or without a solvent, as described by Montet et al. [77]. When a solvent was used, diethylether, diisopropylether, dibutylether, and methyl tertiobutylether or hexane was found suitable. The reaction mixture was composed of free lysine 292 mg, soybean oil 282 mg, Lipozyme 50 mg, and solvent 3 ml. At 65°C, the yield of the reaction seemed to increase when the CHO ratio of the ether used decreased or when the oxygen content of the ether increased. The maximum yield obtained after 24 h of reaction was approximately 35% at 90°C, 15% at 65°C, 0% at 30°C. For synthesis without a solvent, the reactants must be brought into contact with each other and with the lipase, since the two substrates are not absolutely mutually miscible. The quantities of substrates oil/lysine molar ratio is doubled. The yields obtained without solvent are shown in Table 8. The reaction without a solvent gives better yields in general than with a solvent. For the reaction to proceed optimally, the concentration of fatty acids must remain low. In fact, triglycerides may be used as a feedstock to deliver fatty acid as needed. Yields were reported to be low in the presence of oleic acid and high in the presence of triglycerides [77]. A yield of 4% was obtained in the presence of oleic acid alone. Perhaps free fatty acid is somewhat inhibitory due to ionic fixation of the fatty acid onto the Lipozyme's macroporous anion exchange resin, thereby creating a lipophilic macrostruc-

TABLE 8   Yields (%) Obtained Without Solvent

Condition	70°C	90°C
TG/lysine = 2	6.0	7.0
Without catalyst	0	0
Resin only	0	0
Under reduced pressure	18.0	5.0
1 atm with N_2	25.6	36.6
1 atm without N_2	24.0	29.2
2 days	39.4	Not titrated
4 days	39.6	60.7
7 days	Not titrated	73.0
Adding 90 μL of water	Not titrated	16.0

Source: Ref. 77.

ture that prevents the lysine from gaining access to the active site of the lipase. The structure of the product N^ε-oleyllysine was confirmed by the examination of the mass, proton NMR, and microanalysis spectra. The melting points of N^ε-lauryllysine and N^ε-oleyllysine were 263°C and 229°C, respectively, which are reflective of the physical property of the respective fatty acids. It is conceivable that this enzymatic synthesis of N^ε-acyllysine could be scaled up for industrial development.

Homologous series of long-chain N^α-acyl-L-lysine having antimicrobial activity and surface-active properties have been reported [78,62]. The straight-chain length of the fatty acid residue (x), the type of R substituents, or the number of carbons (n) between the ω-basic groups (amine, guanidine) and the asymmetric α-carbon may influence the physicochemical and antimicrobial properties:

$$CH_3-(CH_2)_x-CO-NH-CH-(CH_2)_n - B^+ \cdot Cl^-$$
$$|$$
$$COOR$$

$$B^+: \quad -NH-C=NH \quad \text{or} \quad -NH_3^+$$
$$\qquad\quad |$$
$$\qquad\; NH_3^+$$

Structural modifications are possible to improve these properties, although only those molecules with the blocked carbonyl groups (cationic molecules, where R = Me, Et) showed activity. When the carboxyl group was unprotected (i.e., R = Na$^+$), the resulting amphoteric molecules were inactive. The synthesis of the cationic surfactants type $N^\varepsilon N^\varepsilon N^\varepsilon$-trimethyl N^ε-lauroyl-L-lysine methyl (LLM$_q$), or ethyl (LLE$_q$) esters, as iodide salt, was obtained from their corresponding esters of N^α-lauroyl-L-lysine by the quaternization of the N^ε amine terminal group using the Chen's and Benoiton's method. The sodium salt of N^α-lauroyl-$N^\varepsilon N^\varepsilon N^\varepsilon$-trimethyl-L-lysine iodide (LLS$_q$) was prepared by saponification of LLM$_q$ in a mixture containing MeOH and NaOH for 15 min. The reaction mixture was neutralized by 4 N HCl and then evaporated to dryness and stored in chloroform. Crystallization from chloroform/ether yielded a white solid that corresponded to LLS$_q$.

For the cationic compounds, the water solubility decreased in the following order: N^α-lauroyl-L-lysine esters > N^α-lauroyl-L-arginine esters > N$^\alpha$ lauroyl-$N^\varepsilon N^\varepsilon N^\varepsilon$-trimethyl-L-lysine esters. In contrast, for the amphoteric compounds, water solubility decreased in the following order: sodium salt of N^α-lauroyl-$N^\varepsilon N^\varepsilon N^\varepsilon$-trimethyl-L-lysine > sodium salt of N^α-lauroyl-L-lysine > sodium salt of N^α-lauroyl-L-arginine. It is noteworthy that the amphoteric derivatives of lysine and arginine analogues did not form micelles (i.e., they do not behave as surfactants). However, it is interesting that the sodium salt of the N^α-lauroyl-$N^\varepsilon N^\varepsilon N^\varepsilon$-trimethyl-L-lysine conformed in a way that allowed it to be surface active in solution.

The foaming-power evaluation showed that the higher the hydrophobicity of the radical ester group, the better the foaming power for the lysine or arginine analogue. The foaming power and stability tended to increase when the quaternized derivative was present in a saponified form.

The antimicrobial activity is inhibited when the carboxyl group of the amino acid is unprotected; the amphoteric derivatives LLS are inactive below 128 μm mL, even though the molecule has surface-active properties. All the cationic derivatives (LLM$_q$, LLE$_q$, LAM, and LLE) showed antimicrobial activity against gram-positive and gram-negative microorganisms. The effectiveness of their antimicrobial activity decreased when the cationic molecule contained tetra-alkyl ammonium function as the polar head instead of the amine or guanidine counterparts.

3. Glutamate-Type Surfactants

The glutamic acid is an acidic amino acid with one amino group and two carboxyl groups carrying a negative charge at pH 6–7:

$$HOOC-CH_2-CH_2-\underset{\underset{NH_2}{|}}{CH}-COOH$$

The salts of long-chain N-acylglutamic acids are widely used as good surfactants. They are mild on the skin, with less irritation than the other synthetic surfactants, such as sodium lauryl sulfate and alkyl benzene sulfonate. Takehara et al. [33,34] investigated the physicochemical properties of N-acylglutamate extensively and found that monoequivalently neutralized N-acylglutamate shows weakly acidic pH around 5–6, in spite of its anionic character. Surface activities of N-acylglutamate, such as surface tension and foaming power, vary with the changes of neutralization degree and the chain length of the acyl group [32,33,79–81]. Monosodium N-lauroyl-glutamate shows very good surface tension reduction, foaming power, wetting power, and emulsifying power [33]. Disodium N-palmitoyl or stearoylglutamate was superior in dispersing power to the power of carbon black. Its monosodium salts (acylglutamates, AGS) are less soluble in water, with the exception of monosodium

N-oleoylglutamate, whereas the disodium salts (AGS$_2$) are highly soluble. The AGS generally showed weak acidity in an aqueous solution, the pH of which was almost equal to that of the human skin.

Takehara et al. [34] reported on the preparation and physicochemical of triethanolamine N-acylglutamates (AGTn). AGTn were more water soluble and exhibited good surface activities and less irritation on the skin compared with their corresponding sodium salts. They are often used as an additive in liquid detergents and shampoo. AGTn have shown strong stability to calcium ions and could be used with water having a hardness of 200–300 ppm, calculated as CaCO$_3$ [34].

Recently, several acylglutamates (Amisoft), sodium salt of DL-pyrrolidone carboxylic acid derived from L-glutamic acid (Ajidew), lauryl phytosteryl glutamate (Eldew PS-203), lauryl cholesterol (or 2-octyldodecanol and behenol) (Edeu CL-301) have been developed by Ajinomoto Company (Ajinomoto Co.'s report, 1994–97).

4. Leucine-Type Surfactants

Leucine is a nonpolar neutral amino acid with one each of amino and carboxyl groups; it is also known as α-amino isohexanoic acid:

$$(CH_3)_2CH-CH_2-\underset{\underset{NH_2}{|}}{CH}-COOH$$

Work in this area has been primarily in the use of enzymatic process to improve the surface properties of food proteins. Arai and Watanabe [82 demonstrated the application of a papain-catalyzed one-step process for L-norleucine n-dodecyl ester attachment to succinylate α-s_1-casein to yield a surface-active 20-kDa product. The product had increased emulsifying activity as compared to α-s_1-casein or succinylated α-s_1-casein. More information on the one-step process is found in Chapter 1 in the "Enzyme Modification" section. Watanabe and Arai [83] studied the surface properties of gelatin that had been modified with alkyl esters of L-leucine to increase its amphiphilicity [84]. The modified gelatin was tested for whippability, foam stability, and emulsifying activity [84]. Whippability was greatest for the leucine octyl ester-modified gelatin, whereas incorporation of leucine hexyl esters yielded a modified gelatin with the best foam stabilization property [84]. Improved emulsifying activity correlated well with increasing chain length of the alkyl moiety [84]. The lipophilized gelatin also performed well as an emulsifier in food systems, such as whipped desserts, ice cream, and breads [84].

B.   *C*-Alkylamino Acids and *N*-Alkylamino Acids

C-Alkyl-α-amino acids and *N*-alkylamino acids are derivatives of glycine and have amphoteric structures. Similar to N^ϵ-acyllysine, these molecules make internal salts that reduce the solubility. *N*-Alkyl substitution improves solubility. Table 9 lists the cmc of *N*-alkylamino acids and *C*-alkylglycine with *N*-methyl or hydroxyethyl substitution investigated by Swarbrick and Daluwara [85], Beckett and Woodward [86], Tori and Nakagawa [87], Idem [88–90], and Hidaka [91,92]. As an amphoteric surfactant in which both ionic groups, that is, amino and carboxylic groups, are internally neutralized, these molecules do not have ionic repulsion at the micelle surface and the cmc changes with alkyl chain length similar to nonionic surfactant. Hidaka reported surface activities of ethylene oxide adducts of *N*-(2-hydroxyalkyl)amino acids and found differences in the cmc between optical active and racemic isomers [93], as found in *N*-acylamino acids by Takehara et al. [94].

Other substituted glycine-type amphoteric surfactants containing long-chain alkoxy and methylated amino groups such as *N*-(2-alkoxyethyl) methyl-aminoacetic acids, (*N*-[2-alkoxyethyl]-*N*-[carboxymethyl] dimethylammonium) chloride, or *N*-(*N*-[2-alkoxyethyl]-2-aminoethyl) aminoacetic acids have been synthesized and studied by Abe et al. [95]. They evaluated their surface activities and antimicrobial effect against gram-positive and gram-negative bacilli and some fungi.

The melting point of the hydrochloric acid salts of 2-alkoxyethyl aminoacetic acid decreased in the order C_{12}, C_{13}, C_{11} in the same series [95]. Addition of an aminoethyl group increased the melting point of the compounds. The introduction of a methyl group at the N of the long-chain alkoxy-substituted glycines lowered their melting points.

As for the surface activities, breaking points were obtained at ca. 10^{-3}–10^{-4} mole/L. The presence of an ethylamino group between the alkoxyethyl and ethyl groups in the alkoxyethyl aminoacetic acid influenced surface activity greatly. Moreover, an aqueous solution of these substituted glycine-type amphoteric compounds showed better surface tension reduction potential at pH 4.0 and 10.0 than at neutral pH [95]. The amphoterics were found to be more effective against gram-positive cocci than gram-negative bacilli and fungi, and the introduction of a methyl group further increased the antimicrobial activities against each microbe [96] (Table 10). The introduction of an aminoethyl radical between the alkoxyethyl and amino radicals of the substituted glycine increased their antimicrobial activities.

TABLE 9 Critical Micelle Concentration (cmc) of N-Alkylamino Acids and C-Alkylamino Acids

Structure	Carbon number of alkyl (R), acyl (R'CO)								Method
	8	10	11	12	13	14	15	16	
CH_2COO^- | $R\text{—}^+N(CH_3)_2$	217 (21)[a]	23.27 (25)[b]	7.23 (25)[b]		0.795 (25)[b]		0.056 (25)[b]		L. S.
	170 (23)[c]	18 (23)[c]	6.6 (23)[c]	1.8 (23)[c]		0.18 (23)[c]		0.02 (23)[c]	S. T.
$R\text{—}CHCOO^-$ | $^+N(CH_3)_3$	250 (27)[a] 120 (21)[a,d]	14.4 (21)[d]		1.58 (21)[d]					L. S.
	97 (27)[a,e]	13.1 (27)[e]		1.32 (27)[e]					S. T.
$R'CONH(CH_2)_4CHCOOH$[f] | $N(CH_2)_2$				0.7 (40)				8×10^{-3} (60)	S. T.
$R'CONH(CH_2)_4CHCOO^-$[f] | $^+N(CH_3)_3$				0.89 (40)				0.15×10^{-3} (40)	S. T.
$R\text{—}CHCOOH$ | $NHCH_2CH_2OH$ (pH10)[g]	14.0 (25)	1.2 (25)		0.13 (25)					S. T.
$R\text{—}CHCOOH$ | $N(CH_2CH_2OH)_2$ (pH10)[g]	8.00 (25)	2.20 (25) 0.7 (25) (pH 6)		0.50 (25)		0.18 (25)		0.06 (25)	S. T.
$R\text{—}CHCOOH$ | $NH(CH_2CH_2O)_2H$ (pH10)[h]	25 (30)			5 (30)		0.74 (30)			S. T.

L. S.: light scattering method, S. T.: surface tension method; number in parens ( ): Temperature; Refs.: a: [81], b: [79], c: [80], d: [83], e: [84], f: [58], g: [85], h: [86].
Source: Ref. 85.

TABLE 10 MIC of Glycine-Type Amphoteric Surfactants Against
Microorganisms (μg/mL)

	Compound VIII		Compound XI	
Microorganism	C_{11}	C_{12}	C_{11}	C_{13}
S. aureus 209 P	100	25	12.5	12.5
S. aureus Terashima	100	25	25	12.5
S. epidermidis 10131	100	25	25	12.5
S. faecalis	50	12.5	6.25	12.5
B. subtilis PCI-219	50	12.5	6.25	12.5
E. coli NIHJ	>100	>100	>100	>100
P. aeruginosa A_3	>100	>100	>100	>100
A. niger ATOC-9642	>100	50	>100	>100

Source: Ref. 96.

Commercial glycinates such as Tego Glycinat 818 and potassium cocyl glycinate (Amilite GCK-12) are used in detergents, cosmetics, and toiletries (Ajinimoto Company's 1997 report).

C.   *O*-AcylAmino Acid

O-Acyl-α-amino acids are derivatives of serine and have amphoteric structures. Serine, a polar amino acid without electric charge, is also known as β-hydroxy-aminopropionic acid:

$$OHCH_2CH-COOH$$
$$|$$
$$NH_2$$

Esterfication of the serine hydroxyl group with fatty acids leads to a serine-type surfactant. This type of surfactant is particularly attractive as a food emulsifier, since the synthesis could be achieved enzymatically using naturally occurring amino acids and fatty acids. This approach was used by Nagao and Kito [97] to perform an enzymatic synthesis of *O*-acyl-L-homoserine. Lipase from *Candida cylindracea* (60 u/mg protein) and *Rhizopus delemar* (6000 u/mg protein) were found to catalyze *O*-acylation of L-homoserine or L-serine with free fatty acids or with fatty acids from triacylglycerol. The enzyme reaction was initiated by the addition of fatty acids or oils to the amino acid solution containing 600 units of *Candida cylindracea* lipase. The active pH range of the reaction for *O*-acylation was from 7.5 to 8.5 [97]. The emulsifying activity of O-oleoyl-L-homoserine was higher than that of conventional surfactants [97]

This amphoteric surfactant, O-oleoyl-L-homoserine, could be expected to have some characteristic properties different from those of nonionic food emulsifiers, such as monoglycerides and sugar esters. It may be safe for humans after digestion by pancreatic lipase, for two reasons: (1) O-acyl-L-homoserine is synthesized by an enzymatic process and can be hydrolyzed into fatty acids and L-homoserine by the digestive organs; (2) L-homoserine is a naturally occurring amino acid and is distributed in plants, especially in germinating pea seeds [97].

D. Mixed Amino Acid Surfactants

Most commercial amino acid surfactants are produced from mixed amino acids that are readily obtained from protein hydrolysates due to a cost advantage. A number of readily available and inexpensive sources of raw materials offer interesting prospects for the preparation of hydrolysates. Various plant proteins (e.g., derived from cereal, vegetable, or oilseed) or animal proteins (e.g., derived from milk, whey, blood) or waste proteins could be converted to hydrolysate; hence they are a source of amino acid pool.

Borozdinskaya et al. [98] reported on a process in which amino acids from ''activated sludge biomass'' hydrolysis were acylated with caproyl chloride at 80°C for 1 h (pH 7.5–8.5) to produce surfactants having a surface tension of 33.1–65.1 mN/m, viscosity of 1.08–1.25 cSt, a moisture content of 75.6–98.5%, and a cmc of 0.04–0.21 g/L. Surfactants based on glycine had the highest surface activity and high foaming power [98].

Xia and coworkers [99] have demonstrated the feasibility of converting industrial waste proteins into value-added chemicals, AAS. Such research effort showed the effective conversion of cheap, renewable feedstock to new products of high value. In that work, N-acyl amino acid surfactants were derived from hydrolysates from industrial waste cottonseed protein (CSD), silk residue, silk chrysalis, and glutamic acid mother liquor (GAML). The amino acid composition showed that CSD had the highest amount of polar amino acids, including glutamic acid, aspartic acid, arginine, and nonpolar leucine [99]. Silk chrysalis (SC) was high in glutamic acid and aspartic acid. Silk residue (SR) had more nonpolar amino acids, glycine, alanine, and serine than the other sources.

Two approaches were used for the synthesis of amino acid surfactants. One was to react protein hydrolysates with alkyl acyl chloride, followed by purification and neutralization with alcoholic sodium hydroxide. The other was to react protein hydrolysates with fatty alcohols in organic solvent toluene. The procedure is briefly illustrated here:

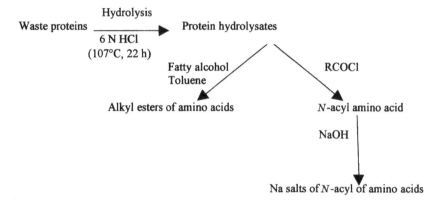

The yield of the purified mixed amino acids as Na salts from waste proteins was ~60–75%, that of amino acid ester from glutamic acid was higher (85–92%) [99]. The active matter contents of most of the synthesized amino acid surfactants were more than 95%. The structural representation of amino acid surfactants from C_{12}Na-SR spectra was identified and confirmed by IR and NMR as follows:

I	II	III
HNCH$_2$COONa	CH$_3$CHCOONa	CH$_2$—CHCOONa
COC$_{11}$H$_{23}$	NHCOC$_{11}$H$_{23}$	OH NHCOC$_{11}$H$_{23}$

The surface tension (γ), cmc, Krafft point, surface adsorption (Γ α 10^{-10} mol/cm^2), and area/molecule (A, Å) of sodium acyl amino acid surfactants as influenced by acyl chain length are shown in Table 11. As can be seen, generally as acyl chain length increased, surface tension of these surfactants increased, the cmc decreased and the Krafft point increased. Essentially, as the number of carbon atoms in the hydrophobic group increased, a corresponding decrease in cmc was seen. This relationship confirmed the expected behavior of homologous series of surfactants. The surface adsorption values of the mixed compounds derived from CSD, SR, and SC were less than those of glutamic acid surfactants, probably due to the steric effect and crosslinking between mixed surfactants and other constituents in the hydrolysates. This view is supported by the larger surface area occupied by the mixed amino acid surfactants compared to glutamic acid surfactants, where the molecules

TABLE 11 Surface Tension (γ), cmc, Γ_∞, A, and Krafft Point of Acyl Amino Acid Surfactants

Amino acid source[a]	Acyl chain length	Surface tension γ (mN m^{-1})	cmc (mM)	Γ_∞ (10^{-10} mol/cm^2)	A (area (Å^2)/ molecule)	Krafft point (°C)
CSD	C_{12}	29.5	2.8	1.6	103	35.0
	C_{14}	33.2	0.8	1.8	93	38.0
	C_{16}	35.2	0.7	1.9	85	42.5
	C_{18}	39.8	0.5	1.6	103	48.4
	$C_{18}{}^=$	34.8	0.6	1.7	100	—
SR	C_{12}	28.0	1.5	1.3	129	31.0
	C_{16}	32.3	0.6	1.4	118	50.0
	C_{18}	34.0	0.4	1.3	125	57.0
	$C_{18}{}^=$	33.2	1.1	1.6	103	—
SC	C_{12}	30.7	1.4	1.2	141	40.0
	C_{16}	35.8	0.5	1.3	129	57.0
	C_{18}	36.3	0.6	1.3	129	65.0
GA	C_{12}	29.2	0.9	2.6	64	37.0
	C_{16}	32.0	0.7	2.9	57	54.0
	C_{18}	33.9	0.6	2.5	65	62.0
	$C_{18}{}^=$	30.3	0.8	2.8	60	—
GAML	C_{12}	29.1	0.82	1.3	128	—
	C_{16}	33.5	0.40	1.2	133	—
	C_{18}	38.9	0.56	1.4	117	—

[a] CSD, cottonseed; SR, silk residue; SC, silk chrysalis; GA, glutamic acid; GAML, glutamic acid mother liquor.
Source: Ref. 99.

TABLE 12 LSDR (g/100 g) of Amino Acid Surfactants as Affected by Acyl Chain Length

Amino acid source	Acyl chain length				
	C_{12}	C_{14}	C_{16}	C_{18}	$C_{18}{}^=$
CSD	6.5	6.1	5.8	5.7	19.0
SR	47.0	42.0	43.0	40.0	56.0
SC	52.0	50.0	45.0	42.0	65.0
GA	5.6	5.2	4.8	4.8	7.0

Literature LSDR values for linear alkyl benzenesulfonate, LAS ($C_{12}PhSO_3Na$) = 40.0 [from Best-line, R. G., et al. J. Am. Oil Chem. Soc. 49:84 (1972)] and 30.0 [from Stirton, A. J., et al. J. Am. Oil Chem. Soc. 42:115 (1965)] for alkyl sulfate, AS ($C_{12}SO_4Na$).
Source: Ref. 99.

of glutamic acid surfactants are more likely to orient at the o/w interface in a close-packed arrangement than the mixed amino acid surfactants.

The foaming power (data not shown) of all the mixed surfactants from waste proteins showed high foaming ability compared to those of glutamic acid surfactants.

Table 12 shows the dispersibility data as "lime soap dispersing requirements" (LSDR) of the mixed surfactants. The LSDR (in grams per 100 grams) define the tolerance of surfactants to calcium and magnesium ions as applicable to hard and soft water. The smaller the LSDR, the better the lime soap dispersing power. The derivatives from cottonseed protein exhibited the best LSDR, 5.7–6.5%. This value was similar to the LSDR of glutamic acid surfactants (4.5–5.6%), but were better than the literature value for alkyl methyl ester sulfonate (9%) and linear alkyl benzenesulfonate (LAS) (40%). The outstanding LSDR of mixed surfactants from cottonseed protein may be attributed to its high glutamic content (25.8%).

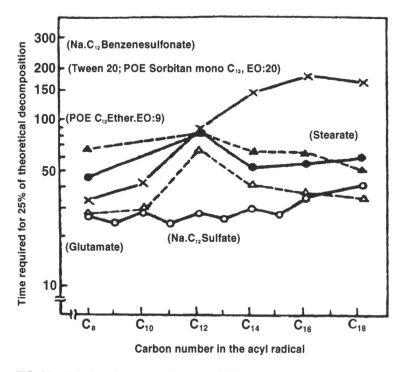

FIG. 22 Relationship between biodegradability and acyl chain length of N-acylamino acids; time required for 25% of theoretical decomposition. (From Ref. 103.)

In sum, the chain length of the acyl group affected surface properties. In particular, the mixed AAS with C_{12} produced good surface properties [99]

E. Biodegradability of Amino Acid Surfactants

Kamimura [100], Shida et al. [101,102], and Kubo et al. [103] conducted extensive studies on biodegradability and found that *N*-acylamino acids are readily biodegradable through decomposition into amino acid and fatty acid. Thus they are desirable surfactants compatible with the environment [Fig. 22].

F. Safety Features and Effect on Skin

Biocompatibility, especially safety and mildness to skin, are very important properties of interest for the application of AAS. Animal tests such as the Draise method for primary eye and skin irritation and the maximization test for sensitization have been widely used in the past. Recently, an alternative or in vitro test was actively developed for the purpose of refining, reducing, and replacing the animal test. Although there are still no perfect alternatives

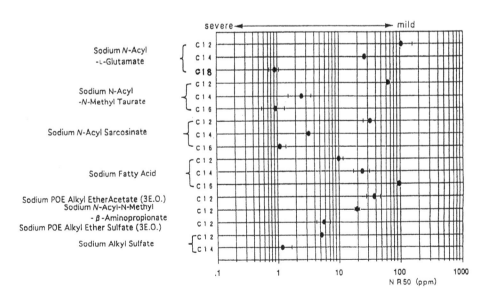

FIG. 23 Cytotoxicity of *N*-acylamino acids and other anionic surfactants for normal human epidermal keratinocyte (HEMA) by neutral red method to detect the end point. (From Ref. 104.)

to the animal test, the advance in in vitro methods is remarkable. In vitro methods are very useful for predicting the safety effect on the newly developed molecule and also for investigating the structure–activity relationship. Further advances have been made in the noninvasive in vivo human test with various bioengineering technologies. Extensive research has been conducted on the safety of *N*-acylamino acid, especially *N*-acylglutamate, with every possible application of the above mentioned testing methodologies. Kanari et al. [104]

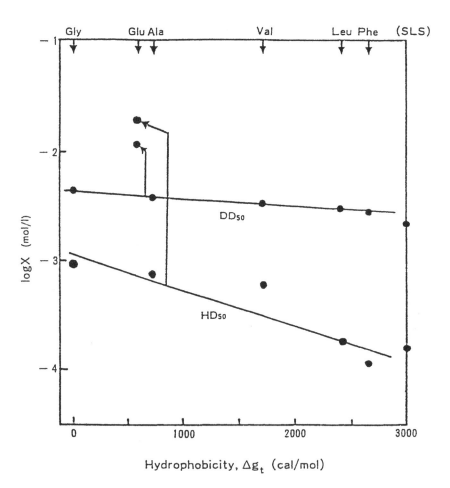

FIG. 24   Relationship between the hydrophobicity and the hemolitiactivity (HD$_{50}$) or denaturation of Cu$_2$-Transferine (DD$_{50}$) of *N*-acylamino acids. (From Ref. 49.)

conducted a cytotoxicity test for various anionic surfactants, and it was found that N-acylamino acids, especially N-acylglutamate, have a low toxicity for cultured human skin cells [Fig. 23]. Miyazawa et al. [105] and Imokawa [106] reported similar results. Sakamoto [49] investigated the relationship between the hydrophobicity of N-lauroylamino acid and several in vitro activities, such as hemolysis and denaturation of protein [Fig. 24].

V. SUMMARY

This chapter has attempted to provide a comprehensive overview of AAS, with an emphasis on their synthesis and physicochemical properties. Some information on their biological properties was discussed. Extensive research was done throughout the past century, along with the commercial development of N-acyl amino acids. The information provided in this chapter has demonstrated that by changing the structure of the hydrophobic hydrocarbon chain or by changing the size, polarity, or charge of the hydrophilic group, the properties of AAS can be altered remarkably.

Biocompatibility and environmental compatibility will continue to be a major consideration in the development and commercialization of new AAS. More work is needed to develop cost-effective biological processes for the synthesis of AAS. Also, further work is needed to better understand the structure–function relationship of existing and new AAS. With advances in their scientific understanding and technological development, AAS will likely find applications in biomedical research, household cleaning products, cosmetics, and pharmaceuticals aids.

REFERENCES

1. M. G. Macfarlane. In: Advances in Lipid Research (Paoletti, R. and Kritchevsky, D, eds.), Academic Press, New York, 1964, Vol. 2, p. 116.
2. S. Cmelik. Z. Physiol. Chem. 299:227 (1955).
3. T. Gender and E. Lederer. Ann. Acad. Sci. Fennicae Ser AII 60:313 (1955).
4. J. Blass. Bull Soc. Chim. Biol. 38:1306 (1956).
5. S. Bondi. Z. Biochem. 17:543 (1909).
6. Y. Kameda and E. Toyoura. Yakugakuzasshi 72:402 (1952).
7. S. Utino, T. Yoneda, and S. Yoshimoto. Ann. Acad. Sci. 607:190 (1957).
8. S. Nagai. J. Biochem. 50:428 (1961).
9. H. Hazama, Kushu. J. Med. Sci. 12:135 (1961).
10. H. Hazama, M. Kitagawa, and Y. Yamamura. J. Biochem. 53:117 (1963).
11. T. Fukui and B. Axelrod. J. Biol. Chem. 236:81 (1961).
12. N. J. Cartwright. Biochem. J. 60:238 (1955).

13. N. J. Cartwright. Biochem. J. 67:663 (1957).
14. H. Fukuda and H. Kimura. Hakkou Kougyou (Fermentation & Industry) 38: 912 (1980).
15. A. Gorchein. Biochem. Biophys. Acta 152:358 (1968).
16. J. Kawanami and H. Otsuka. Chem. Phys. Lipids 3:135 (1969).
17. J. Kawanami. Chem. Phys. Lipids 7:159 (1971).
18. H. W. Knoche and J. M. Shively. J. Biol. Chem. 247:170 (1972).
19. Y. Tahara, M. Kameda, Y. Yamada, and K. Kondo. Agric. Biol. Chem. 40:243 (1976).
20. Y. Tahara and Y. Yamadav. Agric. Bid. Chem. 40:1449 (1976).
21. C. J. Brandy. Biochem. J. 91:105 (1964).
22. A. Van den Oord, H. Danielsson, and R. Ryhage. J. Biol. Chem. 240:2242 (1965).
23. D. A. Holverda and H. J. Vonk. Comp. Biochem. Physiol. 45B:51 (1973).
24. R. W. Hendler. Proc. Natl. Acad. Sci. (U.S.A) 54:1233 (1965).
25. V. R. Wheatley, P. Flesch, E. C. J. Esoda, W. M. Coon and L. Mandol. J. Invest. Dermat. 43:395 (1964).
26. P. Ham and V. R. Wheatley. J. Invest. Dermat. 49:206 (1967).
27. W. Hentrich, H. Keppler, and K. Hintzmann. Patent (Germany) 546,942 (1930), 635,522 (1936); Patent (Britain) 459,039 (1936).
28. H. Staudinger and H. V. Becer. Ber. Deut. Chem. 70:889 (1937).
29. T. Tsubone. Seikagaku (Biology) 35:67 (1963).
30. P. Heitmann. Eur. J. Biochem. 3:346 (1968).
31. P. Heitmann. Eur. J. Biochem. 5:305 (1968).
32. M. Takehara, H. Moriyuki, I. Yoshimura, and R. Yoshida. J. Am. Oil Chem. Soc. 49:143 (1973).
33. M. Takehara, H. Moriyuki, A. Arakawa, and R. Yoshida. J. Am. Oil Chem. Soc. 50:227 (1973).
34. M. Takehara, I. Yoshimura, and R. Yoshida. J. Am. Oil Chem. Soc. 51:419 (1974).
35. M. Takehara, H. Moriyuki, R. Yoshida, and M. Yoshikawa. Cosmet. Toilet. 94: 31 (1979).
36. T. Saitou and M. Takehara. Yushi (Fat & Oils) 31:38 (1978).
37. H. Suyama. Yakugakuzasshi 86:967 (1966).
38. E. B. Kester. U.S. Patent 2,463,779 (1949).
39. E. Jungermann, J. F. Gerechet, and I. J. Krems. J. Am. Oil Chem. Soc. 78:172 (1956).
40. T. Ueda, S. Katou, and S. Toyoshima. Jpn. Patent 9568 (1956).
41. B. Weiss. J. Org. Chem. 24:1367 (1959).
42. R. Yoshida and T. Shishido. Yukagaku 25:546 (1976).
43. T. Paquet. Can. J. Biochem. 58:573 (1979).
44. A. Nagamatsu, T. Okumura, T. Hayashida, and Y. Yamamura. Chem. Pharm. Bull. 16:211 (1968).
45. K. Sakamoto, K. Takizawa, S. Inazuka, and R. Yoshida. Yukagaku 26:110 (1977).

46. M. Naudet. Bull. Soc. Chem. Fr. 358 (1950).
47. Ooki, H. Tokiwa. Bull. Soc. Chem. Fr. 19:897 (1970).
48. K. Sakamoto. Study on the Aggregation States of Sodium Salt of *N*-Acylamino Acids. Ph.D. dissertation, Tohoku University, (1980).
49. K. Sakamoto. Yukagaku 44:256 (1995).
50. C. Tanford. J. Am. Chem. Soc. 84:4240 (1962).
51. M. Takahashi, F. Tanaka, and K. Motomura. Bull. Chem. Soc. Jpn. 57:944 (1984).
52. D. Kaneko and Y. Kawasaki. Fragrance J. 2:43 (1994).
53. K. Sakamoto and M. Hatano. Bull. Chem. Soc. Jpn. 53:339 (1980).
54. H. Yokota, K. Sagawa, C. Eguchi, M. Takehara, K. Ogino, and T. Shibayama. J. Am. Oil Chem. Soc. 62:1716 (1985).
55. K. Ogino, M. Abe, K. Kato, and Y. Sakama. Yukagaku 36:129 (1987).
56. M. Abe, T. Kubota, H. Uchiyama, and K. Ogino. Colloid Polym. Sci. 267:365 (1989).
57. M. Abe, A. Kuwabara, K. Ogino, H. Oshima T. Kondo, and K. M. Ja. J. Sur. Sci. Technol. 8:155 (1992).
58. K. Esumi, S. Yoshida, and K. Meguro. Bull. Chem. Soc. Jpn. 56:2569 (1983).
59. K. Sagawa, H. Yokota, I. Ueno, T. Miyoshi, and M. Takehara. 14th IFSCC, Barcelona, vol. 1, p. 29 (1986).
60. R. Yoshida, K. Baba, T. Saito, and I. Yoshimura. Yukagaku 25:404 (1976).
61. M. R. Infante, J. G. Dominguez, P. Erra, M. R. Julia, and M. Prats. J. Cosmet. Sci. 6:275 (1984).
62. M. R. Infante, J. Molinero, P. Erra, M. R. Julia, and J. G. Dominguez. Fette. Seifen. Anstrichm. 87:309 (1985).
63. J. Xia, Y. Xia, and I. A. Nnanna. J. Agric. Food Chem. 43:867 (1995).
64. K. Yahagi and K. Tsujii. J. Colloid Interface Sci. 117:415 (1987).
65. H. Okabayashi, K. Oshima, H. Etoki, R. Debnath, K. Taga, and T. Yoshida. J. Chem. Soc. Farady Trans. 86:1561 (1990).
66. C. Goddard and R. Felsted. Biochem. J. 253:839 (1988).
67. A. Desai and P. Bahadur. Tenside Surf. Det. 29:425 (1992).
68. H. Okabayashi, K. Taga, T. Yoshida, K. Oshima, E. Etori, T. Uehara, and E. Nishio. Appl. Spectroscopy 45:626 (1991).
69. J. Berger, R. Neumann, and A. Seibt. Tenside Surf. Det. 23:156 (1986).
70. S. Miyagishi, Y. Ishibashi, T. Asakawa, and M. Nishida. J. Colloid Interface Sci. 103:164 (1985).
71. A. Akimova and O. Galakhova. Izv. Vyssh. Uchebn. Zved. Khim. Tekhnol. 30: 114 (1987).
72. S. Mhasker, R. Prasad, and G. Lakshminarayana. J. Am. Oil Chem. Soc. 67: 1015 (1990).
73. S. Miyagishi, M. Higashide, T. Asakwa, and M. Nishida. Langmuir 7:51 (1991).
74. M. Takai, H. Hidaka, and M. Moriya. J. Am. Oil Chem. Soc. 56:537 (1979).
75. T. Izumi, Y. Yagimuma, and M. Haga. J. Am. Oil Chem. Soc. 74:875 (1997).
76. T. Hattori, K. Sano, and H. Yoshihara. Jp. Patent 07061957A to Ajinomoto Co. (1995).

77. D. Montet, F. Servat, M. Pina, J. Graille, P. Galzy, A. Amoud, H. Ledon, and L. Marcou. J. Am. Oil Chem. Soc. 67:771 (1990).
78. M. R. Infante, J. G. Dominguez, P. Erra, and M. R. Julia. Proceeding XII Congress IFSSC, Vol. III, pp. 99–103 (1982); Proceeding XIV Jornadas C.E.D. pp. 267–281 (1983).
79. M. Takehara, M. Moroyuki, I. Yoshimura, and H. Yoshida. J. Am. Oil Chem. Soc. 49:157 (1972).
80. H. Nakayama, A. Hara, and R. Yoshida. Jap. J. Derm. 82:565 (1973).
81. H. Nakayama, Y. Ko, S. Kondo, M. Kamo, and R. Yoshida. Presented at 72nd meeting of the Dermatological Society of Japan, Niigata, May 1973.
82. S. Arai and M. Watanabe. Agric. Biol. Chem. (Japan) 44:1979 (1980).
83. M. Watanabe and S. Arai. In: Development in Food Proteins-6 (B. J. F. Hudson, ed.), Elsevier Applied Science, London, 1988, pp. 179–217.
84. J. McEvily and A. Zaks. In: Biotechnology and Food Ingredients (I. Goldberg and R. Williams, eds.), Van Nostrand Reinhold, New York, 1991, pp. 193–221.
85. J. Swarbrick and J. Daluwara. J. Phys. Chem. 74:1293 (1970).
86. A. H. Beckett and R. J. Woodward. J. Pharm. Pharmcol. 15:422 (1963).
87. K. Tori and T. Nakagawa. Kolloid-Z. u. Z. Polymer 188:47 (1963).
88. Idem. Kolloid-Z. u. Z. Polymer 191:42 (1963).
89. Idem. Kolloid-Z. u. Z. Polymer 191:48 (1963).
90. Idem. Kolloid-Z. u. Z. Polymer 191:50 (1963).
91. H. Hidaka. Yukagaku 27:304 (1978).
92. H. Hidaka, Yukagaku 28:190 (1979).
93. H. Hidaka. 21st Conference of Jpn. Oil Chem. Soc. B37 (1982).
94. M. Takehara, K. Sakamoto, and R. Yoshida. Yukagaku 25:789 (1976).
95. Y. Abe, S. Osanai, and T. Matsushita. J. Am. Oil Chem. Soc. 51:385 (1974).
96. Y. Abe, S. Osanai, and S. Matsumura. J. Am. Oil Chem. Soc. 49:357 (1972).
97. A. Nagao and M. Kito. J. Am. Chem. Oil Chem. Soc. 66:710 (1989).
98. N. K. K. Borozdinskaya, I. A. Kryliv, and M. N. Manakov (USSR). Neftepere-rab. Neftkhim. 3:31 (1984).
99. J. Xia, J. Qian, and I. A. Nnanna. J. Agric. Food Chem. 44:975 (1996).
100. A. Kamimura. Kantou Reg. Agr. Chem. Conf. Jpn., p-3 (1973).
101. T. Shida, Y. Homma, and T. Misato. Agr. Biol. Chem. 37:1027 (1973).
102. T. Shida, Y. Homma, A. Kamimura, and T. Misato. Agr. Biol. Chem. 39:879 (1975).
103. M. Kubo, K. Yamada, and K. Takinami. Hakkoukougyoukaishi (J. Fer. Eng. Jpn), 54:323 (1976).
104. M. Kanari, Y. Kawasaki, and K. Sakamoto. J. Soc. Cosmet. Chem. 27:498 (1993).
105. K. Miyazawa, U. Tamura, Y. Katuura, K. Utikawa, T. Sakamoto, and K. Tomita. Yukagaku 38:34 (1989).
106. G. Imokawa. Fragrance J. 9:57 (1981).

5
Enzyme-Catalyzed Synthesis of Protein-Based Surfactants: Amphoteric Surfactants

YASUKI MATSUMURA and MAKOTO KITO* Kyoto University, Kyoto, Japan

I. INTRODUCTION

Proteins are themselves surface-active compounds with an amphiphilic nature. The interfacial behavior of proteins is different from that of low-molecular-weight amphiphiles with a simple structure, namely, detergents, because proteins are highly complex polymers made up of a combination of 20 different amino acids (this point is described in detail in Chapter 3 of this book). Normally, proteins take on the folded compact structure, in which nonpolar amino acid residues are located in the interior and hydrophilic residues are exposed to molecular surfaces. Since hydrophobic interactions play dominant roles in the adsorption of surfactants to the air–water and oil–water interfaces, such a native structure of proteins should be modified to make full use of the surface activity of proteins [1].

The chemical modification of proteins is not desirable, because of the harsh reaction conditions, the nonspecificity, and the difficulty of removing reagents from the final product [2]. Enzymatic reaction has several advantages, such as the mild reaction conditions, high specificity, and fast reaction rates [3]. Moreover, the use of the enzyme, which occurs naturally in living cells, is acceptable by consumers from the viewpoint of safety. We describe examples of the enzyme-catalyzed synthesis of protein-based surfactants in this chapter.

---

* Professor Emeritus.

Synthesis normally means the construction of larger molecules from small building blocks via covalent linkage. However, we also refer to examples of protein-based surfactants produced via a degradation process such as proteolysis and deamidation, in addition to cases of surfactants produced by the covalent attachment of nonproteinaceous moieties.

II. AMPHIPHILIC STRUCTURE OF PEPTIDES LIBERATED BY PROTEOLYSIS

Proteolysis has been widely used to improve the surface properties of proteins [3,4]. Partial proteolysis may facilitate the unfolding of polypeptides and increase the heterogeneity of the protein, that is, the peptide molecular weight distribution, resulting in improved surface properties. The factors affecting enzymatic hydrolysis as well as the production of protein hydrolysates are described in detail in Chapter 2 of this book. In this section, therefore, we show only examples of the systematic approach on a structural factor relating to amphiphilic or amphoteric properties of peptides produced by proteolysis.

There have been numerous reports on the relationship between proteolysis and changes of functional properties of proteins such as solubility, emulsifying, and foaming activity [4]. Although the control of the degree of hydrolysis (DH) is necessary for optimizing the functional properties of proteins, the systematic data are scanty about the importance of sequential and structural factors of liberated peptides for functional properties, especially amphoteric properties. The Shimizu group studied hydrolysates of some food proteins in terms of emulsifying and oil-binding properties. They found several surface-active peptides that had the following common structural features [5].

At first, surface-active peptides with good emulsifying activity were searched for from hydrolysates of the caseins. Caseins are known to be hydrophobic proteins with a highly flexible structure. Shimizu et al. [6] found that the N-terminal peptide of 23 residues liberated by pepsin digestion from α_{S1}-casein was very surface active. They termed this peptide α_{S1}-CN(f1-23). As shown in Fig. 1, α_{S1}-CN(f1-23) has an amphiphilic structure; i.e., five out of the six basic amino acid residues are located in the N-terminal region (1–10 residues) of the peptide, while the nonpolar amino acid residues are rich in the C-terminal region (11–23 residues). In addition to this sequential amphiphilicity, it should be emphasized that an amphiphilic structure appears when the region corresponding to residues 12–23 form an α-helix (Fig. 2a) [5]. The lipid-binding activity may depend on this amphiphilic helix structure.

Bovine serum albumin (BSA) is a globular protein with a rigid compact structure, but it is also known to have good emulsifying and oil-binding prop-

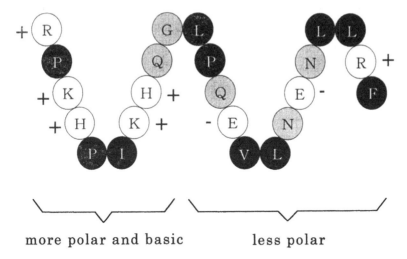

more polar and basic   less polar

FIG. 1 Primary structure of peptides corresponding to the N-terminal of 23 residues of α_{s1}-casein. Hydrophobic amino acids are shown by closed circles, hydrophilic charged amino acids by open circles, and hydrophilic uncharged amino acids by shaded circles. (From Ref. 6.)

erties. Saito et al. [7] found that the third domain of BSA (residues 377–582) produced by limited tryptic hydrolysis was adsorbed more preferentially onto the emulsified oil surface, compared with other peptides, including the intact BSA molecule. Interestingly, this domain also contains two sequences that could provide typical amphiphilic α-helices, as shown in Fig. 2b and c [5], based on crystallographical data [8]. Furthermore, it has been shown that the region of residues 125–143 of β-lactoglobulin, which forms an amphiphilic α-helix (Fig. 2d), is involved in the contact to the oil droplet surfaces [9]. These findings suggest that the amphiphilic helix could be one of the basic conformations in proteins adsorbed onto oil–water interfaces.

In order to examine the possibility of this speculation, an approach using synthetic peptides was made [10]. Three kinds of peptides (H, S, and R) with 16 amino acid residues were synthesized, and their secondary structure and surface properties were investigated to clarify the effects of conformational amphiphilicity. The amino acid compositions of the three peptides were the same (8 Leu and 8 Glu residues), but their sequences were different. The helical wheel structure of peptide H is shown in Fig. 2e [5]. As shown in this structure, peptide H was designed to form an amphiphilic α-helix, whereas the other peptides were designed not to show such amphiphilicity, even when

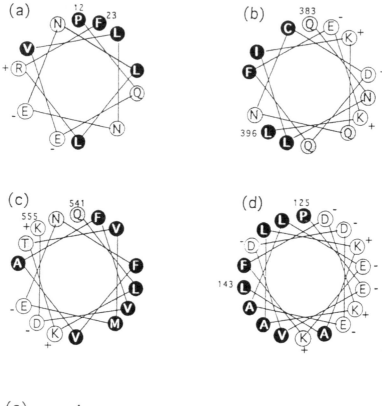

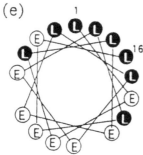

FIG. 2 Helical wheel structures of: (a) α_{s1}-casein residues 12–23; (b) bovine serum albumin residues 383–396; (c) bovine serum albumin residues 541–555; (d) β-lacto-globulin residues 125–143; and (e) synthetic peptide H. Hydrophobic amino acids are shown by closed circles, hydrophilic charged amino acids by open circles, and hydro-philic uncharged amino acids by shaded circles. (From Ref. 5.)

they form α-helices. Circular dichroism (CD) analysis showed that peptide H contained α-helix at pH 5.5 but took a random structure at pH 7.0. The surface pressure of peptide H solution increased very rapidly at pH 5.5, i.e., reached 29.6 mN/m in only 1 min after the start of measurement. The increase of the surface pressure of peptide H at pH 7.0, however, was not so dramatic (13 mN/m). Moreover, peptide H showed higher emulsifying activity at pH 5.5 than at pH 7.0. These results suggest that the amphiphilic α-helical structure is relevant to good surface properties. This conclusion was confirmed by the results of other synthetic peptides, S and R. Peptide R, consisting mainly of random structure (and no propensity for α-helix), showed poor surface properties at both pHs. For peptide S, forming α-helical but not amphiphilic structure at pH 5.5, the surface properties were inferior to those of peptide H. Interestingly, peptide S showed good emulsifying activity and oil-binding properties when the peptide formed an amphiphilic β-sheet structure at pH 7.0.

The importance of amphiphilic the α-helix structure with respect to emulsifying and/or foaming properties of peptides (proteins) was also pointed out in other cases [11]. In biological systems, too, it is well known that the amphiphilic α-helix structure plays important roles in lipid–protein interaction, for instance, the presence of transmembrane α-helices in many integral membrane proteins [12]. Recent experimental evidence has shown that transmembrane regions of membrane-bound proteins such as nicotinic acetylcholine receptor also include the amphiphilic β-sheet structure [13]. Therefore, the amphiphilic α-helix structure and/or amphiphilic β-sheet structure could be general indices for pursuing the good surface activities of peptides or proteins. In other words, a good protein- (peptide-)based surfactant could be developed by incorporating an amphiphilic α-helix or β-sheet structure. This point should be considered when proteolysis of proteins is designed to produce good peptide-based surfactants. For instance, the selection of proteases should be made while considering the substrate specificity of the enzymes, which can leave the peptides with an amphiphilic structure. The numerous data about the three-dimensional structure of proteins obtained from the prediction approach and crystallography must be helpful in the design of the generation of peptide-based surfactants by proteolysis methods.

III. DEAMIDATION

Deamidation of proteins induces the conversion of glutamine and asparagine residues to acid groups, with the concomitant release of ammonia. Deamidation is known to be an important method for improving the functional properties of proteins, for instance, the increase in solubility of cereal prolamins, which contain high contents of glutamine residues and are insoluble in aqueous media, because of the strong hydrogen bonds formed among glutamine residues [3]. The foaming and emulsifying properties of proteins could also be improved by deamidation because of an increase in protein flexibility due to changes in charge balance.

Kato et al. [14,15] reported that proteases such as papain, pronase, and chymotrypsin have the deamidation activity of proteins in an alkaline pH region where proteolytic activity was at the minimum. The optimal pH of deamidation without a high degree of proteolysis was 10, and the optimal temperature was 20°C [14]. Table 1 shows the extent of deamidation and proteolysis in the four proteins after treatment with chymotrypsin [15]. About 20% asparaginyl or glutamynyl residues were deamidated. The extent of proteolysis was very low for ovalbumin and lysozyme, but soy proteins (11S and 7S globulins) were more subject to proteolysis. Table 2 lists the foaming and emulsifying properties of deamidated proteins [15]. The foaming power of demidated proteins was increased about two times. For foam stability, the deamidation effect was more remarkable in the case of ovalbumin and lysozyme than with soy proteins. The emulsifying activity of deamidated forms of all proteins was increased as compared to the nondeamidated forms. These results indicate that proteolytic deamidation by chymotrypsin could be a useful tool for improving the surface properties of proteins.

Although proteolytic deamidation at alkaline pH was shown to be useful for increasing the surface properties of some proteins, as described earlier, proteolysis is not inevitable for other types of proteins. Furthermore, the alkaline pH condition damaged amino acid residues of protein, for instance, racemization or the formation of lysinoalanine. Therefore, it is more desirable to use enzymes that catalyze only the deamidation without touching peptide bonds at neutral pH regions. *Bacillus circulans* peptidoglutaminase was used for the deamidation of proteins. The enzyme was shown to have a very limited

TABLE 1 Deamidation and Proteolysis Percentages of Food Proteins by Treatment with Chymotrypsin at pH 10

Protein	Deamidation, %	Proteolysis, %
Ovalbumin	20	2
Lysozyme	21	0
7S Globulin	24	5
11S Globulin	19	8

Source: Ref. 15.

TABLE 2 Foaming and Emulsifying Properties of Deamidated
Proteins

Protein	Foaming power, $\mu s/cm$	Foam stability, min	Emulsifying activity (OD_{500})
Ovalbumin			
Native	82	2.0	0.220
Deamidated	208	6.5	0.450
Lysozyme			
Native	54	1.0	0.150
Deamidated	96	3.5	0.230
7S globulin			
Native	1250	7.7	0.420
Deamidated	2550	10.2	0.540
11S globulin			
Native	1300	9.8	0.390
Deamidated	2250	11.4	0.600

Source: Ref. 15.

deamidating activity toward intact animal and plant proteins but readily deamidates glutamine in peptides and protein hydrolysates [16]. New enzymes catalyzing the deamidation of seed storage proteins were found in various germinating seeds and were partially purified [17]. It was shown that the enzymes acted on native-form proteins in addition to glutamine and olygopeptides. The structure and function of the enzymes are still unclear, although they are known to differ from transgluaminase. Transglutaminase catalyzes an acyl-transfer reaction between the γ-carboxyamide group of peptide-bound glutamine residues and a variety of primary amines, including the ε-amino group of lysine residues in certain proteins [3,18]. In the absence of amine substrates, transglutaminase can catalyze the deamidation of glutamine residues using water molecules as acyl acceptors.

IV. PHOSPHORYLATION

There are a lot of naturally occurring phosphorylated proteins, including food proteins such as milk casein, egg white albumin, egg yolk phosvitin, and soy bean 7S globulin. The introduction of phosphoryl residues increases the negative charge and hydration, thereby changing the functional properties of proteins.

Phosphoryl residues affect the surface properties of proteins. This can be clearly demonstrated by the effects of "dephosphorylation" on the surface properties of proteins. The effects of dephosphorylation on structure and surface properties are described in more detail in Chapter 3 of this book.

Acknowledgment of the important role of phosphorylation in surface properties urged the investigation of the enzymatic phosphorylation of proteins. Seguro and Motoki [19] tried to phosphorylate soy proteins using cyclic adenosine monophosphate-dependent protein kinase (A-kinase) and found that acidic subunits of 11S globulin were phosphorylated by A-kinase. Emulsifying

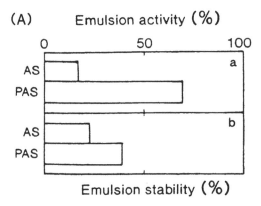

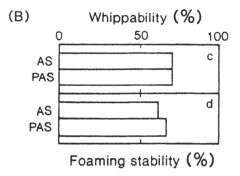

FIG. 3 Effects of phosphorylation on (A) emulsifying properties and (B) foaming properties of acidic subunits of soy bean 11S globulin. a: emulsifying activity; b: emulsion stability; c: whippability; d: foaming stability. AS and PAS indicate native and phosphorylated forms of acidic subunits of 11S globulin, respectively. (From Ref. 20.)

activity and emulsion stability were increased four-fold and two-fold, respectively, by the phosphorylation of 11S acidic subunits (Fig. 3A) [20]. On the other hand, no such changes were observed in whippability and foaming stability upon phosphorylation (Fig. 3B). Campbell et al. [21] also pointed out that emulsifying activity and emulsifying stability were improved by the phosphorylation of soy proteins, but foam stability was lower using the phosphorylated protein. Ralet et al. [22] reported the enzymatic phosphorylation of soy proteins by cAMP-independent type II kinase (CKII) isolated and purified from yeasts. Although A-kinase could not act on 7S globulin, 7S globulin was a better substrate of CKII kinase than was 11S globulin. This result shows that the use of a different type of kinase may promote the application of enzymatic phosphorylation of proteins.

V. GLYCOSYLATION

There have been numerous reports about glycosylation via chemical methods, especially the Maillard reaction between the ε-amino groups in proteins and the reducing-end carbonyl residues in polysaccharides. Kato et al. [23] demonstrated that egg white proteins–galactomannan conjugate produced via the Maillard reaction had excellent emulsifying properties, superior to those of commercial emulsifiers, in acidic pH and high salt concentration. It has also been shown that the conjugation of phosvitin with galactomannan improved both emulsifying activity and emulsion stability [24]. These effects of glycosylation can be attributed to the increase in the amphiphilic nature of conjugated molecule. The hydrophobic residues of protein denatured at the oil–water interface may be anchored to the surface of oil droplets in emulsion, whereas the hydrophilic residues of polysaccharide oriented to water may cover oil droplets to inhibit the coalescence of oil droplets. The polysaccharide moiety should attract water molecules and increase the local surface viscosity surrounding oil droplets.

Examples of enzymatic glycosylation of proteins are scanty. The transglutaminase-catalyzed incorporation of sugar-alkylamine into glutamine residues of proteins was reported [25]. Transglutaminase action resulted in the incorporation of 18 and 57 glycosyl units per mole of β-gliadins and pea legumin, respectively. The corresponding degree of glycosylation of glutamine residues was 15.7% with gliadins and 25.7% with legumin. The surface-active properties of glycosylated proteins by transglutaminase have not been investigated, although the solubility of these neoglycoproteins was markedly increased over that of native proteins in the range of their isoelectric points. In this attempt, sugar-alkylamines were chemically synthesized. The use of natural amino sugars such as glucosamine as donors in the transglutaminase reaction is more desirable. Such an attempt seems to be very difficult, because mammalian transglutaminases shows a high affinity for straight-chain aliphatic amines [26]. However, it is probable that microbial transglutaminases with smaller sizes compared to those of mammalian ones may act on amino sugars under specific conditions [18].

Glycosylated proteins are ubiquitous in biological systems. Glycosyltransferases are responsible for in vivo glycosylation reactions, but so far their in vitro utilization is hampered by the lack of commercially available enzymes capable of directly catalyzing the attachment of carbohydrates or oligosaccharides to sidechains of amino acid residues. Moreover, the necessity of cofactors and the very high specificity of glycosyl transferases toward the glycosyl donor and acceptor would not allow the use of these enzymes as a general method of in vitro glycosylation of proteins. Nakamura et al. [27] attempted to construct polymannnosyl lysozymes in a yeast expression system using genetic engineering methods. The site-directed mutagenesis in complementary DNA encoding hen egg lysozyme was carried out to obtain the Asn-X-Ser/Thr sequence that is the signal for asparagine-linked (N-linked) glycosylation. Namely, two N-linked glycosylation sites (Asn[19]-Try[20]-Thr[21] and Asn[49]-Ser[50]-Thr[51]) were introduced by substituting Arg-21 with Thr and Gly-49 with Asn, respectively. In addition to glycosylated mutants (R21T and G49N), double glycosylated mutant lysozyme (R21T/G49N) was constructed to create the N-linked glycosylation sites at positions 19 and 49 [28]. Table 3 shows the emulsifying properties of various polymannosyl lysozymes. The double-glycosylated lysozyme G49N/R21T exhibited much higher and more stable emulsifying properties than the single lysozyme G49N and R21T [28]. These results suggest that the number of binding sites of the glycosyl chain closely relates to the emulsifying properties of polymannosyl lysozymes. It is possible

TABLE 3 Emulsifying Activity and Emulsion Stability of Polymannosyl Lysozymes

Polymannosyl lysozyme	Emulsifying activity (OD_{500})	Emulsion stability (min)
R21T/G49N	1.635	27.25
G49N	1.472	19.93
R21T	1.218	14.18
Native lysozyme	0.216	<1

Source: Ref. 28.

that the formation of a thick bound water layer adsorbed around the emulsion of R21T/G49N is further enhanced and that coalescence of oil droplets was more effectively inhibited by increasing the oligomannosyl chain.

The glycosylation of proteins by genetic engineering is, so to speak, "enzymatic modification in vivo." The protein engineering technique should be a promising tool for producing neoglycosylated proteins with excellent surface properties in the future, although there are many difficulties for the mass production of modified proteins that are acceptable to consumers at present.

VI. PLASTEIN REACTION

When protein hydrolysates are incubated with a protease at specific conditions, the formation of a precipitate, often gel-like ("plastein"), is observed in many cases. Although this phenomenon has been well known for over 90 years, a Tokyo group reinvestigated the plastein reaction in much more detail, with the object of applying it to the improvement of food proteins [29,30]. Studies of plastein reaction were then developed diversely.

There are at least three parameters affecting the formation of plastein [29,30]. First, the substrate must be a low-molecular-weight peptide mixture; in other words, the hydrolysate used in plastein reaction should have a high DH. Although variety exists for the optimum molecular weights of substrate peptides according to the other conditions, such as the kind of protease, the most favorable size was shown to be a molecular weight range of 451–1450 daltons [31]. Second, the substrate concentration of the reaction medium is an important factor and should be around 20–40%. A lower substrate concentration is more favorable for the hydrolysis reaction because of the lower probability of attack by nucleophile. To avoid hydrolysis by reducing water activity, a water-miscible organic solvent (often acetone is chosen) is added to the reaction mixture. The third factor influencing plastein formation is pH. Optimum pH depends on enzymes, but plastein formation generally takes place at a pH value different from the optimum pH for hydrolysis. This was shown by Adler-Nissen [32], who found that, of the eight different enzymes tested, all but one had a pH optimum for plastein formation 2–3 pH units away from the pH optimum of hydrolysis. Only papain had the same optimum for plastein formation and protein hydrolysis.

The classical method of plastein reaction is based on the use of protein hydrolysates from protease reaction as substrates. Therefore, plastein formation requires two quite different unit processes. The first step is protein hydrolysis, with purification of the resulting hydrolysate, and the second is plastein synthesis with the covalent incorporation of an expected amino acid. Yamashita et al. [33] developed a one-step process that eliminates the first step. In this method, degradation of substrate proteins and the incorporation of amino acid ester into the proteins by proteases occur sequentially in the same reaction mixture.

In order to produce the amphoteric protein–based surfactant, the incorporation of lipophilic amino acid ester was attempted using the one-step method of plastein reaction with papain at pH 9. In a system containing succinylated α_{s1}-casein as a protein substrate and luecine n-dodecyl ester as a lipophile, the peptide bond between Phe[145] and Tyr[146] of casein was first hydrolyzed, and this is followed by the incorporation of luecine n-dodecyl ester at the same position, forming a new C-terminus [34]. The structure of the macropeptide with respect to the distribution of hydrophilic amino acid residues is shown in Fig. 4 [29,34]. Amphiphilic structure consisting of hydrophilic protein portion and lipophilic luecine n-dodecyl ester was clearly demonstrated.

Watanabe et al. [35] tried to use a more hydrophilic and commercially available protein, gelatin, as a substrate instead of α_{s1}-casein. The lypophiles used in this case were luecine n-alkyl esters with alkyl carbon numbers ranging from 2 to 12 in order to produce proteinaceous surfactants with a different hydrophilicity–lipophilicity balance. All of the enzymatically modified gelatin

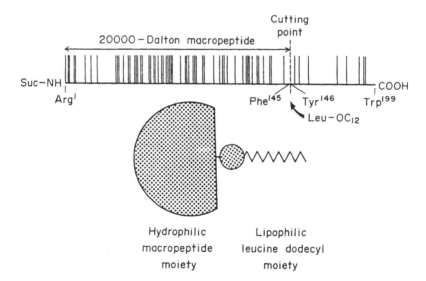

FIG. 4 Amphiphilic structure formed by papain-catalyzed incorporation of L-leucine n-dodecyl ester into succinilated α_{s1}-casein. The position of hydrophilic amino acid residues on the protein molecule are indicated by vertical bars. (From Ref. 34.)

(EMG) products were found to have molecular masses distributed in the range $(2–40) \times 10^3$ daltons, with an average molecular mass of about 7.5×10^3 daltons. Alkyl contents incorporated into EMG were found to lie within a well-controlled narrow range of 1.1–1.2 mol per 7.5 kg. However, surface properties of EMG varied greatly according to the alkyl carbon number of incorporated luecine n-alkyl ester (Fig. 5) [30,35]. The whippability and foam stability showed the maximum values when Leu-OC$_8$ and Leu-OC$_6$, respectively, were incorporated into the gelatin. The emulsifying activity, on the other hand, tended to increase gradually with the carbon chain length of the attached alkyl moiety.

EMG-6 (Leu-OC$_6$ attached) and EMG-12 (Leu-OC$_{12}$ attached), which had been shown to have superior foaming and emulsifying properties, respectively, as shown in Fig. 5, were subjected to further experiments for evaluation of their surface properties. The surface properties of EMG-6 and EMG-12 were compared to those of commercially available surfactants. As shown in Fig. 6,

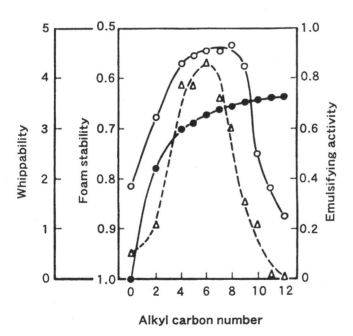

FIG. 5 Functional properties of enzymatically modified gelatin (EMG) products attaching L-leucine n-alkyl esters. Whippability (○), foam stability (△), and emulsifying activity (●) are plotted against the number of carbon atoms in the alkyl chain. (From Ref. 29.)

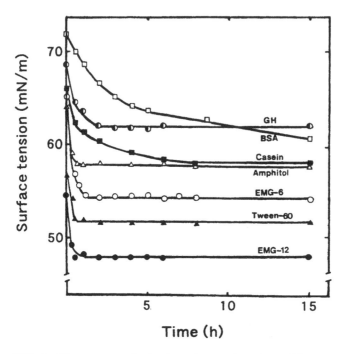

FIG. 6 Time-course reduction in surface tension observed for enzymatically modified proteins and nonmodified proteins at 20°C. EMG-6 and EMG-12 indicate gelatin attaching Leu-OC$_6$ and Leu-OC$_{12}$, respectively. (From Ref. 36.)

when EMG-6 and EMG-12 were added to pure water, the surface tension fell rapidly and reached a plateau value within 15 min [36]. This behavior is similar to that of the synthetic surfactants, Tween-60 (polyoxyethylene sorbitan monostearate) and Amphitol-20BS (N,N'-dimethyl-N-lauryl-carboxy-methyl-ammonium). Under equivalent conditions, the rate of decrease in surface tension for proteins such as casein and bovine serum albumin was substantially slower. These results suggest that EMG-6 and EMG-12 are adsorbed onto the air–water interface without surface denaturation due to their flexible conformations. The critical micelle concentration (cmc) point could be clearly observed in the plot of surface tension versus the logarithm of concentration of EMG-6 and EMG-12, and cmc values were comparable to those of low-molecular-weight surfactants [36].

Many applications of EMG have been demonstrated, e.g., production of a mayonnaise-like emulsion, improvement of baking properties of bread, and preparation of antifreeze emulsions or dispersions [29,30]. EMG-14, an enzymatically modified gelatin product obtained by the attachment of alanine myristyl ester instead of leucine dodecyl ester, has been produced on an industrial scale and applied as a protein-based surfactant for cosmetics. The same method of enzymatic modification with lipophilic amino acid alkyl ester has been applied to other proteins, such as soy protein isolate, fish proteins, casein, and ovalbumin [37]. Although such amphoteric protein–based surfactants have not been available extensively, the attachment of amino acid alkyl ester via plastein reaction still seems to be one of the most important and promising methods. However, such an attempt was not seen in the recent application of the plastein reaction, although there have been many reports about the enrichment of essential amino acids such as methionine to proteins, debittering of peptides, production of gelling materials from hydrolysates, etc. [38].

VII. COVALENT ATTACHMENT OF FATTY ACIDS TO AMINO ACID RESIDUES IN PROTEINS

The covalent attachment of fatty acids to protein molecules is a more direct method for producing amphoteric proteins than other physicochemical and enzymatic modifications. The attached fatty acyl chains fortify the amphiphilic structure, thereby leading to the formation of a polymerlike surfactant. Such proteinaceous surfactants are efficiently adsorbed at the interfaces, because the hydrophobic moiety rapidly penetrates into the hydrophobic layer and anchors the hydrophilic protein portions. Despite the importance of the covalent attachment of hydrophobic groups, enzymatic modifications are extremely more limited than chemical modifications.

The most common method of chemical modification with respect to the attachment of hydrophobic groups is acylation reaction [2,3]. Normally, anhydrides of various carboxylic acids are used to introduce acyl groups into ε-amino groups of lysine residues. Haque and Kito [39] attached several fatty acids with varying alkyl carbon numbers to an α_{S1}-casein and showed that the surface tension decreased with an increase in alkyl number. Magdassi and Toledano [40] reported a similar observation for modified ovalbumin, although they measured interfacial tension at the water–tetradecane interface. In addition to the results on surface tension, the effects of modification were demonstrated for emulsifying properties. Haque and Kito [41] modified α_{S1}-casein by various degrees of palmitoyl groups. The improvement of emulsifying properties was demonstrated when the oil content of emulsions was relatively high (up to nearly 50%). Good emulsifying activity was attained even from a low degree of modification—0.3 groups per casein molecule. But emulsion stability was improved when the degree of modification was above 2.5 groups per molecule. Magdassi and Toledano [40] also pointed out the improvement of emulsifying properties as a result of the attachment of various fatty acyl groups to ovalbumin.

As described earlier, fatty acylation gives rise to the dramatic changes in the surface properties of proteins. If the attachment of fatty acyl groups to proteins is carried out not via chemical reaction but via enzyme-catalyzed reaction, the range of application of fatty acylated proteins should be expanded, especially in the food industry. However, most attempts concerning the fatty acylation of proteins via enzymatic methods have not been successful. On the contrary, enzyme-catalyzed synthesis of acyl-amino acids is still an important challenge worth facing. Several examples of enzymatic acylation of amino acids are discussed next.

Lipase (triacylglycerol acylhydrolase), which hydrolyzes ester bonds of triacylglycerol under biological conditions, has many other catalytic functions, such as the synthesis of esters, transestrification, thiotransesterification, and aminolysis under suitable conditions [42]. There have been several reports about the lipase-catalyzed acylation of amino acids. Nagao and Kito [43] succeeded in O-acylation of L-homoserine by a lipase in a biphasic system and found that the product had excellent emulsifying activity. The enzyme reaction was carried out by mixing fatty acids or oils with amino acid solution containing lipase from *Candida cylyndraceaor* and *Rhizopus delemar*. The synthetic reaction of O-oleoyl-L-homoserine catalyzed by *C. cylyndraceaor* lipase was restricted to the pH range 7.5–8.0. The optimum pH for the hydrolysis of olive oil by *C. cylyndraceaor* lipase was 7.2, and the activity (more than 50% of the maximum) was observed in the pH range 3.0–9.0. This restricted active pH range for synthesis may be due to a pH-dependent change in the ionic state of L-homoserine, the electrical charge of which may have some effect on its affinity to lipase. The reaction proceeded in a linear manner for 4 h after the start of reaction and then gradually reached a plateau, as shown in Fig. 7. After 12 h, fresh lipase (1 mg) was added to the reaction mixture. The amount of O-oleoyl-L-homoserine increased slightly, reaching a plateau after approximately 2 h following the addition of lipase. This fact suggests that the lipase might not be inactivated significantly during the course of the reaction. After 12 h, further addition of buffer (1 mL) to the reaction system caused a rapid decrease in the concentration of O-oleoyl-L-homoserine, plateauing after 8 h. The addition of buffer enhanced the hydrolysis reaction, with the equilibrium being shifted in the direction of hydrolysis. This suggests that equilibrium was reached after a 12-h reaction.

The substrate specificity to the acyl donor was examined (Table 4). Vegeta-

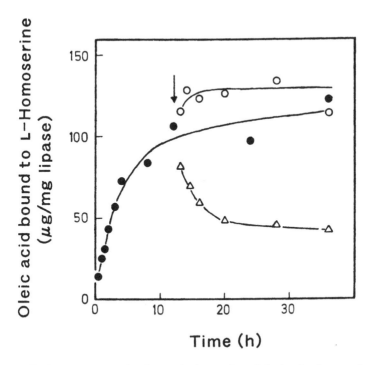

FIG. 7 Time course for the synthetic reaction of O-oleoyl-L-homoserine. The reaction mixture contained 100 mg of oleic acid, 10 mg of lipase, and 3 M L-homoserine in 1 mL of 0.1 M potassium phosphate buffer, pH 7.5 (●). The arrow indicates the time at which 10 mg of fresh lipase (○) or 1 ml of 0.1 M potassium phosphate buffer, pH 7.5 (△), was added to the reaction mixture. (From Ref. 43.)

ble oils and unsaturated fatty acids could be substrates. However, solid fatty acids at the reaction temperature, such as palmitic and stearic acids, were not effective acyl donors. The reactivity of palmitic and stearic acids increased when n-hexane was added to the reaction mixture, whereas the reactivity of oleic acid greatly decreased in the presence of n-hexane.

The emulsifying activity of O-oleoyl-L-homoserine was compared with that of casein, sodium oleate, and other surfactants [43]. O-Oleoyl-L-homoserine formed oil-in-water emulsions at oil-to-water ratios of 0.2, 0.5, and 1.0 (v/v). Its emulsifying activity was higher than that of conventional surfactants and casein, as shown in Fig. 8. O-Acyl-L-homoserine can be synthesized by an enzymatic process and hydrolyzed to fatty acids and L-homoserine by pancreatic lipase in the digestive organs. L-Homoserine is a naturally occurring

TABLE 4 Specificity of Lipase to Acyl Donors

Acyl donor	Relative activity	$(\%)^a$
Myristic acid	10.7	2.5^b
Palmitic acid	0	4.8^b
Stearic acid	0	2.9^b
Palmitoleic acid	106.4	
Oleic acid	100	3.6^b
Linoleic acid	81.7	
Soybean oil	109.8	
Corn oil	108.3	
Olive oil	113.6	

^{a} Relative activity was calculated from the weight of fatty acids bound to L-homoserine. The value for oleic acid was regarded as 100%.
^{b} One mL of n-hexane was added to the reaction mixture.
Source: Ref. 43.

amino acid and is widely distributed in plants, especially in germinating pea seeds. Therefore, O-acyl-L-homoserine may be suitable for human use. Other reports on the lipase-catalyzed synthesis of amino acid–based surfactants, such as N-ε-acyl-lysines [44], N-Lauroyl-β-alanine [45], and O-myristoyl-N-carbobenzyloxy-L-serine [46], have appeared, although these surfactants are thought to be of primary use as pharmaceuticals, biomedicals, and antimicrobial agents.

The use of enzymes other than lipases for the synthesis of amino acid- or peptide-based surfactants has rarely been reported. Although several acyltransferases are found to be involved in the fatty acylation of proteins in biological systems [47], such transferases are not available for commercial use. Recently, the covalent attachment of fatty acid to L-methionine using amino acylase from *Aspergillus* sp. was investigated (Fig. 9) [48]. The reaction was carried out by mixing fatty acids or their ethyl esters with amino acid solution containing the amino acylase at 25°C. Fatty acids are more susceptible to the enzymatic reaction than their ethyl esters. The yield of N-acyl-L-methionine decreased with an increase in the alkyl carbon number of the fatty acids. Maximum yield was obtained in the case of propionic acid (C3), but was still very low (about 5%). The effects on the reaction of adding various organic solvents were investigated. Only the addition of 50% (v/v) glycerol was found to increase the yield, 1.7 times. However, the yield is not satisfactory, and further investigation is necessary to improve the reaction efficiency. The immiscibility

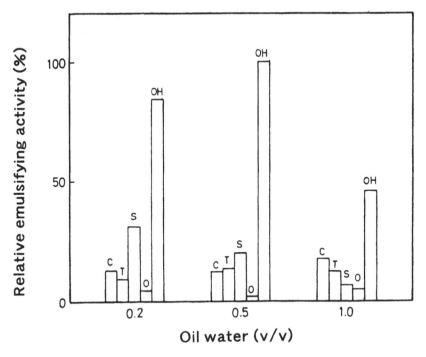

FIG. 8 Relative emulsifying activity of O-oleoyl-L-homoserine and surfactants. Relative emulsifying activity was calculated by regarding the activity of O-oleoyl-L-homoserine to the mixture of the oil-to-water ratio of 0.5 as 100%. C, casein; T, Tween 40; S, span 80; O, sodium oleate; OH, O-oleoyl-L-homoserine. (From Ref. 43.)

of amino acid solution and fatty acids is a serious problem. Furthermore, water content must be reduced, because the equilibrium between hydrolysis and the synthesis reaction depends on water activity. A biphasic reaction system may be effective, because amino acid is highly soluble in water and the fatty acids or their esters are soluble in a water-immiscible solvent [42]. Nakanishi [48] also checked the efficiency of the biphasic reaction system for the synthesis of acyl amino acids, but did not obtain high yields. The low efficiency of synthesis may be due to the use of enzyme soluble in water. Many novel ideas for using enzymes to produce fatty acid derivatives in organic solvents or biphasic systems have been postulated. Immobilization of enzymes onto polymers is one of most important ways to enable enzymes to function in organic solvents [49]. Recently, the new type of organic solvent–soluble enzymes that are covered with synthetic glycoside detergents has been shown to be very

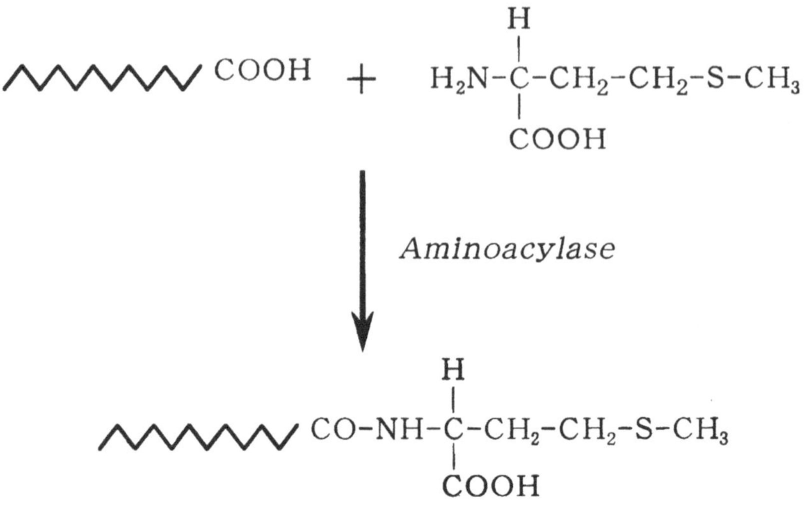

$$\text{\large /\hspace{-2pt}\backslash\hspace{-2pt}/\hspace{-2pt}\backslash\hspace{-2pt}/ COOH} \quad + \quad \underset{\underset{COOH}{|}}{\overset{\overset{H}{|}}{H_2N-C}}-CH_2-CH_2-S-CH_3$$

$$\big\downarrow \quad Aminoacylase$$

$$\text{\large /\hspace{-2pt}\backslash\hspace{-2pt}/\hspace{-2pt}\backslash\hspace{-2pt}/ CO-NH-}\underset{\underset{COOH}{|}}{\overset{\overset{H}{|}}{C}}-CH_2-CH_2-S-CH_3$$

FIG. 9 Covalent attachment of fatty acid to L-methionine catalyzed amino acylase from *Aspergillus* sp.

useful for the catalysis of the covalent attachment of hydrophobic moiety to hydrophilic substrates in organic media as well as in biphasic systems [50–52]. Such a modification of the aminoacylase should be necessary to make a breakthrough in the synthesis of acyl-amino acids in the reaction system, described earlier, of Nakanishi [48]. If this method is used for incorporating fatty acids into peptides, including glutamine, the resultant fatty acylated peptides could be attached to the lysine residues in proteins by a transglutaminase-catalyzed reaction, thereby producing protein-based surfactants. According to Motoki and Seguro [18], transglutaminase can catalyze directly the incorporation of phospholipids into natural proteins using phosophatidyl ethanolamine as acyl donors.

Although the modification of proteins with fatty acids via biochemical methods in vitro is quite difficult, the acylation of some proteins by an autocatalytic process has been reported [53]. Nishimura et al. [54] found that soybean 11S globulin was autoacylated with palmitoyl-CoA. The results on autoacylation of soy proteins in native, denatured, reduced, and both denatured and reduced states are shown in Table 5. The reduced 11S globulin reacted well with palmitoyl-CoA, and both denatured and reduced 11S globulin exhibited twice as much reactivity as the reduced 11S globulin. The native and the denatured 11S globulin did not react with palmitoyl-CoA. These results indicate that a reduction disulfide bond is essential for the reaction with palmitoyl-

TABLE 5 Autoacylation of 11S Globulin and 7S Globulin

	Palmitic acid incorporated (nmol)			
Protein	Native	Denatured	Reduced	Denatured and reduced
7S globulin	0.006			
α-Subunit		ND[a]	ND	0.014
α'-Subunit		ND	ND	0.015
β-Subunit		ND	ND	0.026
11S globulin	0.005			
Acidic polypeptide		<0.005	0.163	0.396
Basic polypeptide		<0.005	0.402	0.690

[a] ND, not determined.
Source: Ref. 54.

CoA. On the other hand, 7S globulin with a low content of cysteine residues did not react with palmitoyl-CoA, even in the denatured and reduced states. Based on the results of the blocking of SH groups and the sensitivity of hydroxylamine treatment, palmitic acid was found to be linked to cysteine residues of 11S globulin via a thioester bond. Emulsifying activity of the acylated 11S globulin was greatly increased [54]. Since it is not thought that 11S globulin itself has catalytic activity of acylation, attachment of acyl-CoA to the protein can be regarded as a sort of chemical reaction. However, it should be noted that acyl-CoA is present in biological systems and autoacylation normally proceeds under mild and physiological-like conditions. Therefore, autoacylation of proteins with acyl-CoA indicates the possibility of biological preparation of fatty acylated proteins in vitro.

We recently have seen a report about lipase-catalyzed lipopholization of soy proteins [55]. The reactions, which were catalyzed by fixed (lipozyme) and free lipases (lipase S of *Rhizopus arrhizus*), were carried out at 60°C with and without solvent. As the authors described, when the reaction was carried out with lipozyme without solvent, 60%, 33%, and 42% of lysine residues were acylated by capric, lauric, and oleic groups, respectively. Such high yields are surprising, judging from the difficulty synthesizing fatty acylated amino acids. Further and more precise investigation is necessary for the analyses of products, since the authors determined the yield of covalent linkage of hydrophobic chains by comparing available lysine residues before and after lipophilization according to the hypothesis that fatty acids acylate only the ε-amino group of lysine residues of proteins. However, this paper is stimulating fundamental and practical interest in the direct incorporation of fatty acids to proteins via enzyme-catalyzed reaction.

Before closing this chapter, we mention the possibility applying protein engineering to the acylation of proteins as well as to the attachment of carbohydrate to proteins. In biological systems, the covalent attachment of the fatty acids occurs cotranslationally or posttranslationally by a specific acyltransferase [56]. One representative form of fatty acid modification of proteins is an amide bond of myristic acid to an amino-terminal glycine (*N*-myristoylation) [57]. Consensus sequence for myristoylation is Met-Gly-X-X-X-Ser (glycine becomes N-terminal amino acid via the aminopeptidase-catalyzed cleavage of methionine prior to myristoylation). The introduction of a consensus sequence to the complementary DNA of proteins was attempted, and the myristoylation of proteins has been demonstrated [58]. Although there are many roadblocks to practical use, protein engineering is a promising tool for the fatty acylation of proteins and for the production of amphoteric protein–based surfactants.

ACKNOWLEDGMENTS

The authors are very grateful to Prof. Nakanishi of Okayama University for providing information and discussions on enzymatic synthesis of acylated amino acids. The experiments on synthesis of amino acid–based surfactants referred to in the text are now under way in collaboration with Prof. Nakanishi, supported partially by a Grant-in-Aid from the Program for Promotion of Basic Research Activities for Innovative Bioscience in Japan. The authors wish to thank Prof. Shimizu of Tokyo University and Prof. Kato of Yamaguchi University for discussions relating to the text.

REFERENCES

1. J. Lefebvre and P. Relkin, in: Surface Activity of Proteins (S. Magdassi, ed.), Marcel Dekker, New York, 1996, pp. 181–236.
2. F. F. Shih, in: Biochemistry of Food Proteins (B. J. F. Hudson, ed.), Elsevier, London, 1992, pp. 235–248.
3. K. D. Schwenke, in: Food Proteins and Their Applications (S. Damodaran and A. Paraf, eds.), Marcel Dekker, New York, 1997, pp. 393–423.
4. P. M. Nielsen, in: Food Proteins and Their Applications (S. Damodaran and A. Paraf, eds.), Marcel Dekker, New York, 1997, pp. 443–472.
5. M. Shimizu and M. Saito, in: Macromolecular Interactions in Food Technology (N. Parris, A. Kato, L. K. Creamer, and J. Pearce, eds.), ACS Symposium Series 650, Washington D.C., 1995, pp. 156–165.
6. M. Shimizu, S. W. Lee, S. Kaminogawa, and K. Yamauchi, J. Food Sci. 51:1248 (1986).

7. M. Saito, K. Chikuni, M. Monma, and M. Shimizu, Biosci. Biotech. Biochem. 57:952 (1993).
8. X. M. He and D. C. Carter, Nature 358:209 (1992).
9. M. Z. Papiz, L. Sawyer, E. E. Eliopoulos, A. C. T. North, J. B. C. Findlay, R. Sivaprasadarao, J. A. Jones, M. E. Newcomer, and P. J. Kraulis, Nature 324:383 (1986).
10. M. Saito, M. Ogasawara, K. Chikuni, and M. Shimizu, Biosci. Biotech. Biochem. 59:388 (1995).
11. S. Damodaran, in: Food Proteins and Their Applications (S. Damodaran and A. Paraf, eds.), Marcel Dekker, New York, 1997, pp. 57–110.
12. C. M. Deber and N. K. Goto, in: Membrane Protein Assembly (G. von Heijne, ed.), Springer, Heidelberg, 1997, pp. 25–37.
13. F. Hucho, U. Gorne-Tschelnokow, and A. Strecker, Trends Biol Sci. 19:383 (1994).
14. A. Kato, A. Tanaka, N. Matsudomi, and K. Kobayashi, J. Agric. Food Chem. 35:224 (1987).
15. A. Kato, A. Tanaka, A. Y. Lee, N. Matsudomi, and K. Kobayashi, J. Agric. Food Chem. 35:285 (1987).
16. J. S. Hamada, J. Am. Oil Chem. Soc. 68:459 (1991).
17. I. A. Vaintraub, L. V. Kotova, and R. Shaha, in: Food Proteins. Structure and Functionality (K. D. Schwenke and R. Mothes, eds.), VCH, Weinheim, Germany, 1993, pp. 187–189.
18. M. Motoki and K. Seguro, Trends Food Sci. Technol. 9:204 (1998).
19. K. Seguro and M. Motoki, Agric. Biol. Chem. 53:3263 (1989).
20. K. Seguro and M. Motoki, Agric. Biol. Chem. 54:1271 (1990).
21. N. F. Campbell, F. F. Shih, and W. E. Marshall, J. Agric. Food Chem. 40:403 (1992).
22. M.-C. Ralet, D. Fouques, T. Chardot, and J.-C. Meunier, J. Agric. Food Chem. 44:69 (1996).
23. A. Kato, K. Mianaki, and K. Kobayashi, J. Agric. Food Chem. 4:540 (1993).
24. S. Nakamura, M. Ogawa, S. Nakai, A. Kato, and D. D. Kitts, J. Agric. Food Chem. 46:3958 (1998).
25. B. Colas, D. Caer, and E. Fournier, J. Agric. Food Chem. 41:1811 (1993).
26. L. Lorand, K. N. Parameswaran, P. Stenberg, Y. S. Tong, P. T. Velasco, N. A. Jonsson, L. Mikiver, and P. Moses, Biochemistry 18:1756 (1979).
27. S. Nakamura, H. Takasaki, K. Kobayashi, and A. Kato, J. Biol. Chem. 268:12706 (1993).
28. Y. Shu, S. Maki, S. Nakamura, and A. Kato, J. Agric. Food Chem. 46:2433 (1998).
29. S. Arai and M. Watanabe, in: Advances in Food Emulsions and Foams (E. Dickinson and G. Stainsby eds.), Elsevier, London, 1988, pp. 189–220.
30. M. Watanabe and S. Arai, in: Biochemistry of Food Proteins (B. J. F. Hudson, ed.), Elsevier, London, 1992, pp. 271–395.
31. V. Sciancalepore and V. Longone, J. Dairy Res. 55:547 (1988).
32. J. Adler-Nissen, J. Agric. Food Chem. 27:1256 (1979).

33. M. Yamashita, S. Arai, Y. Iamaizumi, Y. Amano, and M. Fujimaki, J. Agric. Food Chem. 27:52 (1979).
34. S. Toiguchi, S. Maeda, M. Watanabe, and S. Arai, Agric. Biol. Chem. 46:2945 (1982).
35. M. Watanabe, A. Shimada, and S. Arai, Agric. Biol. Chem. 44:1621 (1981).
36. A. Shimada, I. Yamamoto, H. Sase, Y. Yamazaki, M. Watanabe, and S. Arai, Agric. Food Chem. 48:2681 (1984).
37. M. Watanabe, A. Shimada, E. Yazawa, T. Kato, and S. Arai, J. Food Sci. 46: 1738 (1981).
38. G. Hajos, in: Surface Activity of Proteins (S. Magdassi, ed.), Marcel Dekker, New York, 1996, pp. 131–180.
39. Z. Haque and M. Kito, J. Agric. Food Chem. 32:1392 (1984).
40. S. Magdassi and O. Toledano, in: Surface Activity of Proteins (S. Magdassi, ed.), Marcel Dekker, New York, 1996, pp. 39–60.
41. Z. Haque and M. Kito, J. Agric. Food Chem. 31:1231 (1983).
42. T. Yamane, J. Am. Oil Chem. Soc. 64:1657 (1987).
43. A. Nagao and M. Kito, J. Am. Oil Chem. Soc. 66:710 (1989).
44. D. Monte, F. Servat, M. Pina, J. Graille, P. Galzy, A. Arnaud, H. Leden, and L. Marcous, J. Am. Oil Chem. Soc. 67:771 (1990).
45. T. Izumi, Y. Yagimuma, and M. Haga, J. Am. Oil Chem. Soc. 74:875 (1997).
46. R. Valivety, P. Jauregi, I. Gill, and E. Vulfson, J. Am. Oil Chem. Soc. 67:771 (1990).
47. L. Guttierrez and A. I. Magee, Biochim. Bihophys. Acta 1078:147 (1991).
48. K. Nakanishi (unpublished data)
49. M. Basri, K. Ampon, W. M. Z. Wan Yunus, C. N. A. Razak, and A. B. Salleh, J. Am. Oil Chem. Soc. 72:407 (1995).
50. W. Tsuzuki, Y. Okahata, O. Katayama, and T. Suzuki, J. Chem. Soc. Perkin Trans. 1:1245 (1991).
51. M. Goto, N. Kamiya, M. Miyata, and F. Nakashio, Biotechnol. Prog. 10:263 (1994).
52. W. Tsuzuki, K. Akasaka, S. Kobayashi, and T. Suzuki, J. Am. Oil Chem. Soc. 72:1333 (1995).
53. O. A. Bizzozero, J. F. McGarry, and M. B. Lees, J. Biol. Chem. 262:13550 (1987).
54. T. Nishimura, S. Utsumi, and M. Kito, J. Agric. Food Chem. 37:1266 (1989).
55. C. Roussel, M. Pina, J. Graille, A. Huc, and E. Perrier, OCL. 4:284 (1997).
56. M. Berger and M. F. G. Schmidt, J. Biol. Chem. 259:7245 (1984).
57. M. E. Linder, P. Middleton, J. R. Hepler, R. Taussig, A. G. Gilman, and S. M. Mumby, Proc. Natl. Acad. Sci. USA 90:3675 (1993).
58. A. Kato (unpublished data).

6

Arginine Lipopeptide Surfactants with Antimicrobial Activity

M. R. INFANTE and A. PINAZO Instituto de Investigaciones Químicas y Ambientales de Barcelona, CSIC, Barcelona, Spain

J. MOLINERO The Colomer Group, Barcelona, Spain

J. SEGUER Laboratorios Miret, S.A. (LAMIRSA), Terrassa, Spain

P. VINARDELL Facultad de Farmacia UB, Barcelona, Spain

I. INTRODUCTION

The use of antimicrobial agents for controlling or preventing infection is of major practical importance. Given the adaptability of microorganisms and their tendency to acquire resistance, the development of new antimicrobial agents is a constant challenge. However, antimicrobials carry inherent risks with respect to both environmental and mammalian toxicology [1]. There is accordingly a pressing need for biodegradable antimicrobials with low toxicity profiles.

All forms of life, both plants and animals, produce antimicrobial peptides to prevent and eliminate infection [2]. While exhibiting great structural diversity, they are typically cationic amphiphilic molecules with an excess of basic amino acid residues, particularly lysine (Lys) and arginine (Arg). The presence of a peptide segment capable of forming amphiphilic α-helix or β-sheet elements, surface activity against membrane surfaces, and the highly basic amino acid components are all important requirements for their biological activity [3–5].

One of our main goals is the design and development of new low-toxicity antimicrobial compounds to mimic natural amphiphilic cationic peptides [6–9]. Via acyl, ester, and amide linkages, a simple and effective strategy has been to introduce into the amino acid structure low-molecular-weight amphiphilic structures combining basic amino acids and hydrophobic alkyl chain residues. Thus, since 1985, we have used chemical and enzymatic synthetic technologies [10–12] to explore lipoaminoacids carrying long-chain N^α-acyl, COO^α-ester and N-alkyl amide Lys and Arg derivatives. These amphiphilic molecules, particularly the N^α-acyl arginine alkyl ester derivatives (e.g., N^α-lauroyl arginine methyl ester, LAM) have turned out to be an important class of cationic surface-active compounds with rich self-aggregation properties, bactericidal activity, biodegradability, and low irritation for skin (see Fig. 1 for the structure of N^α-acyl arginine alkyl ester surfactants). These properties make them excellent products for both industrial and basic research applications [11].

We have shown that essential structural factors for the antimicrobial activity of these N^α-acyl arginine alkyl ester surfactants include both the length of the fatty residue (conferring solubility and surface-active properties) and the presence of the protonated guanidine basic function. The literature records a significant group of compounds with excellent antiseptic and pharmacological behavior whose common feature is strongly basic groups of the guanidine type attached to a fairly massive lipophilic molecule [13]. We expected that the condensation of such a long-chain N^α-acyl arginine residue to a second amino acid or peptidic fragment would yield salts of N-acyl arginine-based peptides with interesting biocompatibility and antimicrobial properties.

Because of the simplicity of their structure, good surface properties, mild effect on the skin, and biodegradability, the salts of long-chain N-acyl peptides have been finding increasing utility in detergents, cosmetics, pharmaceuticals, and foods [14–18]. A number of N-acylated peptides with different hydrophobic contents are manufactured commercially from hydrolyzed animal protein

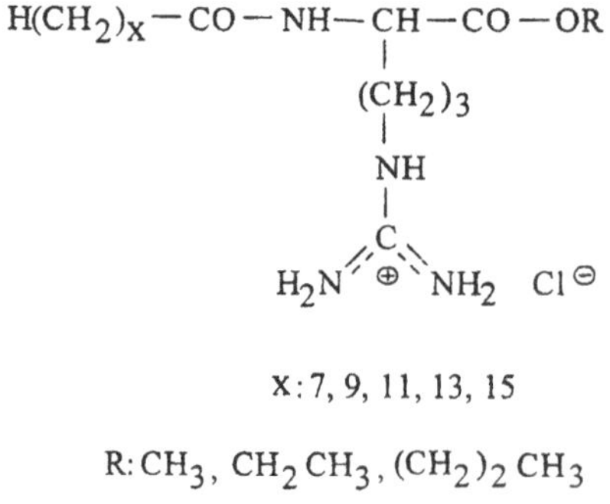

$$H(CH_2)_x - CO - NH - CH - CO - OR$$
$$|$$
$$(CH_2)_3$$
$$|$$
$$NH$$
$$|$$
$$C$$
$$H_2N \quad \oplus \quad NH_2 \quad Cl^\ominus$$

$$x: 7, 9, 11, 13, 15$$

$$R: CH_3, CH_2CH_3, (CH_2)_2CH_3$$

FIG. 1   Structure of N^α-acyl-L-arginine alkyl ester derivatives.

(collagen and keratin) or from plant proteins (soy, wheat). They are claimed to have good foaming behavior, excellent skin and mucosa compatibility, and total biodegradability; however, since they are devoid of antimicrobial activity, formulations containing them require the addition of preservative.

This chapter describes their synthesis on a laboratory scale, some of the properties of noncommercial antimicrobial lipopeptidic surfactants derived from N^α-acyl arginine, and the relationships between antimicrobial function and the chemical structure of the hydrophobic and hydrophilic moieties.

II. STRUCTURE

In the design of these compounds, the ratio between the molecules' hydrophilic and hydrophobic moieties as well as their structures determine antimicrobial properties. Variations include modifying the amino acid composition and/or the fatty acid chain length. Three types of N^α-acyl-L-arginine lipopeptide surfactants have been prepared: (1) pure lipodipeptides from glycine (neutral), phenylalanine (hydrophobic), glutamic acid (acidic), and lysine (basic) amino acids; (2) polydisperse lipodipeptides from an acid-hydrolyzed collagen amino acid mixture; and (3) polydisperse lipopeptides from a partially hydrolyzed collagen peptide mixture of average molecular weight 1000 daltons. The structures and abbreviations are given in Fig. 2 and Table 1, respectively.

Arginine-based lipopeptides can be considered a subgroup of classical N^α-acyl peptidic surfactants [15], which, however, are usually anionic, in contrast to the amphoteric nature of our arginine-based compounds resulting from their positively charged N^δ-guanidine and negatively charged carboxylic groups. Their antimicrobial activity and extreme mildness for skin give the arginine-based compounds a higher added value.

III. METHODS OF PREPARATION

A. Pure Lipodipeptides from Pure Amino Acids (aa)

Ideal chemical conditions for the preparation of pure long-chain N^α-acyl-L-arginine dipeptides would allow rapid and quantitative acyl bond formation under mild conditions and the avoidance of side reactions while maintaining all the adjacent chiral centers. A number of methodologies have been developed to overcome problems related to reactivity and purity.

The simplest and most obvious way of making these materials is based on the introduction of a fatty acid residue as an acid chloride in a strong alkaline aqueous medium that contains the single or mixed amino acids (e.g., from

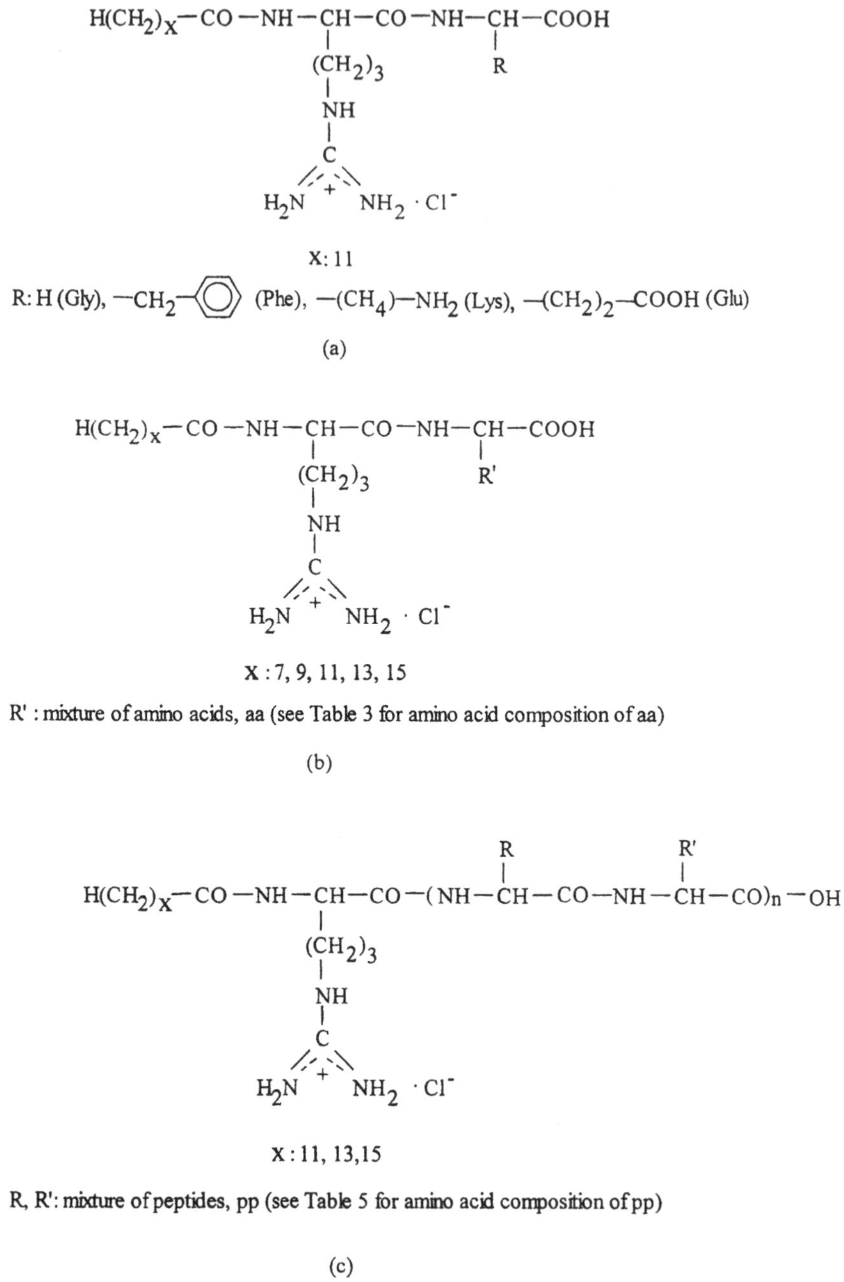

$$H(CH_2)_X-CO-NH-CH-CO-NH-CH-COOH$$

Structure (a):

$$H(CH_2)_X-CO-NH-\underset{\underset{\underset{\underset{\underset{\overset{+}{C}}{\underset{H_2N}{}\ \ NH_2\cdot Cl^-}}{NH}}{(CH_2)_3}}{CH}}-CO-NH-\underset{R}{CH}-COOH$$

X: 11

R: H (Gly), $-CH_2-\bigcirc$ (Phe), $-(CH_4)-NH_2$ (Lys), $-(CH_2)_2-COOH$ (Glu)

(a)

Structure (b):

$$H(CH_2)_X-CO-NH-\underset{\underset{\underset{\underset{\underset{\overset{+}{C}}{\underset{H_2N}{}\ \ NH_2\cdot Cl^-}}{NH}}{(CH_2)_3}}{CH}}-CO-NH-\underset{R'}{CH}-COOH$$

X : 7, 9, 11, 13, 15

R' : mixture of amino acids, aa (see Table 3 for amino acid composition of aa)

(b)

Structure (c):

$$H(CH_2)_X-CO-NH-\underset{\underset{\underset{\underset{\underset{\overset{+}{C}}{\underset{H_2N}{}\ \ NH_2\cdot Cl^-}}{NH}}{(CH_2)_3}}{CH}}-CO-(NH-\underset{R}{CH}-CO-NH-\underset{R'}{CH}-CO)_n-OH$$

X : 11, 13,15

R, R': mixture of peptides, pp (see Table 5 for amino acid composition of pp)

(c)

FIG. 2 Structures of lipopeptidic surfactants synthesized from arginine: (a) pure N^α-lauroyl-L-arginine dipeptides; (b) polydisperse N^α-acyl-L-arginine dipeptides; (c) polydisperse N^α-acyl-L-arginine peptides.

TABLE 1 Structures of and Abbreviations for Lipopeptides Synthesized from Arginine

Compound	x	And/or R, R'	Abbreviation
N^α-Lauroyl-L-arginine-glycine hydrochloride[a]	11	H	C_{12}ArgGly
N^α-Lauroyl-L-arginine-phenylalanine hydrochloride[a]	11	CH_2-Ph	C_{12}ArgPhe
N^α-Lauroyl-L-arginine-glutamic hydrochloride[a]	11	$(CH_2)_2-COOH$	C_{12}ArgGlu
N^α-Lauroyl-L-arginine-lysine dihydrochloride[a]	11	$(CH_2)_4-NH_3Cl$	C_{12}ArgLys
N^α-Octanoyl-L-arginine-dipeptides hydrochloride[b]	7	Mixture of amino acids	C_8Arg aa
N^α-Decanoyl-L-arginine-dipeptides hydrochloride[b]	9	Mixture of amino acids	C_{10}Arg aa
N^α-Lauroyl-L-arginine-dipeptides hydrochloride[b]	11	Mixture of amino acids	C_{12}Arg aa
N^α-Miristoyl-L-arginine-dipeptides hydrochloride[b]	13	Mixture of amino acids	C_{14}Arg aa
N^α-Palmitoyl-L-arginine-dipeptides hydrochloride[b]	15	Mixture of amino acids	C_{16}Arg aa
N^α-Lauroyl-L-arginine-peptides hydrochloride[c]	11	Mixture of peptides[d]	C_{12}Arg pp
N^α-Miristoyl-L-arginine-peptides hydrochloride[c]	13	Mixture of peptides[d]	C_{14}Arg pp
N^α-Palmitoyl-L-arginine-peptides hydrochloride[c]	15	Mixture of peptides[d]	C_{16}Arg pp

[a] See Fig. 2a.
[b] See Fig. 2b.
[c] See Fig. 2c.
[d] Average molecular weight: 1000 daltons.

protein hydrolysates) [15,18,19]. This approach is used by most manufacturers for the preparation of this type of surfactant. A high pH value is necessary to provide a nucleophilic free amino group available for acylation. There are many references describing amino acid–based surfactants and their preparation and properties [11,15,20,21].

The guanidine sidechain of Arg is an extremely strong base that, in the unsubstituted state, remains protonated under normal conditions [22]. Provided the pH is carefully controlled, it is possible to work with an unprotected sidechain, although to obtain very pure products a lateral protection is advisable [23]. The introduction of the strongly electronegative nitro function (NO_2) as a lateral protected group depresses the basic nature of the guanidine function, thus facilitating the incorporation of arginine into peptides by the formation of α-peptidic bonds without competition from the guanidine lateral group [24].

N^α-Lauroyl-L-arginine dipeptides from pure amino acids C_{12}ArgGly, C_{12}ArgPhe, C_{12}ArgGlu and C_{12}ArgLys (Fig. 2a) were prepared by our group on a laboratory scale following the procedure in Scheme 1, which comprised four steps: (1) acylation of N^δ-nitro arginine with lauroyl chloride; (2) peptide coupling; (3) nitrodeprotection; and (4) saponification.

The synthesis used the anhydride method [isobutyl chloroformate (IBClF) in the presence of N-methylmorpholine (NMM) as base] to couple the α-carboxylic group of N^α-lauroyl-L-nitro arginine to the α-amino group of the appropriate methyl ester derivative of the terminal amino acid (Gly, Phe, Glu, or Lys) [25]. Temporary protection of the N^δ-guanidine group of Arg and the N^ε-amino group of Lys was afforded by the nitro and carbobenzoxy (Z) groups, respectively; protection was effective and no undesirable by-products were obtained. Deprotection was carried out by catalytic hydrogenation over palladium catalysts in the presence of formic acid [26]. The final compounds were obtained by subsequent saponification and treatment with methanolic HCl.

The compounds were purified by Flash column chromatography in silica gel to give very hygroscopic solids with a purity of 98.5 + 0.5% as determined by HPLC. All compounds were optically pure, with the absence of racemization being checked by gas chromatography [27]. Table 2 lists the properties of pure N^α-lauroyl-L-arginine dipeptides.

B. Polydisperse Lipodipeptides from a Mixture of Amino Acids

Long-chain N^α-acyl-L-arginine dipeptides from acid-hydrolyzed collagen C_8Arg aa, C_{10}Arg aa, C_{12}Arg aa, C_{14}Arg aa, and C_{16}Arg aa (Fig. 2b) were

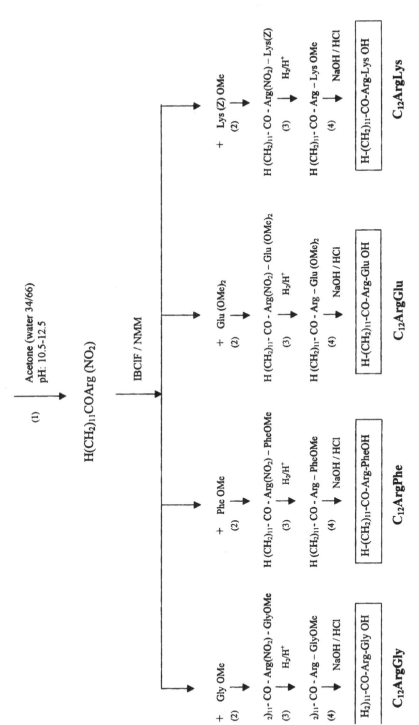

SCHEME 1 Synthetic pathway of pure arginine lipodipeptides.

TABLE 2 Characteristic Properties of Pure N^α-Lauroyl-L-arginine Dipeptides

Compound	Molecular weight	Purity by HPLC[a]	Melting point (°C)	$[\alpha]^{25}$ C = 1%, MeOH	FAB-MS m/e $[M^+H]^+$	aa Analysis molar ratio	Yield[b] (%)
C₁₂ArgGly	449.5	98.0	55–60	−2.65	414	Gly/Arg 1.08	63
C₁₂ArgPhe	539.5	99.0	27–31	−1.60	504	Phe/Arg 0.97	75
C₁₂ArgGlu	521.5	98.5	Hygroscopic	−2.54	486	Glu/Arg 1.004	50
C₁₂ArgLys	557.0	98.0	Hygroscopic	−1.42	485	Lys/Arg 0.97	65

[a] RP 18, λ = 210 nm; acetonitrile-water gradient system.
[b] After column purification.

prepared using a procedure similar to that for the pure N^α-acyl arginine dipeptides. Waste leather trimmings previously hydrolyzed with 6N HCl at 120°C for 24 h were the source of total acid-hydrolyzed collagen amino acids, which were not further purified; the composition of the dry mixture was 93 ± 5% (w/w) amino acids, with the balance as salt. A 10% aqueous solution of the dried hydrolysate gave a pH of 6.5 at room temperature; its amino acid composition expressed in mol (%) is shown in Table 3.

The compounds were synthesized by condensation of the corresponding long-chain N^α-acyl-L-nitro arginine derivatives (C_X Arg[NO_2]) with the collagen hydrolysate. The α-carboxylic group of arginine was coupled to the α-amino groups of the amino acid mixture (yield 54–75%) by aminolysis of the activated nitro-arginine α-carboxylic group with isobutyl chloroformate at low temperature (-15°C) [25]. Deprotection of the NO_2 was carried out by catalytic hydrogenation in the presence of acetic acid, with yields in the range 79–89%. The final products were oils obtained as hydrochloride salts.

TABLE 3 Amino Acid Composition of the Totally Acid-Hydrolyzed Collagen (aa) and Polydisperse N^α-Acyl arginine Dipeptides from Collagen (in mols %)

Amino acid	aa[a]	C_8 Arg aa	C_{10} Arg aa	C_{12} Arg aa	C_{14} Arg aa	C_{16} Arg aa
OH-PRO	10.06	4.91	4.32	6.71	9.17	7.11
ASP	5.15	3.16	2.28	2.29	1.35	2.17
THR	1.17	0.76	0.89	0.59	0.55	0.49
SER	3.64	2.07	1.64	1.44	0.93	1.17
GLU	7.44	6.36	6.79	3.81	4.02	5.17
PRO	12.11	7.11	6.89	6.49	9.58	7.23
GLY	31.43	17.07	18.32	16.07	15.11	18.21
ALA	12.56	7.49	8.28	7.58	7.42	6.17
VAL	1.19	1.01	1.13	0.72	1.17	1.01
MET	0.60	0.15	—	0.59	—	—
ILEU	1.56	1.02	0.83	0.83	—	—
LEU	3.05	1.23	1.61	1.82	0.97	0.53
TYR	0.54	0.11	—	0.18	—	—
PHE	1.25	0.96	—	0.77	0.39	—
HIS	0.53	—	1.55	—	0.24	—
LYS	2.31	1.38	1.47	1.28	1.50	1.44
ARG	4.83	44.18	41.93	47.67	46.71	48.37
OH-LYS	0.57	1.03	1.13	1.17	0.88	0.93

[a] Mixture of amino acids.

TABLE 4 Characteristic Properties of Polydisperse Lipopeptides C_x Arg aa and Their Precursors C_x Arg(NO_2)

Compound	Molecular weight	Melting point (°C)	Yield (%)
C_8 Arg(NO_2)[a]	345	165–168	44
C_{10} Arg(NO_2)[a]	373	175–177	47
C_{12} Arg(NO_2)[a]	401	178–180	46
C_{14} Arg(NO_2)[a]	429	182–184	51
C_{16} Arg(NO_2)[a]	457	188–189	43
C_8 Arg aa[b]	481[c]	Oil	72
C_{10} Arg aa[b]	509[c]	Oil	67
C_{12} Arg aa[b]	537[c]	Oil	75
C_{14} Arg aa[b]	565[c]	Oil	68
C_{16} Arg aa[b]	593[c]	Oil	54

IR: $\nu_{(NH-CO(I))} = 1680$ cm^{-1}; $\nu_{(NH-CO(II))} = 1550$ cm^{-1}; $\nu_{(CH_2)} = 2880-2800$ cm^{-1}.
[a] UV: $\lambda_{(NH-CO)} = 214$ nm; $\lambda_{(NO_2)} = 265$ nm.
[b] UV: $\lambda_{(NH-CO)} = 214$ nm.
[c] Average molecular weight.

The average molecular weight of these polydisperse lipodipeptides was 481–593 g/mol (see Table 4 for these and other characteristics); the amino acid composition of each of the C_x Arg aa compounds is given in Table 3. The high percentage of arginine compared with the other amino acids agreed with the structure of these compounds.

C. Polydisperse Lipopeptides from a Mixture of Peptides

Polydisperse C_{12} Arg pp, C_{14} Arg pp, and C_{16} Arg pp lipopeptides (Fig. 2c) were prepared as described previously but with some modifications. The peptidic fraction was isolated from partially hydrolyzed collagen (Proalan S.A., Spain) by ultrafiltration using an Amicon stirred cell (M8010) supplied with a nominal MW cutoff YM-1000 Diaflo ultrafiltration membrane. The filtration was carried out in aqueous solution at pH 3.5 at room temperature and 3.4 bar. The required peptides were obtained by freeze-drying of the filtrate portion and used without further purification. The final products (obtained as white powders) were polydisperse mixtures of amphiphilic arginine-based peptides with an average molecular weight of about 1000 daltons. Yields were around 60%, and melting points for the three categories of peptide were: C_{12} Arg pp (180–187°C), C_{14} Arg pp (190–192°C), and C_{16} Arg pp (191–194°C).

TABLE 5 Amino Acid Composition of Partially Hydrolyzed Peptides (pp) and Polydisperse N^α-Acyl Arginine Peptides, in mols %

Amino acid	pp^a	C_{12}Arg pp	C_{14}Arg pp	C_{16}Arg pp
OH-PRO	4.40	2.75	2.95	2.35
ASP	2.30	1.15	0.95	0.82
THR	2.15	1.36	1.15	1.07
SER	5.30	2.36	2.50	2.35
GLU	10.05	6.88	6.34	6.53
PRO	16.77	12.29	11.73	10.42
GLY	23.00	36.35	34.40	32.50
ALA	15.58	10.09	11.03	10.20
VAL	2.89	2.71	3.04	2.93
ILE	1.57	1.03	0.98	0.88
LEU	3.22	2.02	2.10	2.02
TYR	0.04	0.04	0.11	—
PHE	1.70	1.03	—	1.15
HIS	1.46	2.44	2.30	2.40
LYS	3.53	2.77	2.50	2.63
ARG	5.96	14.48	14.50	14.34

^{a}Partially acid-hydrolyzed collagen: peptide mixture.

Amino acid analysis, thin-layer chromatography, and nuclear magnetic resonance showed the samples to be devoid of free amino acids, peptides, nitro derivatives, and urethanes. Table 5 gives the amino acid compositions of the starting peptidic fraction and of polydispersed lipopeptides.

IV. RESULTS AND DISCUSSION

A. Physicochemical Properties

1. Solubility

In distilled water at pH in the range 2–9, all three types of lipopeptides displayed good solubility at room temperature (0.5–25%, w/v) except for C_{12}ArgGlu (fully soluble only at pH 2.0) and C_{16}Arg aa, which, because of its very hydrophobic character, was soluble only at pH 9.0. C_{12}ArgGly and C_{12}ArgPhe showed Krafft points around 0°C. All the synthetic lipopeptides were readily soluble in methanol and ethanol at room temperature.

As expected, the solubilities of the compounds decline with increasing alkyl chain length and are enhanced by increasing length of the peptidic moiety.

For instance, while C_{16}Arg aa is poorly soluble in water at pH < 9.0, C_{16}Arg pp has a water solubility of 1% (w/v) at room temperature in the range pH 2–9. Moreover, the polydispersion of the hydrophilic moiety yields greater water solubility compared with the pure analogues with the same alkyl chain length (i.e., at a particular pH, C_{12}Arg aa is more water-soluble than C_{12}ArgGly).

The insolubility of C_{12}ArgGlu at pH 2–9 might be attributable to this molecule's undergoing internal molecular guanidine–carboxylate interactions between the sidechains of Arg and Glu, thereby producing a water-insoluble complex in which only carboxylate anions remain free [28] (Fig. 3).

A similar phenomenon has been observed with the water-insoluble amphoteric derivative N-lauroyl-L-arginine (C_{12}Arg), the insolubility of which is attributed to the formation of guanidinium–carboxylate intermolecular interactions probably stabilized by strong hydrogen bonding (see Fig. 4) [10]. Because of its lack of solubility, this compound is neither effectively surface active nor an antimicrobial. Similar water solubility was described for the N-lauroyl lysine analogue [17]. Since intramolecular guanidine–carboxylate interaction does not take place if the guanidine and carboxylate groups are kept apart, our experience suggests that an interesting strategy for obtaining amphoteric water-soluble lipopeptide compounds is the introduction of a second amino acid or peptide into the amphoteric N^{α}-lauroyl arginine (C_{12}Arg) residue. As we shall shortly see, under these conditions some of the lipopep-

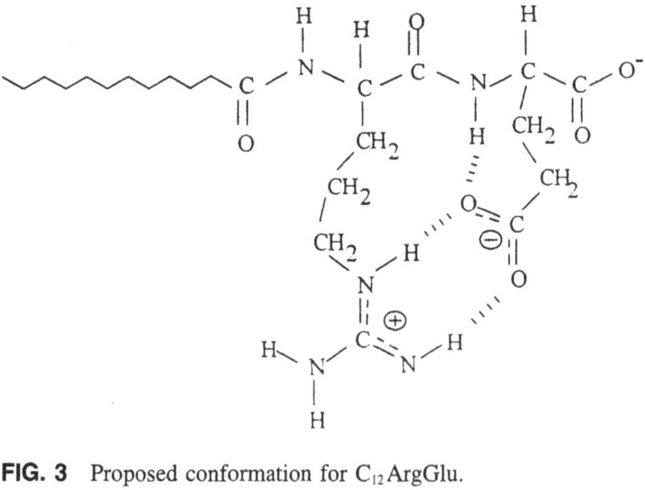

FIG. 3 Proposed conformation for C_{12}ArgGlu.

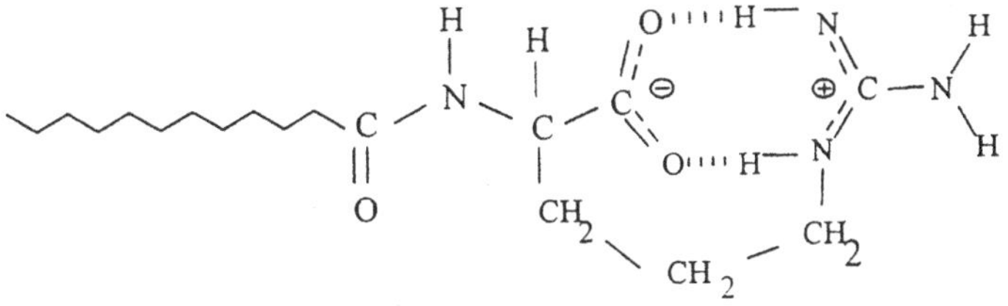

FIG. 4 Proposed conformation for C_{12}Arg.

tides under study show considerable water solubility and are therefore good surface-active compounds replete with antimicrobial activity.

2. Surface-Active Properties

The influences both of the alkyl chain length and of the terminal amino acid sidechain on the physicochemical properties are shown in Table 6. The study was limited to the water-soluble compounds in which the surface tension reduction of the aqueous solutions was determined by the Du Noüy ring technique at room temperature as a function of log (C). Parameters studied were critical micelle concentration (cmc, determined from the break point of the surface tension vs. concentration curves), γ_{cmc} (the surface tension at the cmc), and the minimum area per surfactant molecule at the air/aqueous solution

TABLE 6 Surface-Active Properties of Synthesized Lipopeptides

Compound	γ_{cmc} (mN/m)	cmc (10^{-3} mol/L)	$A_m \times 10^2$ (nm²/molecule)
C_{12}ArgGly	30	1.5	72
C_{12}ArgPhe	30	0.25	79
C_{12}ArgLys	30	1.2	—
C_8Arg aa	35	2.1	70
C_{10}Arg aa	33	0.5	65
C_{12}Arg aa	32	0.2	45
C_{14}Arg aa	28	0.07	41
C_{16}Arg aa[a]	29	0.03	46
C_{12}Arg pp	33.4	0.04% (w/v)	—
C_{14}Arg pp	32	0.01% (w/v)	—
C_{16}Arg pp	32	0.009% (w/v)	—

[a] At pH = 9.

interface (A_{min}, calculated with the Gibbs adsorption equation); values were confirmed by conductivity measurements.

The equilibrium surface tensions of aqueous lipopeptidic solutions decreased gradually with increasing concentration. Above the break, indicating a cmc, the tension curves were flat. In all cases, pure lipodipeptides yielded almost identical decreases in the surface tension of water at the cmc. The hydrophobic nature of the terminal hydrophilic moiety does not produce a substantial change in either the surface tension or the A_{min}. However, the presence of Phe decreases the cmc of the compounds because of increased hydrophobicity. The micellization ability of C_{12}Arg aa (a mixture of lipodipeptides) resembles that of pure C_{12}ArgPhe. This low cmc value for C_{12}Arg aa suggests that the micellization tendency might be enhanced by an effect of synergy resultant from the strength of the interactions between different lipodipeptides in the mixture.

The low surface tension values of solutions of our C_xArg aa (35–28 mN/m) and C_xArg pp (33.4–32 mN/m), together with the existence of a cmc, suggest their utility as surfactants. Surface tension values are similar to those of micellar solutions shown by two commercial surfactants: collagen tripeptide acylated with coconut fatty acid (28 mN/m) and sodium lauryl sulphate (40 mN/m) [29]. As expected, the length of the fatty chain influences the surface activity properties of both polydisperse lipodipeptides (C_xArg aa) and lipopeptides (C_xArg pp). As with a series of conventional surfactants carrying the same polar group, the longer the alkyl chain, the lower the surface tension measurements; and the longer the alkyl chain, the lower the cmc values [30]. A partial phase-behavior study revealed that, at low concentrations (0.1–5%, w/w) and temperatures (1.5–20°C), all pure lipodipeptides exhibit micellar phases except for C_{12}ArgPhe, which self-assembles in a dilute system as liquid crystalline-like phases of high viscosity.

B. Antimicrobial Properties

Antimicrobial activities were evaluated on the basis of minimum inhibitory concentration (MIC) values, defined as the lowest concentration of antibacterial agent inhibiting the development of visible growth after 24 h of incubation at 37°C. The MIC values were determined at pH 7.0 against gram-positive and gram-negative bacteria by the agar plate dilution method [31]; see Tables 7–9 for relevant data. For comparison, Table 7 also gives MIC values for the antimicrobial cationic surfactant LAM [6], determined at the same experimental conditions.

Table 7 shows that the structure of the terminal amino acid has an important

TABLE 7 Minimum Inhibitory Concentration (MIC), in µg/mL of Pure Lipodipeptides and LAM

Microorganism	C_{12}ArgGly	C_{12}ArgPhe	C_{12}ArgLys	C_{12}ArgGlu	LAM[a]
Gram-positive					
Candida albicans (CCM)	32	4	R	R	32
Staphylococcus epidermidis (ATCC 12228)	16	4	R	R	8
Streptococcus faecalis (ATCC 10541)	64	4	R	R	16
Corynebacterium agropyri (CCM)	8	2	32	R	8
Bacillus subtilis (ATCC 6633)	16	4	128	R	8
Bacillus pumilus (CCM)	16	4	128	R	—
Micrococcus luteus (ATCC 10240)	16	2	128	R	16
Micrococcus aurantiacus (ATCC 11731)	16	2	128	R	16
Gram-negative					
Alcaligenes faecalis (ATCC 8750)	R	32	R	R	R
Escherichia coli (ATCC 10536)	64	4	32	R	16
Klebsiella pneumoniae (ATCC 13883)	128	R	R	R	R
Citrobacter freundii (ATCC 22636)	R	R	R	R	R
Serratia marcescens (ATCC 13880)	R	R	R	R	R
Pseudomonas aeruginosa (ATCC 10145)	R	R	R	R	128
Salmonella typhimurium (ATCC 14028)	R	128	R	R	32
Bordetella bronchiseptica (ATCC 4617)	R	32	R	R	16

R: resistant, MIC > 128.

[a] LAM: N^{α}-Lauroyl arginine methyl ester.

TABLE 8 Minimum Inhibitory Concentration (MIC), in µg/mL, of Polydisperse Lipodipeptides

Microorganism	C_8 Arg aa	C_{10} Arg aa	C_{12} Arg aa	C_{14} Arg aa	C_{16} Arg aa
Gram-positive					
Candida albicans (ATCC 10231)	R	R	R	32	R
Staphylococcus epidermidis (ATCC 12228)	R	R	64	32	R
Enterococcus faecalis (ATCC 19433)	R	R	R	32	R
Corynebacterium agropyri (CCM)	R	R	R	32	R
Bacillus subtilis (ATCC 6633)	R	R	128	32	R
Bacillus pumilus (ATCC 7061)	R	R	—	32	R
Micrococcus luteus (ATCC 9341)	R	R	64	32	R
Micrococcus aurantiacus (ATCC 11731)	R	R	128	32	R
Gram-negative					
Alcaligenes faecalis (ATCC 8750)	R	R	—	32	R
Escherichia coli (ATCC 23231)	R	R	R	R	R
Klebsiella pneumoniae (ATCC 13882)	R	R	R	R	R
Citrobacter freundii (ATCC 11606)	R	R	R	32	R
Serratia marcescens (ATCC 13880)	R	R	R	R	R
Pseudomonas aeruginosa (ATCC 27853)	R	R	R	R	R
Salmonella typhimurium (ATCC 14028)	R	R	R	R	R
Bordetella bronchiseptica (ATCC 4617)	R	R	—	R	R

R: resistant, MIC > 128.

TABLE 9 Minimum Inhibitory Concentration (MIC), in µg/mL, of Polydisperse Lipopeptides

Microorganism	C_{12}Arg pp	C_{14}Arg pp	C_{16}Arg pp
Gram-positive			
Candida albicans (CCM)	128	64	64
Staphylococcus epidermidis (ATCC 12228)	128	128	64
Streptococcus faecalis (ATCC 10541)	R	128	R
Corynebacterium agropyri (CCM)	32	32	32
Bacillus subtilis (ATCC 6633)	128	16	32
Bacillus pumilus (CCM)	R	64	16
Micrococcus luteus (ATCC 10240)	32	32	16
Micrococcus aurantiacus (ATCC 11731)	128	32	32
Gram-negative			
Alcaligenes faecalis (ATCC 8750)	R	128	R
Escherichia coli (ATCC 10536)	64	128	128
Klebsiella pneumoniae (ATCC 13883)	128	R	R
Citrobacter freundii (ATCC 22636)	R	R	R
Serratia marcescens (ATCC 13880)	R	R	R
Pseudomonas aeruginosa (ATCC 10145)	R	R	R
Salmonella typhimurium (ATCC 14028)	R	R	R
Bordetella bronchiseptica (ATCC 4617)	R	R	R

R: resistant, MIC > 128.

influence on the antimicrobial activity of pure lipodipeptides. Bacterial growth inhibition comparable to that of LAM was found only when neutral Gly or Phe amino acids were condensed to the N^{α}-lauroyl arginine residue. Note the excellent activity of the C_{12}ArgPhe homologue. More interestingly, although polydisperse compounds are always less effective than those of pure C_{12}ArgGly and C_{12}ArgPhe, condensation of a long-chain N^{α}-acyl-arginine residue to a totally or partially acid-hydrolyzed collagen gave rise in some cases to amphoteric protein-based surfactants with an acceptable antimicrobial activity and potential application as a preservative (Tables 8 and 9, respectively).

Results for polydisperse C_xArg aa and C_xArg pp reported in Tables 8 and 9 also illustrate the effects on MIC values of carbon chain length. It is worth noting that a peak of antimicrobial activity as a function of the alkyl chain length is demonstrated for both kinds of compound. This has also been shown for other homologous series of amino acid–based surfactants with alkyl chain lengths in the range C_8–C_{16} [6,8,10]. According to Ferguson's principle [32], the peaks of antimicrobial activity might result from the combined action of multiple physicochemical (surface activity, adsorption, and solubility) and structural properties (the alkyl chain length and nature of the hydrophilic moiety), possibly explaining why these lipopeptidic compounds are adsorbed at the bacterial interface and show an optimum antimicrobial action.

In all cases, action was specific against gram-positive bacteria; these differ from gram-negatives, in which the outer membrane and lipopolysaccharide resist destabilization by surfactants. The resistance of the gram-negatives may have ecological value for the biodegradation of these antimicrobial materials.

In 1985, at the beginning of these studies, we believed that one essential requirement for the biological activity of amino acid surfactants based on Arg, Lys, and homologues was the cationic character of the molecule [6,10]. While LAM and other cationic surfactant homologues (with their α-carboxyl group protected as alkyl esters) might be good antimicrobial agents, C_{12}Arg and other amphoteric homologues showed no activity up to 256 μg/mL against all the bacteria tested. But when a second amino acid or peptide was incorporated into the monomer C_{12}Arg, α-carboxyl protection was not a requirement for antimicrobial activity so long as it did not interact with a cationic charge on the molecule. The activity of all these amphoteric N^α-acyl arginine lipopeptidic surfactants might be due to the presence of the long-chain N^α-acyl-arginine residue (responsible for both the surface activity and the protonated guanidine function) and to the absence of intramolecular ionic interactions because of the unfavorable guanidinium–carboxylate configuration. The free guanidine group of these compounds, together with the surface activity (dependent on their alkyl chain length), might interact with the polyanionic components of the cell surface, thereby triggering their biocidal effects.

C. Ocular Irritation Properties

Published results on the toxicological properties of different lipoaminoacids suggest that amino acid–based surfactants are excellent nontoxic surfactants for a great variety of potential applications [33–36]. Ocular irritation by C_{12}ArgGly, C_{12}Arg pp, LAM, and lauryl dimethyl amino betaine (LDAB), a commercial amphoteric surfactant, has been evaluated in male albino rabbits by a double-blind method. The eyes were examined after 24 h of administration and then each day for a week. The ocular damage was evaluated in accordance with the criteria of Draize [37].

Table 10 gives the ocular irritation index for each product, along with its classification. The differences between the products are marked: While LAM and LDAB were irritants at 10% (w/v), lipopeptides C_{12}ArgGly and C_{12}Arg pp were quite nonirritant, perhaps because their chemical compositions resem-

TABLE 10 Mean Values, Standard Error, and Classification of Selected Compounds in Water at pH 7.0 and 10% Concentration (w/v)

Compound	Mean value	Standard error	Classification
C_{12}ArgGly	67.2	5.5	Nonirritant
C_{12}Arg pp	57.0	1.0	Nonirritant
LAM	2.2	1.0	Irritant
LDAB	7.5	2.2	Irritant

ble that of corneal collagen. It is interesting to note that while LAM is a cationic surfactant, it shows an ocular irritation similar to LDAB, a substance which is generally considered an acceptable soft surfactant. Moreover, the progress of ocular irritation induced by these compounds after 7 days shows that the effects of LAM and LDAB are reversible.

V. CONCLUSION

Water-soluble antimicrobial lipopeptides have been prepared by the condensation of amino acids or peptides to N^{α}-acyl arginine residues with suitable lipophilic contents. Our data suggest that such arginine lipopeptidic surfactants have value as soft preservatives in cosmetic, food, and dermopharmaceutical formulations, as well as being tools for fundamental research.

ACKNOWLEDGMENTS

The authors gratefully acknowledge Prof. Vivian Moses for his valuable advice in the final version of this chapter. We are also indebted to Amalia Vilchez for her experimental assistance.

REFERENCES

1. T. J. Franklin and G. A. Snow. In: Biochemistry of Antimicrobial Action. 3rd ed. Chapman & Hall, New York, 1981.
2. R. E. W. Hancock and R. Lehrer. Trends Biotechnol. 16:82 (1998).
3. M. Zasloff. Proc. Natl. Acad. Sci. U.S.A. 84:5449 (1987).
4. D. Wade, D. Andreu, S. A. Mitchell, A. M. V. Silveira, A. Boman, H. G. Boman, R. B. Merrifield. Int. J. Pept. Protein Res. 40:429 (1992) and references therein.
5. R. E. W. Hancock. Lancet 349:418 (1997).
6. M. R. Infante, J. J. García Dominguez, P. Erra, M. R. Juliá, M. Prats. Int. J. Cosm. Sci. 6:275 (1984).

7. J. Molinero, P. Erra, M. R. Juliá, M. Robert, M. R. Infante. J. Am. Oil Chem. Soc. 65:975 (1988); M. R. Infante, J. Molinero, P. Bosch, M. R. Juliá, P. Erra. J. Am. Oil Chem. Soc. 66:1835 (1989); M. R. Infante, J. Molinero, P. Erra. J. Am. Oil Chem. Soc. 69:647 (1992).

8. L. Pérez, J. L. Torres, A. Manresa, C. Solans, M. R. Infante. Langmuir 12:5296 (1996).

9. E. Piera, P. Erra, M. R. Infante. J. Chem. Soc. Perkin 2 Trans. 335 (1998).

10. M. R. Infante, J. Molinero, P. Erra, M. R. Juliá, J. J. García Dominguez. Fett. Wiss. Tech. 87:309 (1985).

11. M. R. Infante, A. Pinazo, J. Seguer. Colloids Surfaces A 123–124:49 (1997) and references therein.

12. P. Clapés, C. Morán, M. R. Infante. Biotech. Bioeng. 63:333 (1999).

13. H. Gibson, J. T. Holah. In: Preservation of Surfactant Formulations (F. F. Morpeth, ed.). Blackie Academic and Professional, Glasgow, 1995, Ch. 3.

14. K. Takizawa, R. Yoshida. US Patent 3,897,466 to Ajinomoto Co. (1975).

15. J. D. Spivak. In: Anionic Surfactants II (W. M. Linfield, ed.). Marcel Dekker, New York, 1976, Ch. 16.

16. P. Than. In: Surfactants in Cosmetics (M. M. Rieger, ed.). Marcel Dekker, New York, 1985.

17. M. Takehara. Colloids Surfaces 38:149 (1989).

18. S. Matsumura, Y. Kawamura, S. Yoshikawa, K. Kawada, T. Uchibori. J. Am. Oil Chem. Soc. 70:17 (1993).

19. R. J. Sims, J. A. Fioriti. J. Am. Oil Chem. Soc. 52:144 (1975).

20. E. Jungermann, J. F. Gerecht, I. J. Krems. J. Am. Chem. Soc. 78:172 (1956).

21. M. Takehara, I. Yoshimura, K. Takizawa, R. Yoshida. J. Am. Oil Chem. Soc. 49:157 (1972).

22. R. Geiger, W. Köning. In: The Peptides, Vol. 3 (E. Gross, J. Meienhofer, eds.). Academic Press, London, 1981, Ch. 1.

23. R. Yoshida, K. Babak, T. Saito, I. Yoshimura. J. Jpn Oil Chem. Soc. 25:404 (1976).

24. T. Hayakawa, F. Fujiwara, J. Noguchi. Bull. Chem. Soc. Japan 40:1205 (1967).

25. J. R. Vaughan. J. Am. Chem. Soc. 73:3547 (1951).

26. J. Mery, B. Calas. Int. J. Pept. Prot. Res. 31:412 (1988).

27. F. E. Kaiser, Ch. W. Gehrke, R. W. Zumwalt, K. C. Kuo. J. Chromatog. 94:113 (1974).

28. G. Lancelot, R. Mayer, Cl. Hélène. J. Am. Chem. Soc. 101:1569 (1979).

29. P. Sokol. J. Am. Oil Chem. Soc. 52:101 (1974).

30. M. J. Rosen. In: Surfactants and Interfacial Phenomena. 2nd ed. Wiley, New York, 1989.

31. J. A. Washington, V. L. Sutter. In: Clinical Microbiology (E. H. Lenette, A. Balows, W. J. Hausler, J. P. Truant, eds.). American Society for Microbiology. Washington, DC, 1980.

32. J. Ferguson. Proc. Roy. Soc. Serv. B 127:387 (1939).

33. J. Seguer, M. Allouch, P. Vinardell, M. R. Infante, L. Mansuy, C. Selve. New J. Chem. 18:765 (1994).

34. M. Macian, J. Seguer, M. R. Infante, C. Selve, P. Vinardell. Toxicology 106:1 (1996).

35. M. A. Vives, M. Macian, J. Seguer, M. R. Infante, P. Vinardell. Toxicol. in vitro 11:779 (1997).

36. M. A. Vives, M. R. Infante, E. García, C. Selve, P. Vinardell. Chem-Biol. Interact. 118:1 (1999).

37. J. H. Draize. In: Appraisal of the Safety of Chemicals in Foods, Drugs and Cosmetics. Assoc. Food & Drug Officials of the U.S.A., Austin, TX, 1959.

7

Essentially Fluorinated Synthetic Surfactants Based on Amino Acids or Oligopeptides

CLAUDE SELVE, CHRISTINE GÉRARDIN, and LUDWIG RODEHÜSER   Université Henri Poincaré—Nancy I, Vandoeuvre-les-Nancy, France

I. INTRODUCTION

Surfactants are amphipathic molecules, generally characterized by the presence of two distinctly different regions in the same molecule: a lipophilic part (restrictively called hydrophobic) and a lipophobic or hydrophilic portion. The existence in the same molecule of two moieties, one with an affinity for the solvent and the other antipathetic to it, is termed *amphiphily* or *amphipathy*. Surface-active agents constitute a versatile class of natural or synthetic compounds. They may contain a large variety of polar ionic or nonionic parts (called *head*) and apolar moieties (named *tail*), generally consisting of long hydrocarbon chains [1–5]. This dual nature is responsible for the phenomena of surface activity and of micellization, the formation of molecular organized systems [6] (lamellar phases, liquid crystals, vesicles, etc.), and their capacity of solubilization by forming emulsions and microemulsions [1–6].

The polar head group may dissociate into ions in polar solvents, with the extent of ionization depending on the pH of the medium. The apolar part may be constituted of a perhydrogenated carbon chain (majority of the surfactants encountered), but also of perfluorinated chains or substituents containing other atoms, such as silicon (silicone-type derivatives).

The amphiphilic products are a very important class of compounds for practical applications: nearly all human activities imply the use of surfactants because of their extraordinary amphipathic properties. This chapter will be restricted to surfactants with essentially perfluorinated hydrophobic chains.

The preparation of surfactant molecules in the laboratory is often achieved by the method of modular synthesis [7,8]. This means that the molecule is considered as an assembly of subunits called *moduli*. In this way we can distinguish in natural surfactants **bi-** or **trimodular** structures, as shown in Scheme 1. Based on this principle, the synthetic approaches and strategies are variable. The amino acid by itself may or may not constitute the hydrophilic modulus; alternatively, it may play the role of the junction modulus or that of an anchor unit for the hydrophobic group attached to it via a C–C bond, depending on its specific structure.

The choice of perfluorinated chains as lipophilic moduli is due to their particular properties. Distinctly more hydrophobic than their hydrogenated analogues, these fluorinated substituents enhance the amphiphilic character of the resulting surfactant, increasing significantly its hydrophobic interactions and surface activity. Fluorinated surfactants exhibit stronger tendencies to self-

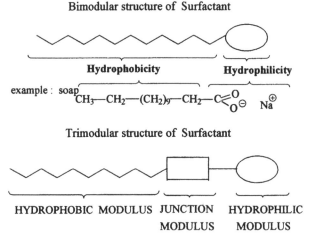

SCHEME 1   Modular structures of natural surfactants.

assemble at the interfaces or to self-associate into organized molecular systems. These surfactants are both more effective and more efficient. They display stronger surface-tension-lowering effects at lower concentrations. The dispersions and phases they form exhibit unique properties [9,10]. A large number of fluorinated surfactants have been developed for many different industrial purposes.

II. AMINO ACID–BASED SURFACTANTS

Many compounds containing a perfluorinated or hydrogenated hydrophobic chain and an amino acid as either polar head or junction modulus have been synthesized. Among the different structures that have been prepared, a variety of amino acids are encountered, but α-amino acids are the most common; however, β-alanine or its derivatives play an important role, too. Moreover, molecules based on perfluoroalkyl-amino acids, prepared from different functionalized perfluoroalkyl starting compounds, have been synthesized. It should be stressed that the different types of surfactants based on amino acids encountered in this family of products may be classified as ionic, nonionic, or zwitterionic. In all the structures obtained, the concept of modularity is important, since the amino acid—with or without prior chemical transformation—represents either the polar head group or a pivot modulus linking one or more lipophilic chains to different polar heads.

A. Simple Amino Acids

Derivatives of *N*-acylated amino acids find a large variety of applications as cosmetics and pharmaceutical products. Their surfactant [11] and antimicrobial properties [12] have been well established. They are obtained generally via the condensation of a fatty acid derivative with an amino acid [12,13]. Numerous syntheses of compounds containing perfluorinated hydrophobic groups have been reported. Generally, they follow synthetic pathways analogous to those used for the preparation of hydrogenated compounds. Frequently, the products are obtained by reaction between a perfluoroalkylester and an ester of an amino acid bearing a free amine function [14]. Subsequent hydrolysis leads to the *N*-acylated acid **1** (Scheme 2).

The same type of reaction using a fatty acid chloride instead of the ester leads to better yields [15] for structures like **2** (Scheme 3), especially with sarcosine.

Activated esters of fatty acids have also been recommended as a starting material [16]. They give satisfactory results when hydrogenated compounds

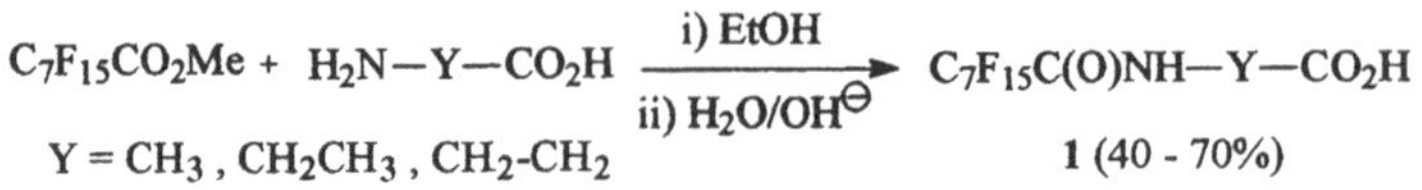

$$C_7F_{15}CO_2Me + H_2N-Y-CO_2H \xrightarrow[\text{ii) H}_2\text{O/OH}^\ominus]{\text{i) EtOH}} C_7F_{15}C(O)NH-Y-CO_2H$$

$$Y = CH_3 , CH_2CH_3 , CH_2\text{-}CH_2 \qquad\qquad \mathbf{1}\ (40\text{ - }70\%)$$

SCHEME 2 Perfluoroalkyl-acyl-amino acids of type **1**.

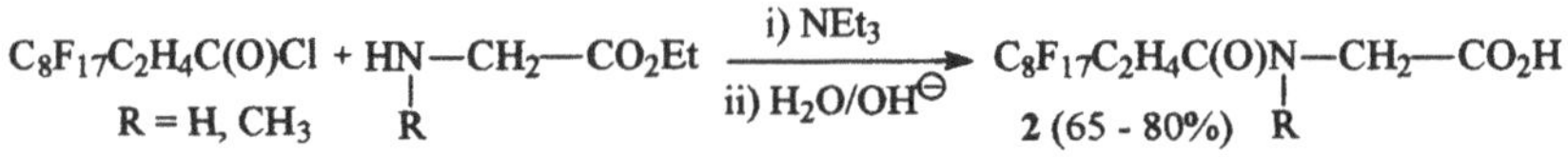

$$C_8F_{17}C_2H_4C(O)Cl + \underset{R}{HN}-CH_2-CO_2Et \xrightarrow[\text{ii) H}_2\text{O/OH}^\ominus]{\text{i) NEt}_3} C_8F_{17}C_2H_4C(O)\underset{R}{N}-CH_2-CO_2H$$

$$R = H, CH_3 \qquad\qquad\qquad\qquad\qquad \mathbf{2}\ (65\text{ - }80\%)$$

SCHEME 3 Perfluoroalkyl-acyl-amino acids of type **2**.

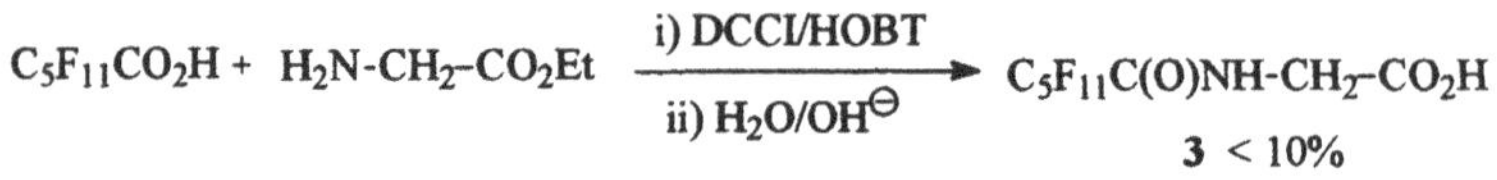

$$C_5F_{11}CO_2H + H_2N\text{-}CH_2\text{-}CO_2Et \xrightarrow[\text{ii) H}_2\text{O/OH}^\ominus]{\text{i) DCCI/HOBT}} C_5F_{11}C(O)NH\text{-}CH_2\text{-}CO_2H$$

$$\mathbf{3}\ < 10\%$$

SCHEME 4 Perfluoroalkyl-acyl-amino acids of type **3**.

$$C_7F_{15}CO_2H + H_2N\text{-}CH_2\text{-}CO_2Et \xrightarrow[\text{ii) H}_2\text{O/OH}^\ominus]{\text{i) BOP}} C_7F_{15}C(O)NH\text{-}CH_2\text{-}CO_2H$$

$$\mathbf{3}\ \sim 20\%$$

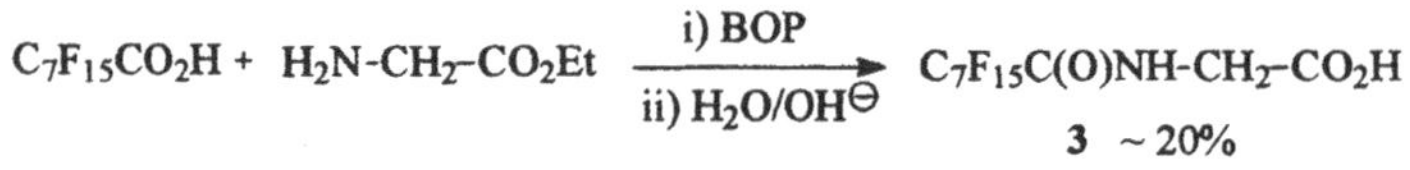

SCHEME 5 Alternative synthesis of perfluoroalkyl-acyl-amino acids of type **3**.

are used, but our experience with perfluorinated acids (Scheme 4) shows that the yields are rather poor [17].

The amidation reaction carried out with DCCI/N-hydroxy-succinimide [18,19] or DCCI/hydroxy-benzotriazole [20,21] as coupling agents yields less than 15% of the corresponding amides. The utilization of BOP [22,23] as activating agent for the fluorinated acid increases the yields to 20–30%. The hydrolysis step is also not very efficient, so the overall yield of compound **3** (Scheme 5) is less than 10%. Moreover, the products are poorly soluble in water ($<10^{-6}$ mol · L^{-1}), even at high pH values [24].

Many N-acyl-aminoacids contain substituents with enhanced polarity, such as sulfonates, quaternary ammonium groups, and zwitterionic moieties, that endow them with interesting physicochemical properties, such as the ability to form vesicles [26] or globular aggregates [27]. In a similar way (see Scheme

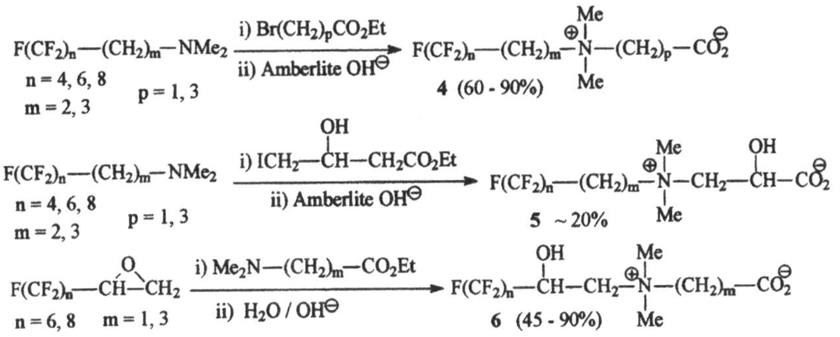

$$F(CF_2)_n—(CH_2)_m—NMe_2 \xrightarrow[\text{ii) Amberlite OH}^{\ominus}]{\text{i) Br(CH}_2)_p\text{CO}_2\text{Et}} F(CF_2)_n—(CH_2)_m—\overset{\overset{\displaystyle Me}{\oplus|}}{N}—(CH_2)_p—C\overset{\ominus}{\underset{\displaystyle Me}{\overset{\displaystyle O}{O}}}$$

$$n = 4, 6, 8 \qquad p = 1, 3$$
$$m = 2, 3$$
$$\mathbf{4}\ (60 - 90\%)$$

$$F(CF_2)_n—(CH_2)_m—NMe_2 \xrightarrow[\text{ii) Amberlite OH}^{\ominus}]{\text{i) ICH}_2—\overset{\displaystyle OH}{\overset{|}{C}H}—CH_2CO_2Et} F(CF_2)_n—(CH_2)_m—\overset{\overset{\displaystyle Me}{\oplus|}}{N}—CH_2—\overset{\displaystyle OH}{\overset{|}{C}H}—C\overset{\ominus}{\overset{\displaystyle O}{O}}$$

$$n = 4, 6, 8 \qquad p = 1, 3$$
$$m = 2, 3$$
$$\mathbf{5}\ \sim 20\% \qquad Me$$

$$F(CF_2)_n—\overset{O}{\overset{\diagup\diagdown}{CH—CH_2}} \xrightarrow[\text{ii) H}_2\text{O / OH}^{\ominus}]{\text{i) Me}_2\text{N—(CH}_2)_m—CO_2Et} F(CF_2)_n—\overset{\displaystyle OH}{\overset{|}{C}H}—CH_2—\overset{\overset{\displaystyle Me}{\oplus|}}{N}—(CH_2)_m—C\overset{\ominus}{\overset{\displaystyle O}{O}}$$

$$n = 6, 8 \quad m = 1, 3$$
$$\mathbf{6}\ (45 - 90\%) \qquad Me$$

SCHEME 6   Perfluoroalkyl-acyl-amino acids of types **4**, **5**, and **6**.

6), zwitterionic derivatives with perfluorinated sidechains (**4**) have been prepared [28]. Additionally they may contain a hydroxyl function either in the aminoacid modulus (derivatives **5**) or on the perfluorinated substituent (derivatives **6**) [28,29].

An important number of esters of long-chain alcohols and amino acids have been described in the literature. The permethylation of the amino function (Scheme 7) leads to the ammonium salts (**7**) showing generally interesting surfactant properties [30]. We have chosen a similar pathway by reacting a perfluorinated fatty alcohol with chloroacetic acid; the subsequent substitution of the chlorine atom by trimethylamine leads to the corresponding ammonium salt (**8**) [31]. The latter is water soluble, but the hydrolysis of the ester function is rather rapid (<6 h), which renders this type of compound less interesting for possible applications.

B.   Multifunctional Amino Acids

A large variety of compounds have been prepared starting from basic or acid trifunctional α-amino acids, in particular, lysine, aspartic acid, and glutamic acid. Lysine has been used as polar head group in structures containing a fatty acid attached to the nitrogen in the ε-position via an amido bond (*N*-ε-acyl-

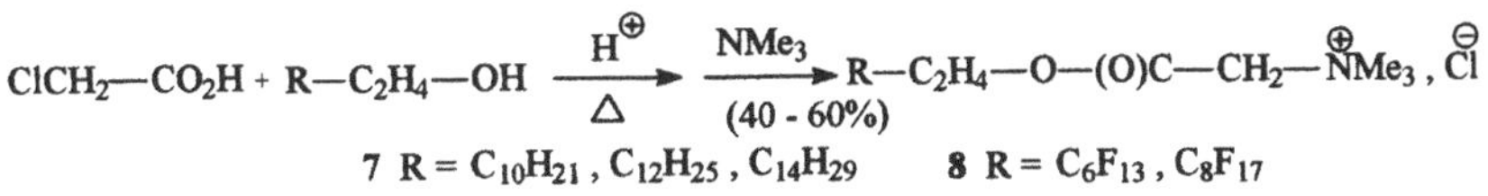

$$ClCH_2—CO_2H + R—C_2H_4—OH \xrightarrow[\Delta]{\overset{\oplus}{H}} \xrightarrow{NMe_3} R—C_2H_4—O—(O)C—CH_2—\overset{\oplus}{N}Me_3, \overset{\ominus}{Cl}$$
$$(40 - 60\%)$$

$$\mathbf{7}\ R = C_{10}H_{21}, C_{12}H_{25}, C_{14}H_{29} \qquad \mathbf{8}\ R = C_6F_{13}, C_8F_{17}$$

SCHEME 7   Perfluoroalkyl-esters of amino acids of types **7** and **8**.

lysines). The syntheses of these molecules follow purely chemical or enzymatic pathways. The first chemical acylations have been realized by reacting a paranitrophenyl ester [32] or a fatty acid halogenide [33] with the copper(II) complex of lysine in which the α-amino group is protected by coordination to the copper ion. The synthesis of N-ε-acyl-lysine derivatives and of other dibasic amino acids, such as homolysine and ornithine, is equally possible without protection when the reaction is carried out at pH 11. In this case, the selective acylation is achieved by using paranitrophenyl esters of the different fatty acids [34]. However, the most interesting method to obtain these compounds seems to be the thermal dehydration of the salts formed between lysine and a fatty acid [35]. These N-ε-acyl-lysines are poorly soluble in water. The quaternization by permethylation of the α-amino group yields products (**9**) (Scheme 8) that show good surfactant properties [36].

Perfluoroalkylated derivatives of these compounds have been synthesized [37] by following basically the same method with modifications suitable for perfluoroalkylated fatty acids [15] (Scheme 9).

It should be noted, however, that the yields for the quaternization of products (**10**) leading to the betaines (**11**) (Scheme 10) are rather modest (∼16%) [15], whereas the quaternized N-ε-acyl-lysines (**9**) are generally obtained in good yields.

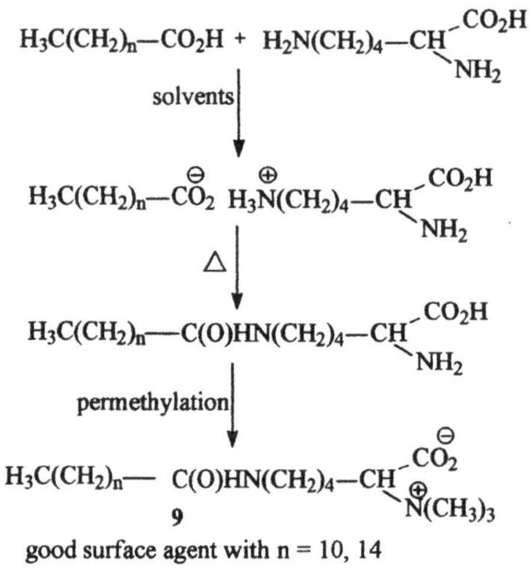

$$H_3C(CH_2)_n-CO_2H \; + \; H_2N(CH_2)_4-CH\overset{CO_2H}{\underset{NH_2}{<}}$$

solvents $\downarrow$

$$H_3C(CH_2)_n-\overset{\ominus}{C}O_2 \; \overset{\oplus}{H_3}N(CH_2)_4-CH\overset{CO_2H}{\underset{NH_2}{<}}$$

$\Delta \downarrow$

$$H_3C(CH_2)_n-C(O)HN(CH_2)_4-CH\overset{CO_2H}{\underset{NH_2}{<}}$$

permethylation $\downarrow$

$$H_3C(CH_2)_n- \; C(O)HN(CH_2)_4-CH\overset{\overset{\ominus}{CO_2}}{\underset{\overset{\oplus}{N}(CH_3)_3}{<}}$$

9

good surface agent with n = 10, 14

SCHEME 8 Synthesis of zwitterionic compounds by quaternization of N-ε-acyl-lysines **9**.

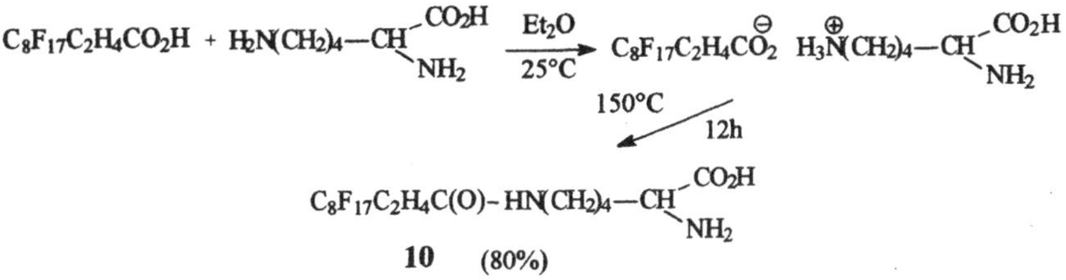

$$C_8F_{17}C_2H_4CO_2H \ + \ H_2N(CH_2)_4-CH\overset{CO_2H}{\underset{NH_2}{\diagup}} \ \xrightarrow[25°C]{Et_2O} \ C_8F_{17}C_2H_4C\overset{\ominus}{O_2} \ \ H_3\overset{\oplus}{N}(CH_2)_4-CH\overset{CO_2H}{\underset{NH_2}{\diagup}}$$

$$\xleftarrow[12h]{150°C}$$

$$C_8F_{17}C_2H_4C(O)-HN(CH_2)_4-CH\overset{CO_2H}{\underset{NH_2}{\diagup}}$$

10   (80%)

SCHEME 9   Synthesis of *N*-ε-perfluoroalkyl-acyl-lysines **10**.

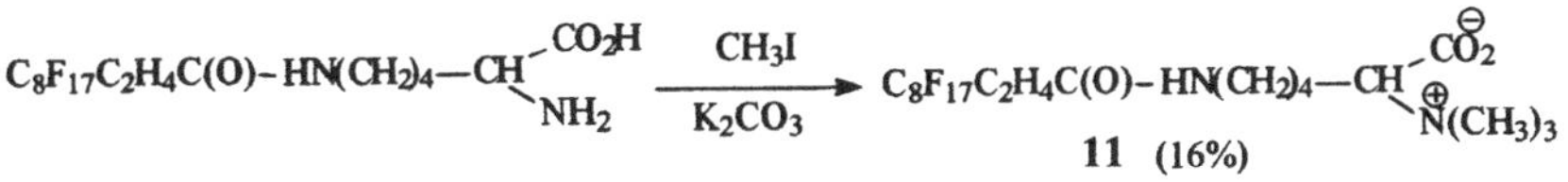

$$C_8F_{17}C_2H_4C(O)-HN(CH_2)_4-CH\overset{CO_2H}{\underset{NH_2}{\diagup}} \ \xrightarrow[K_2CO_3]{CH_3I} \ C_8F_{17}C_2H_4C(O)-HN(CH_2)_4-CH\overset{\overset{\ominus}{CO_2}}{\underset{\overset{\oplus}{N}(CH_3)_3}{\diagup}}$$

11 (16%)

SCHEME 10   Quaternization of *N*-ε-perfluoroalkyl-acyl-lysines **11**.

The acylation of the α-amino group of lysine is another strategy for the preparation of surfactants and biocides [38]. It is achieved mostly by the classical methods of peptide synthesis. They have been realized exclusively starting from perhydrogenated DCCI activated fatty acids by condensation with the ethyl ester of lysine, the second amino group of which had been protected with the benzyloxycarboyl group. Subsequent hydrolysis and catalytic hydrogenation yields α-acyl-lysines (of type **12**) (Scheme 11). Quaternization with methyliodide leads to the surface-active derivatives (**13**) [38–40].

The direct diacylation of the unprotected lysine with fatty acid chlorides [41] furnishes anionic surfactants (of type **14**). In the same way, lysine plays the role of junction modulus in the synthesis and study of the properties of lecithin analogues [42]. The latter are of the nonionic type, one of the series

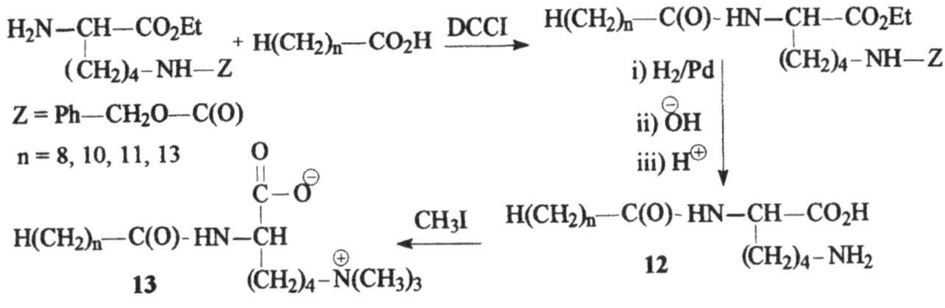

$$H_2N-\underset{(CH_2)_4-NH-Z}{\overset{|}{CH}}-CO_2Et \ + \ H(CH_2)_n-CO_2H \ \xrightarrow{DCCI} \ H(CH_2)_n-C(O)-HN-\underset{(CH_2)_4-NH-Z}{\overset{|}{CH}}-CO_2Et$$

Z = Ph—CH₂O—C(O)

n = 8, 10, 11, 13

i) H₂/Pd

ii) $\overset{\ominus}{O}H$

iii) H$^{\oplus}$

$$H(CH_2)_n-C(O)-HN-\underset{\underset{(CH_2)_4-\overset{\oplus}{N}(CH_3)_3}{}}{\overset{|}{CH}}\overset{\overset{O}{\parallel}}{\underset{}{C-O}}^{\ominus} \ \xleftarrow{CH_3I} \ H(CH_2)_n-C(O)-HN-\underset{(CH_2)_4-NH_2}{\overset{|}{CH}}-CO_2H$$

13   **12**

SCHEME 11   Synthesis of zwitterionic *N*-α-acyl-lysines **13**.

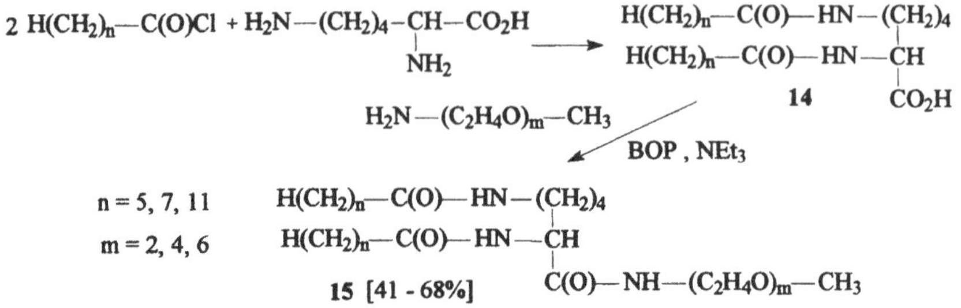

$$2\ H(CH_2)_n\!-\!C(O)Cl + H_2N\!-\!(CH_2)_4\!-\!\underset{\underset{NH_2}{|}}{CH}\!-\!CO_2H \longrightarrow$$

$$\begin{array}{l} H(CH_2)_n\!-\!C(O)\!-\!HN\!-\!(CH_2)_4 \\ H(CH_2)_n\!-\!C(O)\!-\!HN\!-\!\underset{\underset{CO_2H}{|}}{CH} \quad \mathbf{14} \end{array}$$

$$H_2N\!-\!(C_2H_4O)_m\!-\!CH_3$$

$$\text{BOP , NEt}_3$$

$$n = 5, 7, 11$$
$$m = 2, 4, 6$$

$$\begin{array}{l} H(CH_2)_n\!-\!C(O)\!-\!HN\!-\!(CH_2)_4 \\ H(CH_2)_n\!-\!C(O)\!-\!HN\!-\!\underset{\underset{C(O)\!-\!NH\!-\!(C_2H_4O)_m\!-\!CH_3}{|}}{CH} \end{array}$$

$$\mathbf{15}\ [41 - 68\%]$$

SCHEME 12 Monocephalic nonionic analogues of lecithins **15**, based on lysine.

of products bearing a polar head group with one polyethyleneglycol chain (monocephalic compounds) with structure **15** shown in Scheme 12. The other series [43] contains two polyethyleneglycol chains in its head group (bicephalic compounds) of structural type **16** [41,43], as represented in Scheme 13.

Recently, by following analogous pathways, we have prepared lysine derivatives with two hydrophobic chains stemming from a long-chain alcohol and a fatty acid, respectively. The latter is linked to the ε-nitrogen, whereas the alcohol forms an ester with the acid function of lysine. The hydrophilic part of the molecule is a sugar substituent [44] introduced as shown in Scheme 14. The compounds (**17**) obtained in this manner show very good surfactant properties, a relatively high solubility in water, even for those containing 16 methylene groups, leading to a decrease in surface tension of ca. 30 mN·m^{-1}, and self-organization to form lamellar or cubic phases. They have a pro-

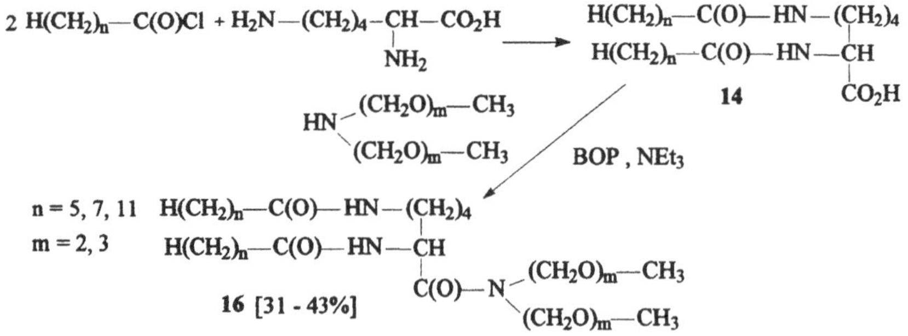

$$2\ H(CH_2)_n\!-\!C(O)Cl + H_2N\!-\!(CH_2)_4\!-\!\underset{\underset{NH_2}{|}}{CH}\!-\!CO_2H \longrightarrow$$

$$\begin{array}{l} H(CH_2)_n\!-\!C(O)\!-\!HN\!-\!(CH_2)_4 \\ H(CH_2)_n\!-\!C(O)\!-\!HN\!-\!\underset{\underset{CO_2H}{|}}{CH} \quad \mathbf{14} \end{array}$$

$$HN\!\!\!\begin{array}{l} ^{(CH_2O)_m\!-\!CH_3} \\ _{(CH_2O)_m\!-\!CH_3} \end{array}$$

$$\text{BOP , NEt}_3$$

$$n = 5, 7, 11$$
$$m = 2, 3$$

$$\begin{array}{l} H(CH_2)_n\!-\!C(O)\!-\!HN\!-\!(CH_2)_4 \\ H(CH_2)_n\!-\!C(O)\!-\!HN\!-\!\underset{\underset{C(O)\!-\!N\begin{array}{l} ^{(CH_2O)_m\!-\!CH_3} \\ _{(CH_2O)_m\!-\!CH_3} \end{array}}{|}}{CH} \end{array}$$

$$\mathbf{16}\ [31 - 43\%]$$

SCHEME 13 Bicephalic nonionic analogues of lecithins **16**, based on lysine.

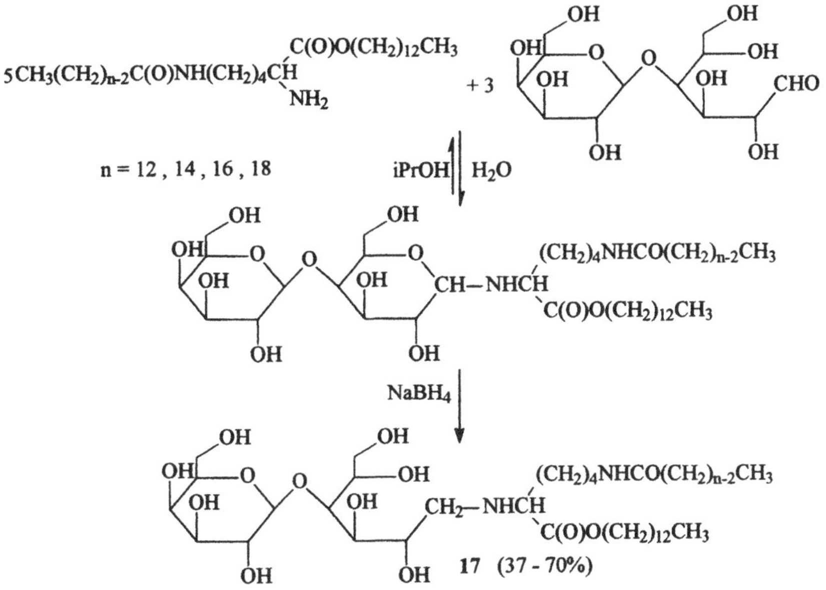

$$5\,CH_3(CH_2)_{n-2}C(O)NH(CH_2)_4\overset{\displaystyle C(O)O(CH_2)_{12}CH_3}{\underset{\displaystyle NH_2}{CH}}$$

$$+3$$

$$n = 12, 14, 16, 18$$

iPrOH $\|$ H$_2$O

$(CH_2)_4NHCO(CH_2)_{n-2}CH_3$

$CH{-}NHCH$

$C(O)O(CH_2)_{12}CH_3$

NaBH$_4$

$(CH_2)_4NHCO(CH_2)_{n-2}CH_3$

$CH_2{-}NHCH$

$C(O)O(CH_2)_{12}CH_3$

17 (37 - 70%)

SCHEME 14   Synthesis of ε-alcanoylamido-α-alkylester-α-lactylamino-lysine **17**.

nounced effect on the development of yeast cells; part of the products stimulate the growth of the microorganisms, whereas another part inhibits it [45].

Based on lysine as before, single chain derivatives (**18**) and bicatenar compounds (**19**) have been prepared by a strategy the inverse of that described earlier [46]. The products (**18** and **19**) are obtained with higher overall yields than **17**. As before, no protection or deprotection steps are needed in this case [46,47]. The synthesis consists of a reaction between two moduli prepared beforehand, one representing the polar head group in the final product and the other the junction modulus. The preparation of this intermediate starts from aldoses or acids derived from them that are coupled with lysine in its basic form (Scheme 15). The yields for this step are quantitative. A subsequent amidation reaction with a hydrogenated or perfluorinated fatty acid (Scheme 16) leads to the monosubstituted compounds (**18**). Esterification of the free acid function of lysine (Scheme 17) yields in a final step the bicatenar structures (**19**).

Finally the synthesis of "biosurfactants" following enzymatic pathways [48] should be mentioned; this has been described in particular for the preparation of *N*-acyl-lysine. The overall yields are modest, but the method is nonethe-

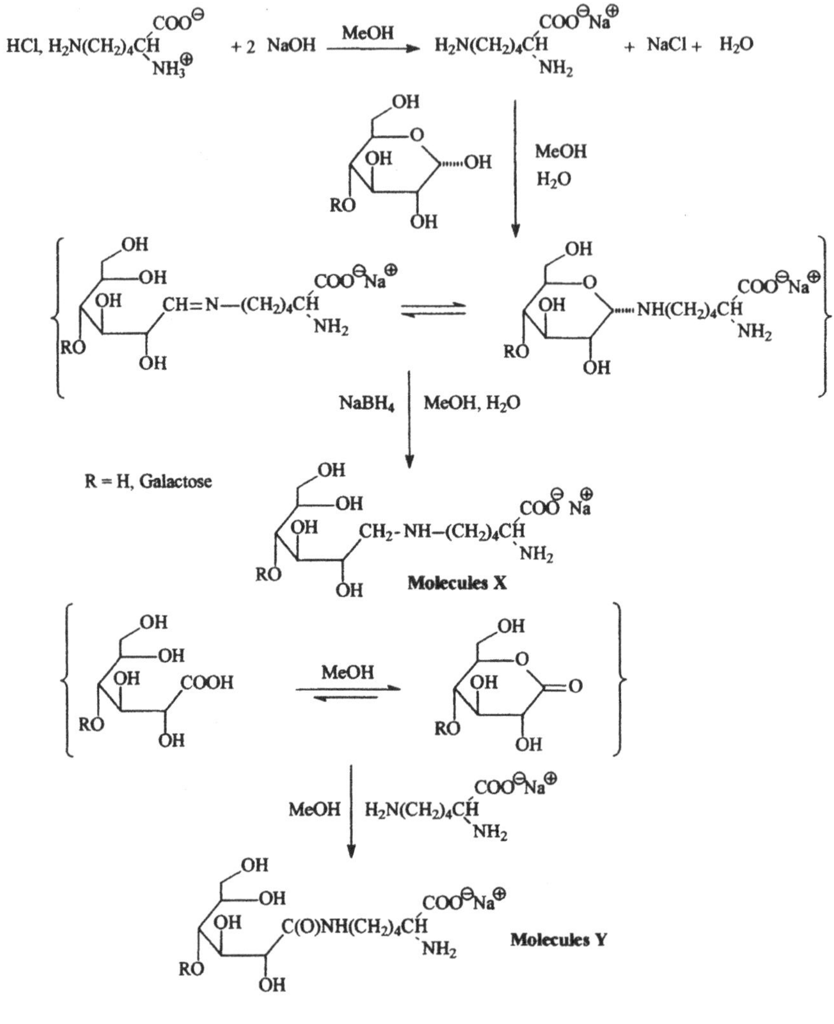

SCHEME 15 Preparation of hydrophilic synthons "Lys-sugar": **molecules X** and **Y**.

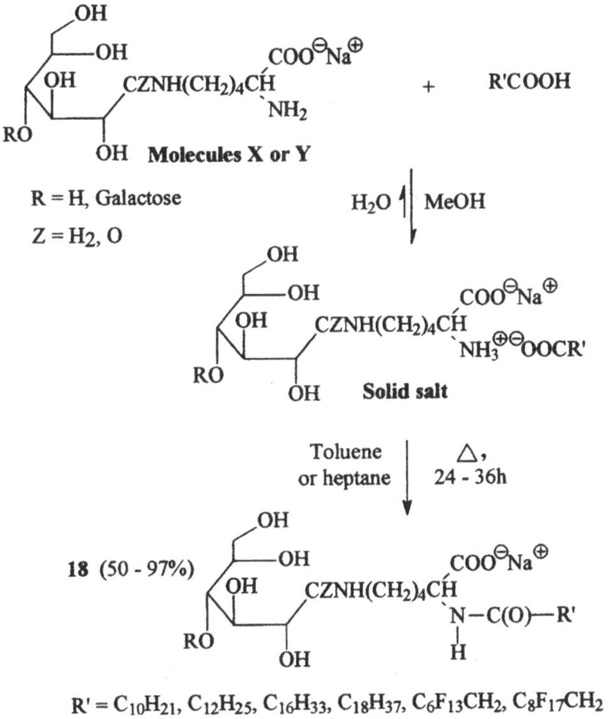

SCHEME 16 Synthesis of monocatenar surfactants **18**.

less of interest, since it allows the utilization of natural soybean oil and lysine as starting material, opening access to N-ε-oleyl-lysines via a biotechnological process [50].

Aspartic and glutamic acid have also been used frequently as starting material for the preparation of surfactants. The diesters of glutamic acid, obtained by reaction with a long-chain alcohol, have been rendered surface active by grafting a quaternary ammonium group on the α-amino function via an acylation reaction [51] (structure **20**, Scheme 18).

These molecules (**20**) show an interesting phase behavior [51] and form helicoidal superstructures [52]. Nonionic derivatives of aspartic and glutamic acid with structures analogous to lecithins may contain perhydrogenated or perfluorinated apolar chains [53]. The N,N-diacylated compounds are synthesized in two steps (Scheme 19). The hydrophilicity of the molecules (**21**) is due to the presence of an acyl-polyethyleneglycol-methylether [54].

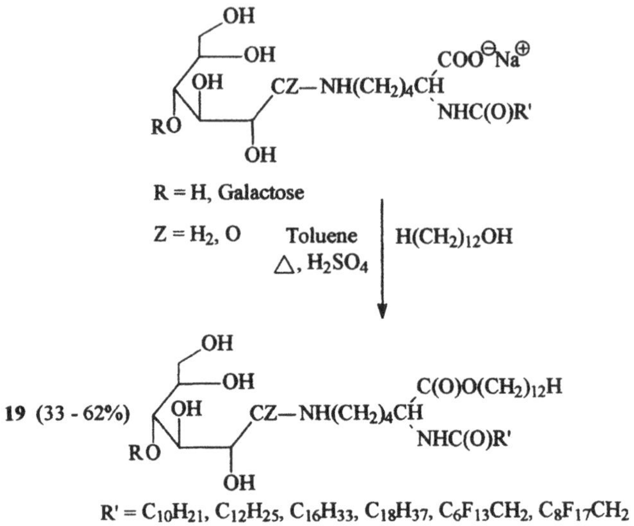

R = H, Galactose

Z = H$_2$, O   Toluene | H(CH$_2$)$_{12}$OH
                      $\triangle$, H$_2$SO$_4$

19 (33 - 62%)

R' = C$_{10}$H$_{21}$, C$_{12}$H$_{25}$, C$_{16}$H$_{33}$, C$_{18}$H$_{37}$, C$_6$F$_{13}$CH$_2$, C$_8$F$_{17}$CH$_2$

SCHEME 17 Synthesis of bicatenar surfactants **19**.

The hydrogenated products (**21**) generally show good surface activity and a phase behavior close to that of short-chain lecithins. The corresponding perfluorinated compounds (**22**) are not very soluble in aqueous media and the introduction of a sugar moiety as polar group (Scheme 20) leading to structure **23** favors good surfactant properties [55].

The synthesis of monoesters based on perfluorinated alcohols and aspartic or glutamic acid has been reported, and the corresponding betaines (**24**) have been prepared (Scheme 21). The yields for these reactions are universally poor, however [15].

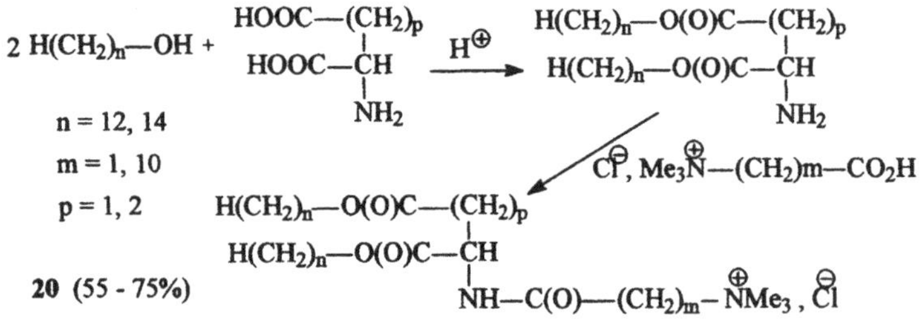

n = 12, 14
m = 1, 10
p = 1, 2

20 (55 - 75%)

SCHEME 18 Synthesis of diesters **20**.

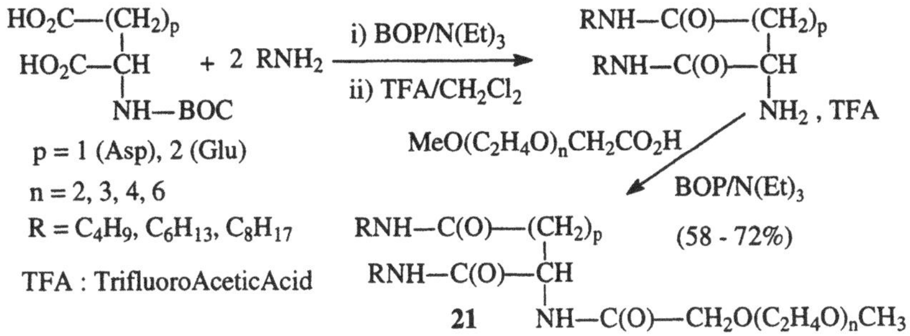

$$\text{HO}_2\text{C}-(\text{CH}_2)_p$$
$$\text{HO}_2\text{C}-\overset{|}{\text{CH}} \quad + \ 2 \ \text{RNH}_2 \xrightarrow[\text{ii) TFA/CH}_2\text{Cl}_2]{\text{i) BOP/N(Et)}_3}$$
$$\overset{|}{\text{NH}-\text{BOC}}$$

$$\text{RNH}-\text{C(O)}-(\text{CH}_2)_p$$
$$\text{RNH}-\text{C(O)}-\overset{|}{\text{CH}}$$
$$\overset{|}{\text{NH}_2 \text{, TFA}}$$

p = 1 (Asp), 2 (Glu)

n = 2, 3, 4, 6

R = C$_4$H$_9$, C$_6$H$_{13}$, C$_8$H$_{17}$

TFA : TrifluoroAceticAcid

$\text{MeO}(\text{C}_2\text{H}_4\text{O})_n\text{CH}_2\text{CO}_2\text{H}$

$$\text{RNH}-\text{C(O)}-(\text{CH}_2)_p$$
$$\text{RNH}-\text{C(O)}-\overset{|}{\text{CH}}$$
$$\mathbf{21} \quad \overset{|}{\text{NH}}-\text{C(O)}-\text{CH}_2\text{O}(\text{C}_2\text{H}_4\text{O})_n\text{CH}_3$$

BOP/N(Et)$_3$

(58 - 72%)

SCHEME 19 Synthesis of diamides **21**.

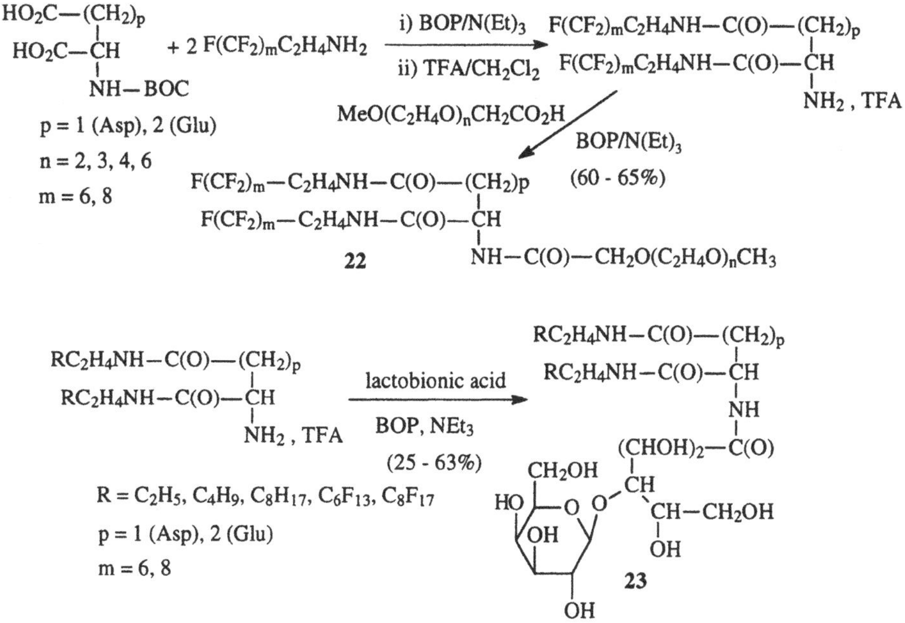

$$\text{HO}_2\text{C}-(\text{CH}_2)_p$$
$$\text{HO}_2\text{C}-\overset{|}{\text{CH}} \quad + \ 2 \ \text{F(CF}_2)_m\text{C}_2\text{H}_4\text{NH}_2 \xrightarrow[\text{ii) TFA/CH}_2\text{Cl}_2]{\text{i) BOP/N(Et)}_3}$$
$$\overset{|}{\text{NH}-\text{BOC}}$$

$$\text{F(CF}_2)_m\text{C}_2\text{H}_4\text{NH}-\text{C(O)}-(\text{CH}_2)_p$$
$$\text{F(CF}_2)_m\text{C}_2\text{H}_4\text{NH}-\text{C(O)}-\overset{|}{\text{CH}}$$
$$\overset{|}{\text{NH}_2 \text{, TFA}}$$

p = 1 (Asp), 2 (Glu)

n = 2, 3, 4, 6

m = 6, 8

$\text{MeO}(\text{C}_2\text{H}_4\text{O})_n\text{CH}_2\text{CO}_2\text{H}$

BOP/N(Et)$_3$

(60 - 65%)

$$\text{F(CF}_2)_m-\text{C}_2\text{H}_4\text{NH}-\text{C(O)}-(\text{CH}_2)p$$
$$\text{F(CF}_2)_m-\text{C}_2\text{H}_4\text{NH}-\text{C(O)}-\overset{|}{\text{CH}}$$
$$\mathbf{22} \quad \overset{|}{\text{NH}}-\text{C(O)}-\text{CH}_2\text{O}(\text{C}_2\text{H}_4\text{O})_n\text{CH}_3$$

$$\text{RC}_2\text{H}_4\text{NH}-\text{C(O)}-(\text{CH}_2)_p$$
$$\text{RC}_2\text{H}_4\text{NH}-\text{C(O)}-\overset{|}{\text{CH}} \xrightarrow[\substack{\text{BOP, NEt}_3 \\ (25 - 63\%)}]{\text{lactobionic acid}}$$
$$\overset{|}{\text{NH}_2 \text{, TFA}}$$

R = C$_2$H$_5$, C$_4$H$_9$, C$_8$H$_{17}$, C$_6$F$_{13}$, C$_8$F$_{17}$

p = 1 (Asp), 2 (Glu)

m = 6, 8

$$\text{RC}_2\text{H}_4\text{NH}-\text{C(O)}-(\text{CH}_2)_p$$
$$\text{RC}_2\text{H}_4\text{NH}-\text{C(O)}-\overset{|}{\text{CH}}$$
$$\overset{|}{\text{NH}}$$
$$\overset{|}{(\text{CHOH})_2-\text{C(O)}}$$

$$\text{CH}_2\text{OH} \quad \overset{|}{\text{CH}}$$
$$\text{HO}\diagdown\!\!\diagup\text{O} \ \ \text{O}\diagdown\!\text{CH}-\text{CH}_2\text{OH}$$
$$\diagup\text{OH} \qquad \overset{|}{\text{OH}}$$
$$\overset{|}{\text{OH}} \qquad \mathbf{23}$$

SCHEME 20 Synthesis of hydrophilic compounds **22** and **23**.

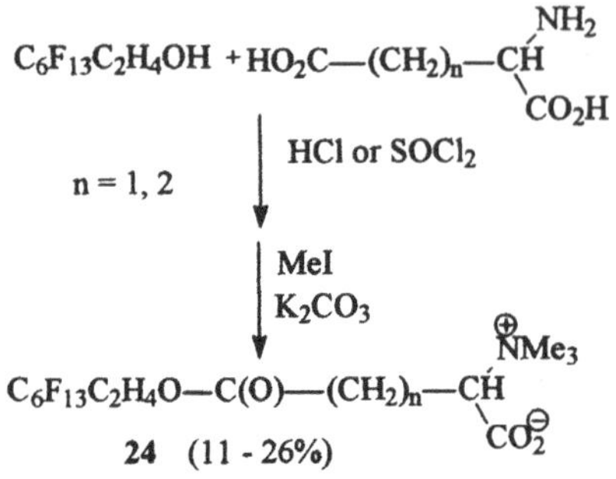

$C_6F_{13}C_2H_4OH$ + HO_2C—$(CH_2)_n$—$\overset{\displaystyle NH_2}{\underset{\displaystyle CO_2H}{CH}}$

$\downarrow$ HCl or SOCl$_2$

n = 1, 2

$\downarrow$ MeI
K$_2$CO$_3$

$C_6F_{13}C_2H_4O$—$C(O)$—$(CH_2)_n$—$\overset{\displaystyle \overset{\oplus}{N}Me_3}{\underset{\displaystyle C\overset{\ominus}{O_2}}{CH}}$

24   (11 - 26%)

SCHEME 21   Synthesis of perfluoroalkylesters of the lateral acid group of Asp or Glu **24**.

N-Acyl-glutamates are well known, and their synthesis has been realized by various methods [56]. We have used one of them to prepare the perfluoro-alkylated glutamic acid **25** (Scheme 22), which has been obtained with a yield of 18% [57].

The synthesis of disubstituted derivatives of glutamic acid leaving one of the acid functions free for further coupling has been attempted (Scheme 23) by amidation of the carboxyl function in position 1 of the amino acid with a long-chain amine, followed by acylation of the α-amino group with a fatty acid chloride [58]. The intermediate disubstituted product (**26**), however, re-acts further by intramolecular nucleophilic attack of the acylated nitrogen atom on the second acid function to finally yield the cyclic structure **27**. The mono-amides with $n > 9$ show surface activity, whereas the cyclic compound is too little soluble in water to determine its cmc. The properties of a monomolecular film on water, however, have been studied by means of a Langmuir balance [59]. Syntheses of the fluorinated analogues of these compounds are in progress.

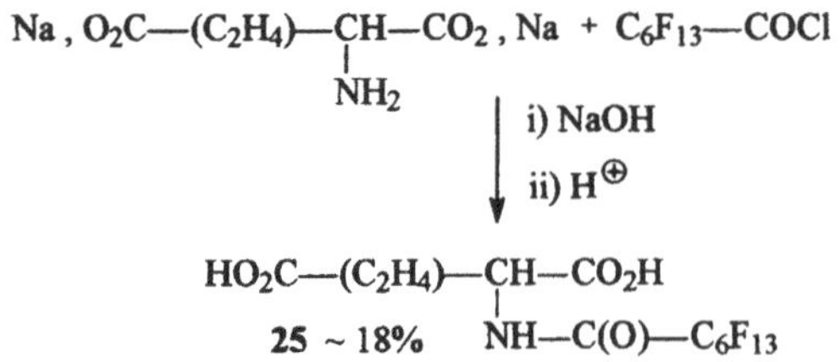

Na , O_2C—(C_2H_4)—$\underset{\displaystyle NH_2}{CH}$—$CO_2$, Na + C_6F_{13}—COCl

$\downarrow$ i) NaOH
ii) H$^{\oplus}$

HO_2C—(C_2H_4)—$\underset{\displaystyle NH-C(O)-C_6F_{13}}{CH}$—$CO_2H$

25 ~ 18%

SCHEME 22   Synthesis of perfluoro-acyl-glutamic acid **25**.

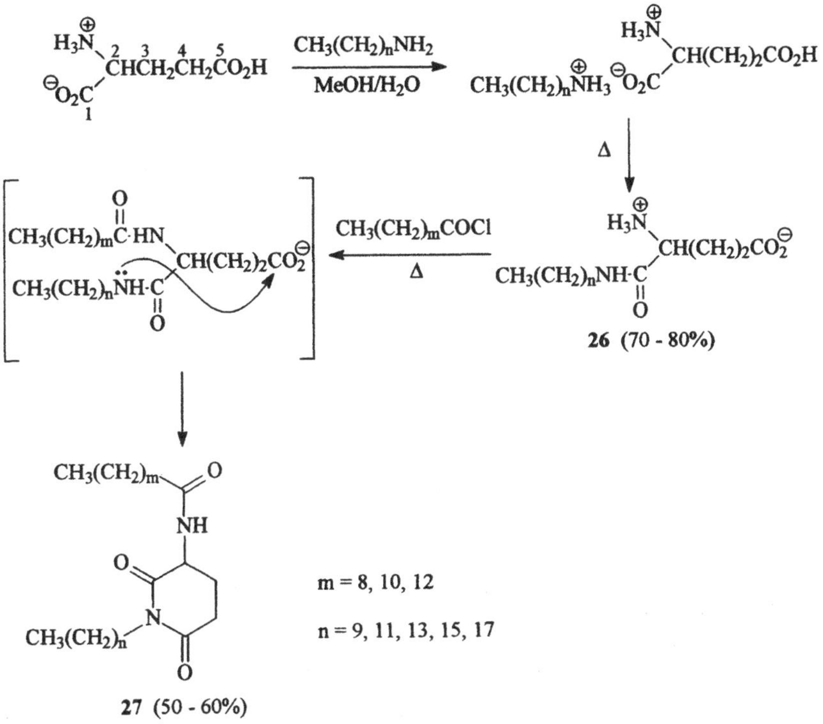

SCHEME 23 Synthesis of glutamic amide derivatives **26** and **27**.

Carnitine is at the origin of two types of fluorinated surfactants. The perfluoroalkyl-acylcarnitines (**28**) (Scheme 24), obtained with satisfactory yields (44–80%), show good surfactant characteristics [60]. The synthesis of N-perfluoroalkyl carnitine (**29**) (Scheme 25) is less efficient, and its physico-chemical properties have not been evaluated [61]. The betaines of perfluoro-alkyl α,ω-amino acids (**30**) (Scheme 26) are accessible in a similar manner [62]. They are good surface agents [63].

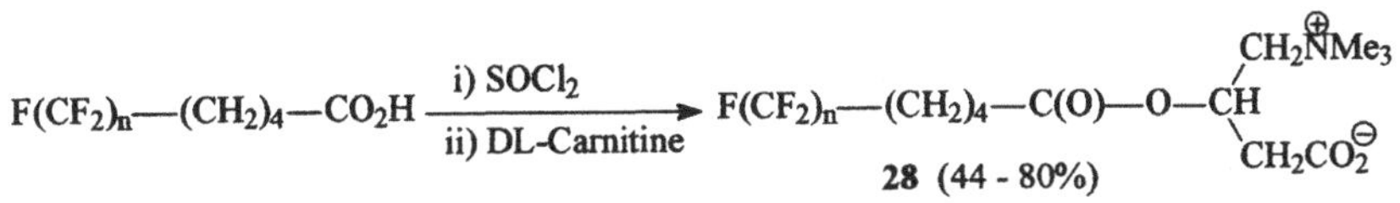

SCHEME 24 Synthesis of perfluoroalkyl-acyl-carnitine **28**.

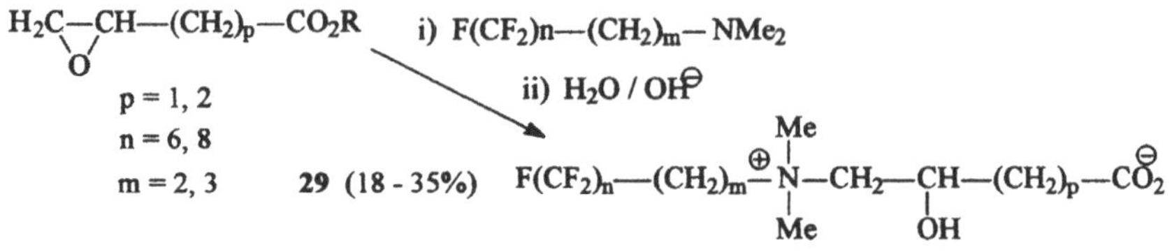

$$H_2C\overset{\diagdown}{\underset{O}{-}}CH\!-\!(CH_2)_p\!-\!CO_2R$$

p = 1, 2
n = 6, 8
m = 2, 3   **29** (18 - 35%)

i) F(CF₂)n—(CH₂)ₘ—NMe₂

ii) H₂O / OH⊖

$$F(CF_2)_n\!-\!(CH_2)_m\!-\!\overset{Me}{\underset{Me}{\overset{\oplus}{N}}}\!-\!CH_2\!-\!\underset{OH}{\overset{|}{CH}}\!-\!(CH_2)_p\!-\!C\overset{\ominus}{O_2}$$

SCHEME 25   Synthesis of *N*-perfluoroalkyl-carnitine **29**.

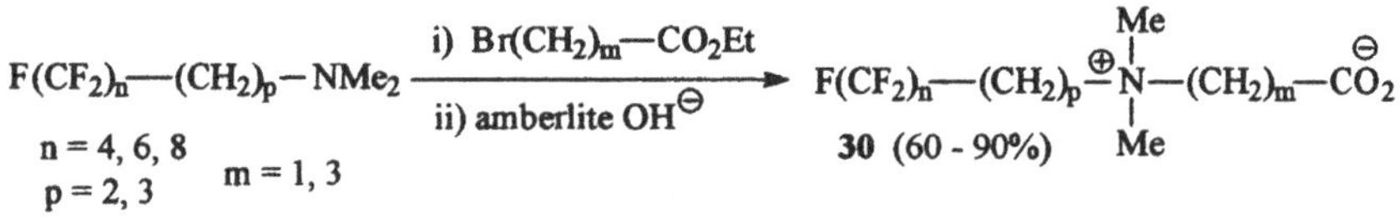

$$F(CF_2)_n\!-\!(CH_2)_p\!-\!NMe_2$$

n = 4, 6, 8
p = 2, 3   m = 1, 3

i) Br(CH₂)ₘ—CO₂Et
$\xrightarrow{\qquad\qquad}$
ii) amberlite OH⊖

$$F(CF_2)_n\!-\!(CH_2)_p\!-\!\overset{Me}{\underset{Me}{\overset{\oplus}{N}}}\!-\!(CH_2)_m\!-\!C\overset{\ominus}{O_2}$$

30 (60 - 90%)

SCHEME 26   Synthesis of betaine **30**.

Among the numerous approaches described to graft a perfluoroalkyl chain on an aminoacid skeleton, the methods allowing one to create a C–C bond between the fluorinated part and the amino acid are of special interest. The access to this less frequently encountered type of compound uses classical methods of synthesis, such as that leading to the perfluoroalkyl-glycine **31** (Scheme 27). The compounds **31** are barely soluble, except for the betaine derivatives **32**, which are obtained with very poor yields, however [64]. The perfluoroalkyl-glycines have been condensed on cholic acid (Scheme 28) to give products **33**, also with unsatisfactory yields [37].

A synthesis of β-perfluoroalkyl-β-alanine (**34**) (Scheme 29) has been de-scribed [65,66] and extended to the *N*-alkyl derivatives (**35**) (Scheme 30) [67]. The compounds **34** show a relatively poor solubility in water, but their sodium

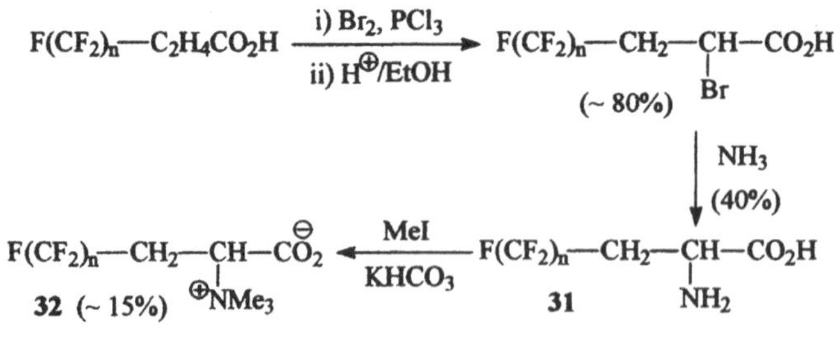

$$F(CF_2)_n\!-\!C_2H_4CO_2H \xrightarrow[\text{ii) } H^{\oplus}/EtOH]{\text{i) Br}_2,\ PCl_3} F(CF_2)_n\!-\!CH_2\!-\!\underset{Br}{\overset{|}{CH}}\!-\!CO_2H$$

(~ 80%)

$\downarrow$ NH₃
(40%)

$$F(CF_2)_n\!-\!CH_2\!-\!\underset{\underset{NMe_3}{\oplus}}{\overset{\ominus}{C}}O_2 \xleftarrow[\text{KHCO}_3]{\text{MeI}} F(CF_2)_n\!-\!CH_2\!-\!\underset{NH_2}{\overset{|}{CH}}\!-\!CO_2H$$

32 (~ 15%)   **31**

SCHEME 27   Synthesis of α-perfluoroalkyl-Gly **31** and corresponding betaine **32**.

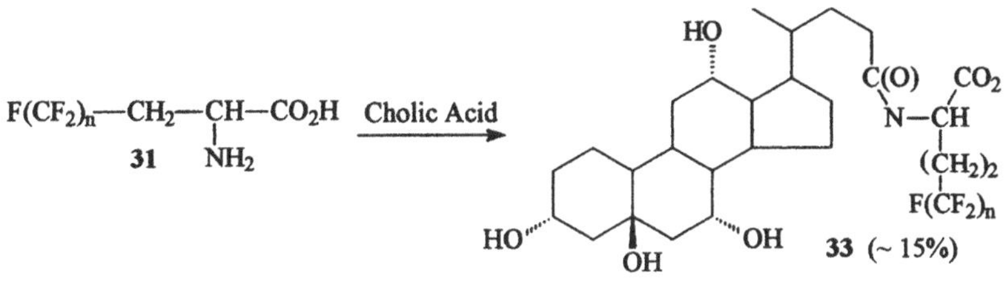

$$F(CF_2)_n-CH_2-CH-CO_2H \quad \xrightarrow{\text{Cholic Acid}}$$
$$\overset{|}{\underset{\textbf{31} \quad NH_2}{}}$$

SCHEME 28 Synthesis of structures of type **33**.

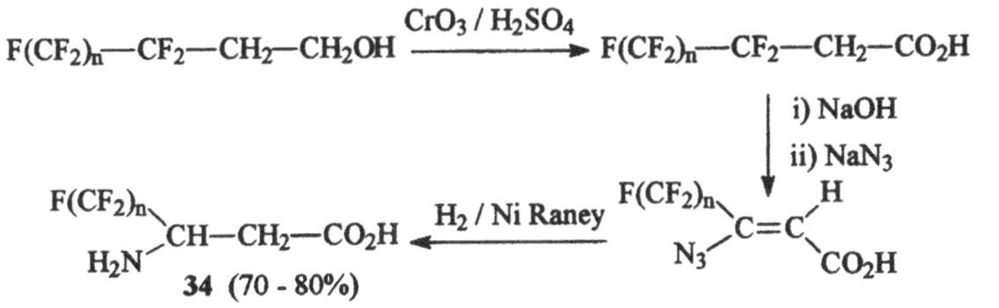

$$F(CF_2)_n-CF_2-CH_2-CH_2OH \xrightarrow{CrO_3 / H_2SO_4} F(CF_2)_n-CF_2-CH_2-CO_2H$$

i) NaOH
ii) NaN_3

$$\underset{H_2N}{\overset{F(CF_2)_n}{\diagdown}}CH-CH_2-CO_2H \xleftarrow{H_2 / Ni \ Raney} \underset{N_3}{\overset{F(CF_2)_n}{\diagdown}}C=C\overset{H}{\underset{CO_2H}{\diagup}}$$
34 (70 - 80%)

SCHEME 29 Synthesis of β-perfluoroalkyl-β-alanine **34**.

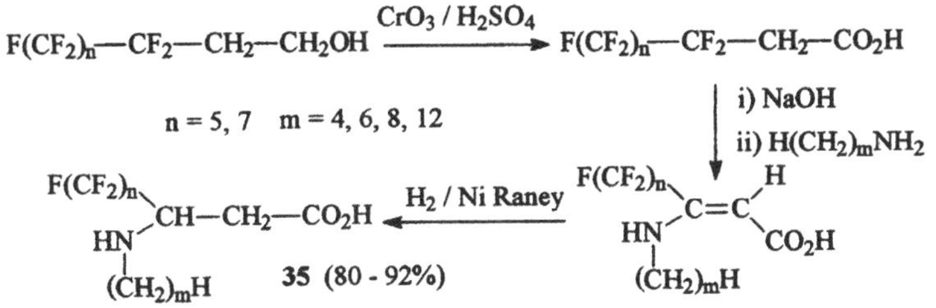

$$F(CF_2)_n-CF_2-CH_2-CH_2OH \xrightarrow{CrO_3 / H_2SO_4} F(CF_2)_n-CF_2-CH_2-CO_2H$$

n = 5, 7   m = 4, 6, 8, 12

i) NaOH
ii) H(CH_2)_mNH_2

$$\underset{\underset{(CH_2)_mH}{\overset{|}{HN}}}{\overset{F(CF_2)_n}{\diagdown}}CH-CH_2-CO_2H \xleftarrow{H_2 / Ni \ Raney} \underset{\underset{(CH_2)_mH}{\overset{|}{HN}}}{\overset{F(CF_2)_n}{\diagdown}}C=C\overset{H}{\underset{CO_2H}{\diagup}}$$
35 (80 - 92%)

SCHEME 30 Preparation of N-alkyl-β-perfluoroalkyl-N-alkyl-β-alanine **35**.

salts are soluble and display the characteristic properties of anionic surfactants. The products **35** are hardly soluble in water, even in the form of their salts [68]. They are of interest for the preparation of bicatenar amphiphilic molecules forming molecularly organized systems such as bilayers, lamellar phases, and vesicles.

To conclude this chapter one may say that fluorinated surfactant molecules based on amino acids have been the subject of numerous studies. This is due, on one hand, to the particular properties conferred to them by the perfluoroalkyl chains and, on the other, to their potential applications in the field of biology, e.g., as oxygenation agents. Synthetic fluorinated surfactants are particularly useful in the form of emulsions [69] or microemulsions [70,71] for the transport of respiratory gases. A large number of formulations to obtain this type of system have been elaborated [72–74].

III. PEPTIDE-BASED SURFACTANTS

Analogously to amino acids, oligopeptides can play the role of pivot or polar modulus in synthetic compounds [75–77] and biosurfactants [78]. An important number of molecules containing a peptide as polar part and a perhydrogenated apolar chain have been described, but fluorinated analogues are much less frequently encountered. Generally, these compounds are prepared with peptides of biological interest having a suitable hydrophilicity together with specific properties in the physiological domain. To this category belongs a family of surfactants based either on carnosine [79–81] or on carcinine [82–84] containing perhydrogenated or perfluorinated hydrophobic substituents [85]. These compounds have been obtained by modular methodology; their structures (**36** and **37**) are shown in Scheme 31.

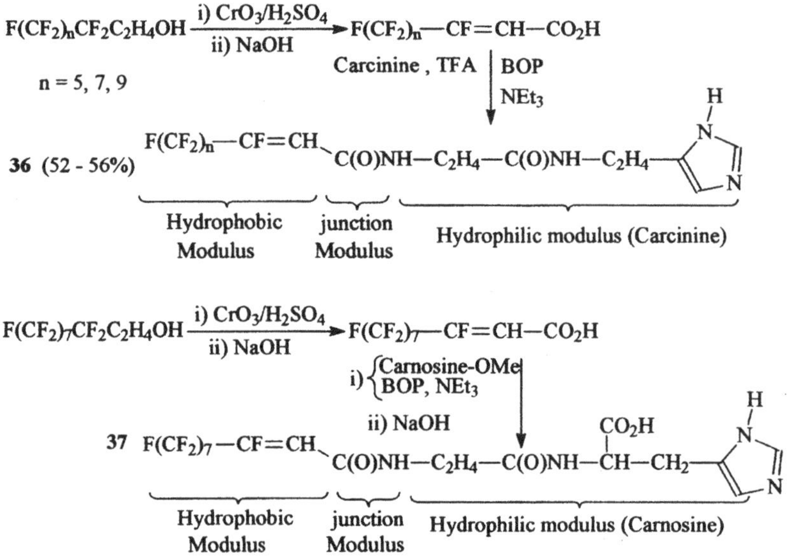

SCHEME 31 Synthesis of perfluoroacyl-peptidoamines **36** and **37**.

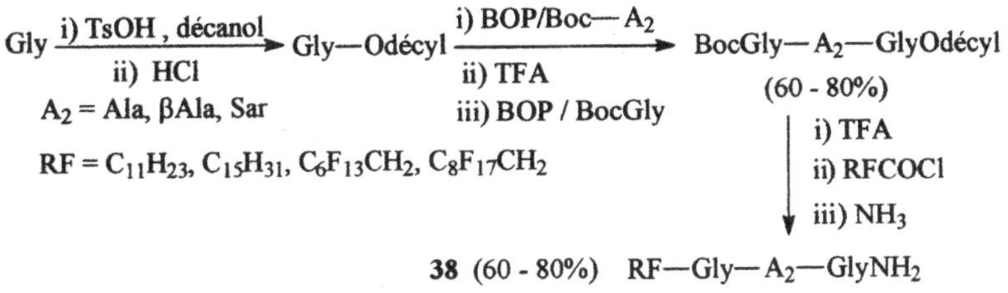

Gly $\xrightarrow[\text{ii) HCl}]{\text{i) TsOH , décanol}}$ Gly—Odécyl $\xrightarrow[\text{ii) TFA}]{\text{i) BOP/Boc—A}_2}$ BocGly—A$_2$—GlyOdécyl

A$_2$ = Ala, βAla, Sar   iii) BOP / BocGly   (60 - 80%)

RF = C$_{11}$H$_{23}$, C$_{15}$H$_{31}$, C$_6$F$_{13}$CH$_2$, C$_8$F$_{17}$CH$_2$

i) TFA
ii) RFCOCl
iii) NH$_3$

38 (60 - 80%)   RF—Gly—A$_2$—GlyNH$_2$

SCHEME 32   Synthesis of fluorolipopeptides **38**.

The physicochemical properties (surface tension, cmc) of these surfactants in binary mixtures with water are comparable to other nonionic surface-active agents. They have no hemolytic activity toward red blood cells. As to the cellular viability of hybridoma cell cultures, compounds **36** and **37** show a behavior strictly identical to that of commercial biocompatible surfactants [85].

A synthesis of fluorolipopeptides by the modular methodology (Scheme 32) leads to the nonionic amphiphilic structures **38** [86,87]. Because derivatives **38** are poorly soluble in water, their surfactant properties have been determined in formamide [88]. Hemolytic tests carried out with some of these compounds indicate no activity toward red blood cells; they show, however, a marked toxicity for hybridoma cells [89].

Another family of compounds containing proline [39] derived from the products of preceding work [90] is obtained in a similar way (Scheme 33) and shows properties typical of hydrophobic surfactants. Their ability to form molecularly organized systems in polar solvents such as DMSO has been studied [91].

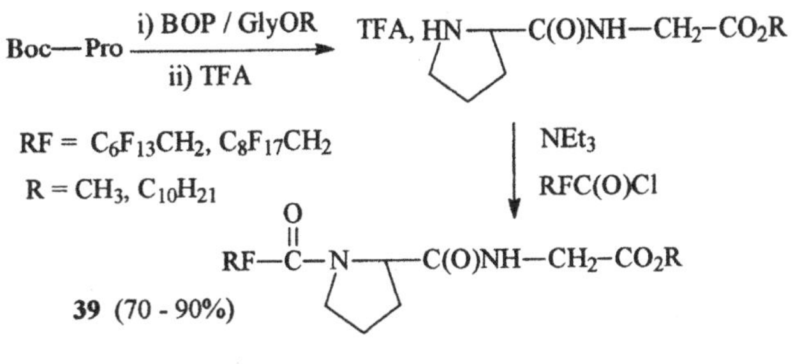

Boc—Pro $\xrightarrow[\text{ii) TFA}]{\text{i) BOP / GlyOR}}$ TFA, HN⟨—C(O)NH—CH$_2$—CO$_2$R

RF = C$_6$F$_{13}$CH$_2$, C$_8$F$_{17}$CH$_2$

R = CH$_3$, C$_{10}$H$_{21}$

NEt$_3$
RFC(O)Cl

$$RF—\overset{O}{\overset{\|}{C}}—N⟨—C(O)NH—CH_2—CO_2R$$

39 (70 - 90%)

SCHEME 33   Synthesis of proline-containing surfactants **39**.

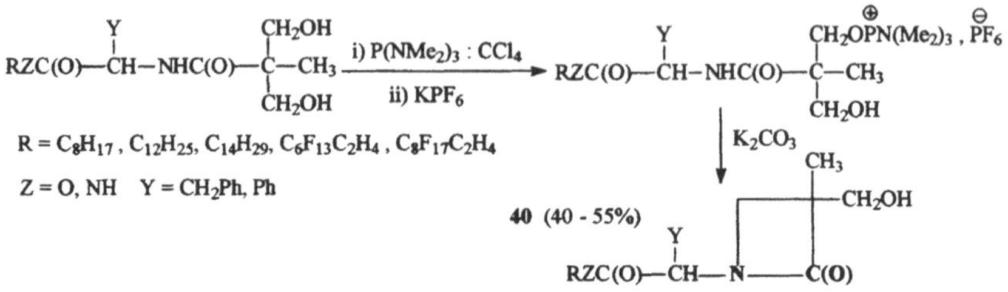

SCHEME 34 Synthesis of β-lactam surfactants **40**.

The synthetic pathways to this type of molecule follow the classical methods of peptide synthesis [92], with certain improvements to increase their efficiency [87,88,93]. Nevertheless, examples of amphiphilic molecules with perfluorinated substituents based on oligopeptides are still rare, probably due to the greater difficulties inherent in these methodologies. In this context a synthetic approach to obtain the perfluoroalkyl-phenylalanyl-β-lactams **40** with satisfactory yields should be mentioned [94]. In these compounds the dipeptide structure plays the role of a junction modulus (Scheme 34). It confers a polar character to the molecule and introduces simultaneously, via the benzyl side-chain of phenylalanine, a lipophilic part. These derivatives show good surfactant properties. The products with short lipophilic chains form L_1 phases, whereas those containing longer chains give rise to lamellar phases. Generally speaking, their behavior is like that of hydrophobic surfactants [95].

By the addition of a sugar group (Scheme 35), the compounds **41** are ob-

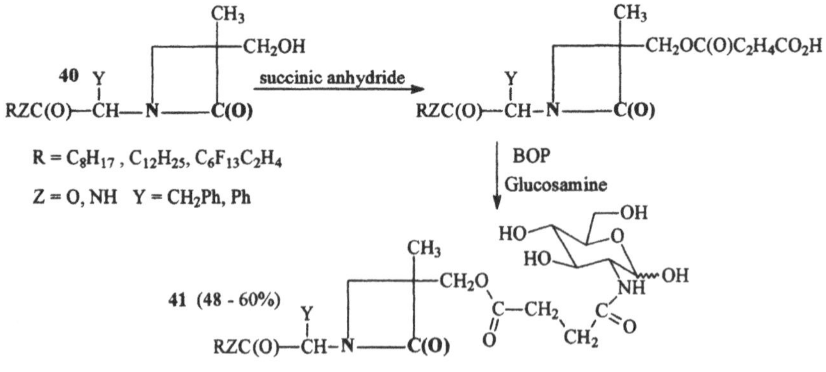

SCHEME 35 Synthesis of β-lactam surfactants **41** containing a sugar group.

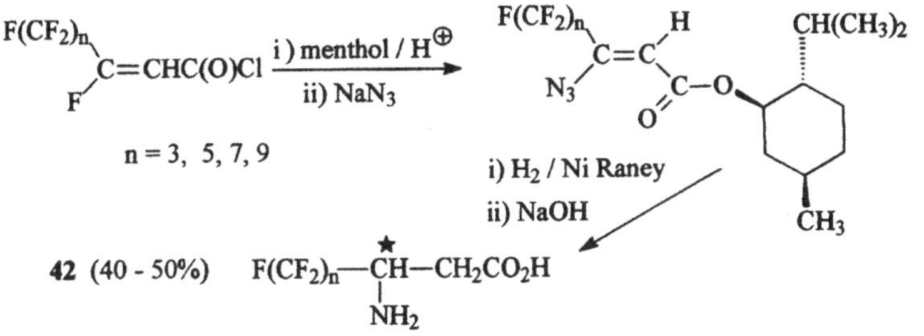

SCHEME 36 Enantioselective synthesis of perfluoroalkyl-β-alanine **42**.

tained, showing the characteristics of hydrophilic surface agents [96]. Derivatives **40** and **41** are of reduced toxicity toward cell cultures and have no hemolytic effect; they have, however, a certain antibiotic activity [97].

In an analogous series, surfactants derived from 6-aminopenicillanic acid but bearing only hydrogenated substituents [98] have been synthesized, with yields of ca. 50%. One of the aims of these preparations is to access amphiphilic substances useful for chiral recognition [99,100]. We have recently attempted the enantioselective synthesis of the β-perfluoroalkyl-β-alanines (**42**). The enantiomeric excess achieved is rather modest for the moment. The approach is described in Scheme 36, with the enantioselective step being the hydrogenation of the chiral menthol ester.

The products (**42**) are surface active. The corresponding racemic mixtures (**34**) have been coupled either with histidine, to yield the amphiphilic carnosine analogues **43** [80], or with histamine, leading to the carcinine derivatives [84]. Structures **43** and **44** [67], containing a racemic β-perfluoroalkyl-β-alanine (Schemes 37 and 38), have been obtained with yields of the order of 40% [101]. The preparation of derivatives based on optically active compounds **42**

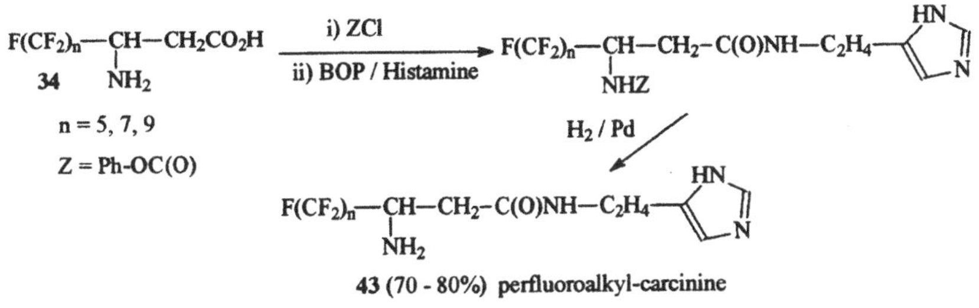

43 (70 – 80%) perfluoroalkyl-carcinine

SCHEME 37 Synthesis of DL-perfluoroalkyl-β-alanyl-histidine **43**.

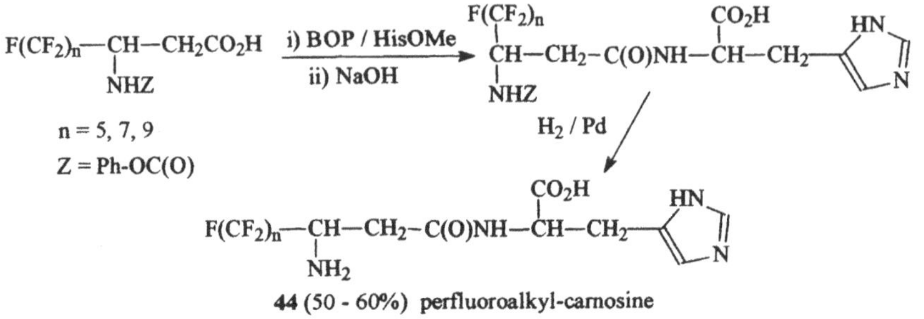

SCHEME 38   Synthesis of DL-perfluoroalkyl-β-alanyl-histamine **44**.

is under study. Substances **43** and **44** are water soluble. They lower the surface tension of aqueous solutions to 25 mN·m^{-1} and complex copper(II) ions, analogous to peptidoamines [102–105].

To complete this section on oligopeptide-based compounds, the hydrophilic polyamino acids should be mentioned. They are frequently encountered in amphiphilic hydrogenated surfactants [106,107]. To our knowledge no fluorinated counterparts have been prepared so far, although the methods described earlier should allow their synthesis. The organized systems obtained with the hydrogenated molecules are quite original [108–110], and the fluorinated analogues might show an interesting phase behavior.

IV. CONCLUSION

A large number of amphipathic molecules derived from amino acids or peptides have been described up to date. The possibilities of structure modulation are nearly without limits, allowing one to adjust their HLB at will and to synthesize derivatives with special properties à la carte. In particular, it is possible to prepare, in a relatively simple way, compounds able to form vesicles, liposomes, bilayers, and other specific organized systems. The presence of an asymmetrical carbon atom in the naturally occurring amino acids opens the way to further applications, e.g., chiral recognition. So far the majority of these possibilities have been explored for hydrogenated surfactants; for compounds containing perfluorinated substituents, the progress in synthetic methodology during the past 20 years has led to numerous new surface-active substances. The particularly interesting properties of these amphiphiles open the way to a large variety of original applications [111]. The field of investigations in the biological domain appears especially vast for these fluorinated amphiphatic products [112]. These perspectives should encourage chemists concerned with amino acid– and peptide-based surfactants of this type to continue their efforts. There is no doubt that they will be rewarded with a surprising wealth of new results in the physicochemical and biological domains.

REFERENCES

1. C. Larpent, Technique de l'Ingénieur, traité constantes physico-chimiques, K342, pp. 1–15 (1995).
2. M.G. Rosen, Surfactants and Interfacial Phenomena, Wiley, New York (1978).
3. M.J. Schick, ed., Surfactant Science Series, Marcel Dekker, New York, an ongoing series.
4. K.L. Mittal, ed., Surfactants in Solution, Plenum Press, New York, an ongoing series.
5. Th. F. Tadros, ed., Surfactants, Academic Press, London (1984).
6. D. Attwood, A.T. Florence, Surfactant Systems, Chapman and Hall, New York (1983).
7. V. Emmanouil, M. El Ghoul, C. André-Barres, B. Guidetti, I. Rico-Lattes, A. Lattes, Langmuir, 1998, 14(19):5389–5395.
8. I. Rico-Lattes, A. Lattes, International Festschrift for Pr S.E. Friberg, Colloids Surf., 1997, 123–124:37–48, and references cited therein.
9. H.C. Fielding, in: Organofluorine Chemicals and Their Industrial Applications, R.E. Banks, ed., Elis Horwood, Chichester, England (1978).
10. T. Kunitake, Angew. Chem. Int. Ed. Engl., 31, 709–716 (1992).
11. B.R. Bluestein, R.D. Cowell, M.L. Dausner, J. Am. Oil Chem. Soc., 58:173 (1981).
12. J.D. Spinach, in: Anionic surfactants, W.M. Linfield, ed. Marcel Dekker, New York, Surfactants Sciences Series, Vol. 7(II), Chap. 16 (1976).
13. J.P. Greenstein, M. Winitz, in: Chemistry of the Amino Acids, Wiley, New York (1961).
14. C. Kimura, K. Kashigawaya, M. Kobayashi, K. Murai, Yukagaku, 33(12):838–842 (1984).
15. C. Blaignon, Ph.D. dissertation, University of Nice, Nice, France (1987).
16. A. Paquet, Canadian J. Chem., 54:733–737 (1976), and references cited therein.
17. T. Gartiser, Ph.D. dissertation, University H. Poincaré of Nancy 1, France (1982).
18. J.P. Greenstein, M. Winitz, in: Chemistry of the Aminoacids, Vol. 2, Wiley, New York, p. 1027 (1961).
19. G.W. Anderson, J.E. Zimmerman, F.M. Callahan, J. Am. Chem. Soc. 85:3039 (1963).
20. W. König, R. Geiger, Chem. Ber., 103:788–798 (1970).
21. W. König, R. Geiger, Chem. Ber., 103:2024–2030 (1970).

22. B. Castro, J.R. Dormoy, J.R. Evin, C. Selve, Tetrahedron Lett., pp. 1219–1222 (1975).

23. L. Jiang, A. Davison, G. Tennant, R. Ramage, Tetrahedron, 54:14233–14254 (1998).

24. T. Gartiser, C. Selve, E.M. Moummi, unpublished results.

25. Y. Murakami, A. Nakano, K. Fukuya, J. Am. Chem. Soc. 102:4253 (1980).

26. Y. Murakami, J. Kikuchi, T. Takahi, K. Uchimura, J. Am. Chem. Soc. 107: 3373–3374 (1985).

27. Y. Murakami, A. Nakano, H. Ikeda, J. Org. Chem. 47:2137–2144 (1982).

28. J.B. Nivet, M. le Blanc, M. Haddach, M. Lanier, R. Pastor, J.G. Riess, New J. Chem. 18:861–869 (1994).

29. J.B. Nivet, R. Bervelin, M. le Blanc, J.G. Riess, Eur. J. Med. Chem., 27:891–899 (1992).

30. B. Bosycha-Raszah, S. Pryestalski, S. Witk, J. Colloid Interf. Sci. 125(1):80–85 (1988).

31. M. Allouch, C. Selve, unpublished results.

32. L. Benoiton, J. Leclerc, Canadian J. Chem. 43:991 (1965).

33. K. Vogler, P. Lanz, P. Quitt, Helv. Chim. Acta 47(2):526–544 (1964).

34. J. Leclerc, L. Benoiton, Canadian J. Chem. 46:1047–1050 (1968).

35. K. Akizawa, R. Yoshida, U.S. Patent #3,897,466 (1975).

36. H. Yokota, K. Sagawa, C. Eguchi, J. Am. Oil Chem. Soc. 62(12):1716–1719 (1985).

37. C. Blaignon, M. le Blanc, J.G. Riess, Proceedings of the 2nd World Surfactant Congress, ed. ASPA, Paris, 2:137–142 (1988).

38. M.R. Infante, J. Molinero, P. Erra, M.R. Julia, Fette Seifen Anstrichmittel 87(8): 309–313 (1985).

39. M.R. Infante, P. Erra, M.R. Julia, Proceedings of the XIV Jornadas C.E.D., pp. 267–281 (1983).

40. J.J. Garcia-Dominguez, M.R. Infante, Anales de Quimica, 82:413–417 (1986).

41. J. Seguer, M. Allouch, P. Vinardell, M.R. Infante, L. Mansuy, C. Selve, New J. Chem., 18:765 (1994).

42. J. Seguer, Ph.D. dissertation, University of Barcelona, Spain (1993).

43. J. Seguer, C. Selve, M. Allouch, M.R. Infante, J. Am. Oil. Chem. Soc. 73(1): 79–86 (1996).

44. V. Alliot, C. Gérardin, C. Selve, unpublished results.

45. V. Alliot, Ph.D. dissertation, University H. Poincaré of Nancy I, France (1999).

46. V. Alliot, B. Naji, C. Gérardin, C. Selve, unpublished results.

47. V. Alliot, A. Meyer, C. Gérardin-Charbonnier, J. Amos, C. Selve, Amino Acids, 13(1):77 (1997).

48. A. Fichter, Trends Biotechnol. 10:208–217 (1992).

49. D.B. Sarney, E.N. Vulson, Trends Biotechnol., 13:164–172 (1995).

50. D. Montet, F. Servat, M. Pina, J. Graille, P. Galzy, A. Arnaud, H. Leda, L. Marcon, J. Am. Oil Chem. Soc. 67(11):771–774 (1990).

51. N. Nakashima, S. Asakuma, J.M. Kim, T. Kunitake, Chem. Lett., pp. 1709–1702 (1984).
52. N. Nakashima, S. Asakuma, T. Kunitake, J. Am. Chem. Soc., 107:509–510 (1985).
53. M. Allouch, Ph.D. dissertation, University H. Poincaré of Nancy 1, France (1994).
54. M. Allouch, M.R. Infante, J. Seguer, M.J. Stébé, C. Selve, J. Am. Oil Chem. Soc. 73(1):87–95 (1996).
55. M. Allouch, M.R. Infante, L. Mansuy, J. Seguer, C. Selve, J. Com. Esp. Det. 25:443–458 (1994).
56. M. Takehara, I. Yoskimura, K. Takizawa, R. Yoshida, J. Am. Oil Chem. Soc. 49:157–161 (1972).
57. E.M. Moumni, C. Selve, unpublished results.
58. H. Chaumette, Rapport de DEA, University H. Poincaré of Nancy 1, France (1998).
59. L. Rodehüser, H. Chaumette, A. Neyers, E. Rogalska, Ch. Gérardin, C. Selve, Amino Acids, 18/1:89–100 (2000).
60. J.B. Nivet, M. le Blanc, J.G. Riess, Eur. J. Med. Chem., 26:953–960 (1991).
61. J.B. Nivet, M. le Blanc, M. Haddach, M. Lanier, R. Pastor, J.G. Riess, New J. Chem. 18:861–869 (1994).
62. J.B. Nivet, R. Bervelin, M. le Blanc, J.G. Riess, Eur. J. Med. Chem., 27:891–899 (1992).
63. J.B. Nivet, M. le Blanc, J.G. Riess, J. Disp. Sci. Technol., 13:627–646 (1992).
64. R. Bervelin, M. le Blanc, J.G. Riess, Abstract, Int. Chem. Cong. Pacific Basin Soc., Org. 662 Honolulu (1989).
65. S. Thiébaut, Ph.D. dissertation, University H. Poincaré of Nancy 1, France (1995).
66. M. Özer, S. Thiébaut, C. Gérardin, C. Selve, Synthetic Com. 28(13):2429–2441 (1998).
67. M. Özer, S. Thiébaut, C. Gérardin-Charbonnier, L. Rodehüser, C. Selve, Amino Acids, 16/3–4:381–389 (1999).
68. S. Cosgun, C. Gérardin, S. Thiébaut, C. Selve, unpublished results (1999).
69. T.M.S. Chang, R.P. Geyer, Blood Substitutes, Marcel Dekker, New York, pp. 365–623 (1989).
70. G. Mathis, P. Leempoel, J.C. Ravey, C. Selve, J.J. Delpuech, J. Am. Chem. Soc., 106:6162–6171 (1984).
71. J.J. Delpuech, G. Mathis, J.C. Ravey, C. Selve, G. Serratrice, M.J. Stébé, Bull. Soc. Chim. Fr., 4:578–581 (1985).
72. J.G. Riess, New J. Chem., 19:891–909 (1995).
73. B. Castro, J.J. Delpuech, T. Gartiser, G. Mathis, A. Robert, C. Selve, G. Serratrice, M.J. Stébé, C. Tondre, Médecine Armée, 12(2):103–106 (1984).
74. J.J. Delpuech, J.C. Ravey, Images de la Recherche: De la Matière au Vivant, Les Systèmes Moléculaires Organisés, CNRS ed., pp. 51–54 (1994).

75. H. Goldschiniedt, I. Lubowe, Am. Cosm. Perf. 87:35 (1972).
76. J. Molinero, M.R. Julia, P. Erra, M. Robert, M.R. Infante, J. Am. Oil Chem. Soc. 65/6:971–978 (1988).
77. M.R. Infante, A. Pinazo, J. Seguer, Colloids Surf., 126/1:49–58 (1997).
78. K. Jenuy, V. Deltrieu, O. Käppeli, Biosurfactants, N. Kosaric, ed., Marcel Dekker, New York, pp. 135–156 (1993).
79. J.M. Arnould, Can. J. Physiol. Pharmacol., 65:70–74 (1987).
80. M.A. Babizhayev, Biochem. Biophys. Acta, 1004:363–371 (1989).
81. R. Kohen, Y. Yamamoto, K.C. Cundy, B.N. Ames, Proc. Natl. Acad. Sci. USA, 85:3175–3179 (1988).
82. J.M. Arnould, R. Frentz, Comp. Biochem. Physiol., 50C:59–65 (1975).
83. J.M. Arnould, Biochem. Systématics Ecol., 14/4:431–433 (1986).
84. J.M. Arnould, J. Neurochem., 48/4:1316–1324 (1987), and references cited therein.
85. F. Bichat, F. Poirson, B. Henry, M. Maugras, C. Selve, Amino Acids, 5/1:145–146 (1993).
86. F. Hamdoune, Ph.D. dissertation, University H. Poincaré of Nancy 1, France (1990).
87. C. Selve, F. Hamdoune, Tetrahedron Lett., 30/42:5755–5758 (1989).
88. C. Selve, F. Hamdoune, L. Mansuy, M. Allouch, J. Chem. Res. (S), pp. 22–23 (1922); (M) pp. 401–426 (1992).
89. C. Selve, F. Hamdoune, J.C. Ravey, J.J. Delpuech, M. Maugras, S. Ramzaoui, B. Lakhssassi, J.F. Stoltz, Inov. Tech. Biol. Med., 11/6:685–704 (1990).
90. J. Chrisment, J.J. Delpuech, W. Rajerison, C. Selve, Tetrahedron, 42/17:4743–4756 (1986).
91. J. Chrisment, J.J. Delpuech, F. Hamdoune, J.C. Ravey, C. Selve, M.J. Stébé, J. Chem. Soc. Faraday Trans., 89/6:927–934 (1993).
92. M. Bodanszky, Principes of Peptide Synthesis, Springer Verlag, Berlin (1984).
93. T. Schmittberger, A. Cotté, H. Waldmann, Chem. Com., pp. 937–938 (1998).
94. L. Billaut-Molina, Ph.D. dissertation, University H. Poincaré of Nancy 1, France (1995).
95. L. Molina, A. Perrani, M.R. Infante, M. Maugras, M.J. Stébé, C. Selve, Progr. Colloid, Polym. Sci., 100:290–295 (1996).
96. L. Molina, C. Gerardin-Charbonnier, C. Selve, M.J. Stébé, M. Maugras, M.R. Infante, J.L. Torres, M-A. Manresa, P. Vinardell, New J. Chem., 21:1027–1035 (1997).
97. C. Gerardin-Charbonnier, S. Auberger, L. Molina, S. Achilefu, M-A. Manresa, P. Vinardell, M.R. Infante, C. Selve, Prep. Biol. Biotechnol., 29:17 (1999).
98. M. Bouzige, G. Okafo, P. Dhanals, P. Camilleri, Chem. Com., pp. 671–672 (1996).
99. S. Terabe, K. Otsuka, T. Ano, Anal. Chem., 57:834–841 (1985).
100. V. de Biasi, J. Senior, J.A. Zukowski, R.C. Haltmanger, D.S. Eggleston, P. Camilleri, Chem. Com., pp. 1575–1576 (1995).
101. M. Özer, S. Thiébaut, C. Gérardin, C. Selve, unpublished results.

102. T. Gajda, B. Henry, J.J. Delpuech, J. Chem. Soc. Dalton Trans., pp. 2313–2319 (1992).
103. T. Gajda, B. Henry, J.J. Delpuech, J. Chem. Soc. Dalton Trans., pp. 1301–1306 (1993).
104. T. Gajda, B. Henry, A. Aubry, J.J. Delpuech, Inorg. Chem., 35:586–593 (1996); T. Gajda, B. Henry, J.J. Delpuech, J. Chem. Soc. Dalton Trans., pp. 2313–2319 (1992).
105. B. Henry, M. Özer, C. Gerardin, C. Selve, to be published.
106. H. Ihara, T. Fukumoto, C. Hirayama, K. Yamada, Polymer Com., 27:282–285 (1986).
107. H. Ihara, H. Hachisako, C. Hirayama, K. Yamada, Chem. Com., pp. 1244–1245 (1992).
108. M. Kogisco, T. Hanada, K. Yase, T. Shimizu, Chem. Com., pp. 1791–1792 (1998).
109. B. Shade, J.H. Furhop, New J. Chem. 22/2:97–104 (1998).
110. T. Shimizu, M. Mori, H. Minamikawa, M. Mato, Chem. Lett., pp. 1341–1344 (1989).
111. E. Kissa, Fluorinated Surfactants, Marcel Dekker, New York (1993).
112. J. Greiner, J.G. Riess, P. Vierling, Organofluorine Compounds in Medicinal Chemistry and Biomedical Applications, Elsevier, New York, Vol. 48, pp. 339–380 (1993).

8

Interactions of Amino Acid–Based Surfactants with Other Compounds

YUN-PENG ZHU Crompton Corporation, Dublin, Ohio

I. INTRODUCTION

A great number of surfactant-based products, such as shampoos, detergents, cosmetics, paints and coatings, and pharmaceutical products, contain one or more surfactants and other materials in one formulation. As a matter of fact, surfactant-based formulations in practical applications are almost always mixtures, consisting of surfactants, polymers, and other ingredients. These systems provide us with the opportunity to obtain unique properties and performance that are superior to those of the individual ones. These unique properties and the performance of the mixture system are associated with the interaction of surfactants and other ingredients. For instance, polymer in shampoos is used as a viscosity and moisturizing modifier; and in the formulation of detergents, polymer is an effective antireprecipitating agent as well as a viscosity modifier and chelating agent.

Interactions of surfactants are of great interest for theoretical and practical purposes. Physicochemical studies of the interaction of the surfactants with other materials can provide useful information helping us understand the applications regarding aggregated structures, adsorption at the surface, and interaction characteristic parameters. Technical-grade surfactants are mixtures of isomers/different alkyl chain length that interact to produce unique functionality. In simple and complex emulsions and dispersions, surfactants and other materials such as polymers interact to provide stability. In such systems, surfactants provide emulsifying properties, interfacial tension controls, and colloidal stability, whereas polymers stabilize an emulsion and provide other rheological features.

There are numerous studies on mixtures of surfactants [1–3] and surfactants/polymers [4–7]. As would be expected, the interfacial and colloidal properties of such a mixture can be quite different from those of a solution of the individual components. There are interactions among the components through mutual electrostatic attraction/repulsion, van der Waals attraction of hydrophobic groups, formation of hydrogen bonds, and dipolar force.

Studies on a mixture composed of amino acid–based surfactants have not been extensively made as compared to other surfactants, such as sodium dodecyl sulfate (SDS), and nonionics. To the best of our knowledge, there are few publications dealing with mixtures where amino acid–based surfactants, especially amino acid–based surfactants/polymers, are involved. With this in mind, this chapter first discusses methods for examining a mixture containing amino acid surfactants, essentially to provide prospective methods to study surfactant mixture systems. In addition, interactions of amino acid–based surfactants with other components, including inorganic and organic molecules, are reviewed.

II. METHODS TO INVESTIGATE SURFACTANT MIXTURE SYSTEMS

Several methods are available to investigate surfactant mixture systems, ranging from the classical surface tension method to the modern instrumental methods. An extensive review on the methods of investigation of surfactant solutions has been published [8].

A. Surface Tension Method

This is one of the classical, still most widely used, methods for studying the surfactant mixture systems. The simplicity, availability of apparatus, and advances in thermodynamic understanding have made this old method very attractive and practical. As a result, there are a large number of recent studies on the mixed surfactant systems by use of this method [1–3].

Since Rubingh [9] proposed the so-called β^M parameter to evaluate interactions between surfactants in a mixed micelle, Rosen and co-workers have done extensive work on the interaction between surfactants in a binary-surfactant system and introduced a new parameter, β^δ [10] (β_{LL}^δ for liquid–liquid interface [11], β_{LS}^δ for liquid–solid interface [12]) to measure the interaction between surfactants in mixed monolayer at the surface of an aqueous solution.

The evaluation of the interaction parameters is based on the following equations using nonideal or regular solution approximates.

1. The molecular interaction parameter for the mixed micelle formation (β^M) [13] as derived from the following equation:

$$\frac{(X_1^M)^2 \ln(\alpha C_{12}^M / X_1^M C_1^M)}{(1 - X_1^M)^2 \ln[(1 - \alpha) C_{12}^M / (1 - X_1^M) C_2^M]} = 1$$

$$\beta^M = \frac{\ln(\alpha C_{12}^M / X_1^M C_1^M)}{(1 - X_1^M)^2}$$

This equation is solved numerically for X_1^M, and the substitution of X_1^M yields the values of β^M.

2. The molecular interaction parameter for the mixed monolayer formation at the interface of an aqueous solution β^δ [13]. In the same way, X_1 can be found:

$$\frac{X_1^2 \ln(\alpha C_{12} / X_1 C_1^0)}{(1 - X_1)^2 \ln[(1 - \alpha) C_{12} / (1 - X_1) C_2^0]} = 1$$

$$\beta^\sigma = \frac{\ln(\alpha C_{12} / X_1 C_1^0)}{(1 - X_1)^2}$$

To determine the values of β^δ and β^M experimentally, three surface-tension values versus $-\log$ concentration curves are required, as shown in Fig. 1. Two

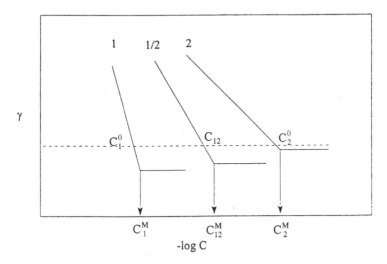

FIG. 1 Experimental determination of parameters β^δ, β^M, X_1, and X_1^M. 1 and 2 represent individual surfactants 1 and 2, respectively; 1/2 represents a mixture of surfactants 1 and 2 at a certain molar ratio α. C_1^0, C_{12}, and C_2^0 should be taken as the concentrations at which the surface tensions are very close to the γ_{cmc} values of each curve.

are for each of the pure individual components, and one is for a mixture of them at a certain molar fraction α. The more negative the value of the two parameters, the greater the interaction between the two surfactants compared to the reference states of the individual surfactants by themselves. From the analysis of these quantities, C_1^M, β^M, C_1, and β^δ, one can determine the following useful properties of a mixed surfactant system: (1) the composition of mixed micelle and the mixed monolayer; (2) the ratio of the two surfactants at which the maximum synergistic effect can be achieved.

In the case of a mixture of surfactant and polymer, the surface tension method has become popular and has had a significant impact in this area since 1967, when Jones published his work [14] on mixtures of polyethylene oxide (PEO)/SDS by using the surface tension method. From the plot of surface tension versus −log concentration at a certain concentration of polymer, one can evaluate whether and where the surfactant and polymer associate (some call it clusters) in the bulk solution and at the air/water interface.

It must be mentioned that although the surface tension method is facile and readily applicable to the study of a mixture of surfactant/polymer, it also has its limitation. Generally, the polymer used should be less surface active than the surfactant itself. Otherwise, interpretation of the data may be difficult in terms of T_1 (cac, critical aggregation concentration) and T_2 (apparent cmc) in spite of a clear indication from the curve of the interaction in such a system. For example, the surface activity of polypropylene glycol makes it difficult to interpret the surface-tension results [15]. Similar to a system of mixed surfactants, the composition of polymer is also a factor in the surfactant/polymer system. The case of poly(vinyl alcohol) (PVA) is more complicated than that of other synthetic polymers, since PVA is known to be prepared by hydrolysis of the acetate group of poly(vinyl acetate). The hydrolysis is only partially done; hence acetate and hydroxyl groups are present. Therefore, consideration of the compositions of polymer is important for a reliable interpretation of experimental data.

B. Electrical Conductivity Method

The electrical conductivity method can be applied to a system consisting of ionic surfactants and ionic/nonionic surfactant mixtures. Also, this method is suitable for a mixture of ionic surfactant and polymer.

Advances in the technologies of the ion-specific electrode make the ion-specific electrode very attractive for studying the amount of surfactant bound to micelle or polymer by measuring the activity of the surfactant monomeric species in solution. Ionic surfactant (RX; R: surfactant ions, and X: denoting a counterion), as an electrolyte in aqueous solution, should follow the Nernst equation:

$$E = E_0 + 2.303RT \log \alpha_s/zF$$

where E_0 is a constant depending on the electrode system, z is the valence of the surfactant ion, and α_s is the surfactant ion activity. From the foregoing, it is clear that the ion-electrode method can provide a direct way for determining the concentration of the monomeric surfactant ion in the solution. This method is widely used in the study of surfactant binding. Advances in this technology have made the electrode stable even in a micellar solution (at a concentration greater than the cmc) [16]. However, in the study of the surfactant/polymer system, the problem is the possible polymer adsorption onto the electrode surface, yielding an incorrect response. As a matter of fact, in the system of ionic surfactant and polymer, the use of ion-specific electrodes to study the binding of surfactant ions to the polymer has attracted much attention [17,18]. On the other hand, a less direct but very popular method to obtain information about the binding of surfactant is to measure the change in counterions (Na$^+$, K$^+$, Ca^{2+}, Cl$^-$, Br$^-$) by the use of counterion-specific electrodes [19–21]. Francois et al. [22] used a sodium ion electrode to study the interaction of SDS and PEO 20,000, and reported that the values of the degree of ionic dissociation α were 0.2 and 0.3 for SDS micelles and its complex with PEO, respectively. Zana [23] studied the mixture of SDS with poly(vinylpyrrolidone) (PVP) and PEO and reported a pronounced increase, 0.85 and 0.65, respectively, in the value of α of SDS, compared with a value of $\alpha = 0.35$ for regular micelles. The values obtained by Francois et al. [22] are different from those obtained by Zana, but the results show a similar tendency, indicating the formation of a cluster due to the surfactant molecules bound to the unit of polymer.

C. Fluorescent Probes

Fluorescent probes are widely used to study the microstructure of micelles and microemulsion, providing information on microstructural aspects of the aggregates. Zana has recently reviewed static and dynamic fluorescence techniques and pointed out the advantages and disadvantages of these techniques [24]. For the application of photophysical and photochemical techniques in the study of surfactants, the reader is also referred to the review [25].

   Fluorescence probe and fluorescent label are two approaches that are generally applied to the study of surfactant-containing systems, yielding information on the properties of the micro-domain, such as the mean aggregation number and the mobility or fluidity of the aggregates. Pyrene is one of the popular probes used in this method because of its high hydrophobicity and relatively long lifetime in the excited state. From steady-state emission spectra of pyrene, the ratio of I_1/I_3 (the first and third vibronic peaks) can be used to estimate the polarity of the microenvironment of an aggregate such as micelles. Turro and co-workers investigated the surfactant/polymer systems of PVP/SDS and PEO/SDS using pyrene fluorescence spectra [26]. Pyrene in the excited state can interact with the pyrene molecule in the ground state to form excimers (generally dimers). The extent of formation of the excimers is inversely proportional to the viscosity of the medium. Therefore, the ratio of I_e/I_m from its emission spectra can yield insight into the microviscosity of the aggregate. Bis(1-pyryenylmethyl) ether spectra can provide information on both the polarity and the microviscosity of a microenvironment.

Static fluorescence quenching has been used to measure the mean aggregation number of surfactant-related systems [27–29]. This method was originally developed by Turro and Yekta for anionic surfactant micelles [30]. This method works well without significant errors when $k_q/k \gg 1$ and no quencher redistribution has taken place. On the other hand, fluorescence-decay experiment, namely, time-resolved fluorescence quenching (TRFQ), has been extensively used to determine the micelle aggregation number in a surfactant system. The reader is referred to the review [24] on this subject.

The fluorescence labeling method can be more informative and direct, because the data report on the phenomena of the subjects studied. It is applied in surfactant/polymer systems [31] and the study of lipid membrane [32], but is rarely used in surfactant/surfactant systems. Also, there is always a question of whether the fluorescent-labeled molecule interferes with the system under study.

D. Electron Spin Resonance Spectroscopy

Electron spin resonance (ESR) is one of powerful techniques for studying the microstructure of micelles formed by a single surfactant and its mixtures [33]. This method can provide information on the microviscosity of aggregates. When the concentration of a surfactant solution is beyond its cmc, the spin probe molecule is believed to be located predominantly in a micellar phase, owing to solubilization. Moreover, a spin probe nitroxide radical can be designed to attach to a specific position on an alkyl chain according to the demands of investigation to obtain the information at different positions. The hyperfine splitting of the ESR is an index of the microviscosity of the environment around the probe. The rotational correlation time τ_c can be estimated from the ESR linewidth by using the following equation [33,34]:

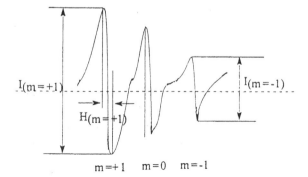

FIG. 2 Representative ESR spectrum in micellar solution.

$$\tau_c = 6.6 \times 10^{-10} \times \Delta H_{(m=+1)} \left[\left(\frac{I_{(m=+1)}}{I_{(m=-1)}} \right)^{1/2} - 1 \right] s$$

A typical ESR spectrum is shown in Fig. 2. As described, τ_c can be calculated from the spectra. Shah and co-workers [35] used ESR to study the structure of microemulsion, and Rosen et al. [36] used ESR and found the microviscosity gradient in the middle phase of three-phase micellar systems. Shirhama et al. [37] and Witte et al. [38] have used ESR to study the interaction of SDS with PEO and PVP, yielding information on the structure of complex from the polymers and SDS. But the influence of the spin-label molecule on the microstructure should be considered.

E. Nuclear Magnetic Resonance

Nuclear magnetic resonance (NMR) spectroscopy is a powerful technique for determining molecular structure and conformation. Since the breakthrough 2-D NMR technique was developed, NMR has been able to give us more detailed information about structure and conformation at the molecular level. In the study of surfactant-involved systems, the main methods used involve the measurement of chemical shifts, relaxation times (T_1 and T_2), and the self-diffusion coefficient.

Chemical-shift measurement has proven to be useful in determining cmc, surfactant chain conformation, and the extent of water penetration into micelle. Menger [39] found that as many as three methylene groups of an alkyl chain near the head group are hydrated in a micelle. Relaxation times are related to translational and rotational molecular motions. These also can give us information on the nature of the micellar core and the change in the organization of alkyl chain from the micelle surface to the micellar core. Tokiwa and Tsujii [40] studied mixed surfactant systems via chemical-shift measurement. The 2-D NMR method [41] has been used to study the mixed system of SDS and PEO nonionics.

The self-diffusion coefficient method was developed originally by Lindman and co-workers [42]. This method is by far the most powerful and successful application of NMR in the study of surfactant systems. The self-diffusion method permits a full characterization of the system under investigation, since it can yield the values of the free monomeric surfactant concentration, the degree of counterion binding, and the amount of water bound (hydration). Moreover, these data can help one deduce the information on micelle size, shape, and composition from the monomer surfactant concentration in a surfactant mixture:

$$D_{i,\exp} = \frac{C_M}{C_i} D_M + \frac{C_{i,\text{free}}}{C_i} D_{i,\text{free}}$$

where $D_{i,\exp}$ is the average self-diffusion coefficient of species i, C_i is the total concentration of species i, C_M is the critical micellar concentration of species i, $C_{i,\text{free}}$ is the concentration of monomeric species i, D_M is the self-diffusion coefficient of species i in the micellar phase, and $D_{i,\text{free}}$ is the self-diffusion coefficient of species i in the monomeric state or in the unmicellar state.

F. Light Scattering

Light scattering has been widely applied in polymer solution to get information on the molecular weight and conformation of polymer from the Rayleigh ratio and Yamakawa theory [43]. On the other hand, light-scattering techniques have contributed greatly to the study of surfactant systems, providing information on the size of micelle and aggregate growth. Dubin and co-workers [44,45] pioneered the study of surfactant and polymer systems via light scattering. From the decay of the autocorrelation function, they concluded that there is a coexistence of free micelles of polymer-bound micelles. With the light-scattering method, changes in micelle size of alcohol ethoxylates and SDS by the addition of chemicals such as oil and electrolyte was reported by Chiu et al. [46]. They also related the rate of solubilization to the change in surfactant aggregate size.

G. Viscosity Measurement

At a dilute concentration, a solution of ionic and nonionic surfactants usually behaves as Newtonian liquid. The viscosity of these solutions can be utilized to gain information on the size, the shape, and the hydration of micelles. However, a concentrated surfactant solution shows a more complicated rheological behavior. The rheological properties of surfactant solution are of most interest in the handling and application of such a system.

Saito [47] observed a considerable increase in the viscosity of methylcellulose and PVC after the addition of SDS. Also, Jones [14] and Lange [48] reported a steady increase in the viscosity of PEO and PVP solution, respectively, upon adding SDS. These results clearly indicate that a change in polymer conformation, which is an expansion of the polymer coil, depends on the association with charged surfactant, as reported by Takagi et al. [49] and Prud'homme and Uhl [50]. Chang and Rosano [51] used the rheology method to study the interaction of long-chain dimethylamine oxide (LDAO)/SDS and reported an increase in the viscosity, approaching a maximum value with the addition of SDS to LSAO. The mixture is non-Newtonian in nature. Ogino et al. [52] reported a noticeable increase in the viscosity of SDS/$C_{18}POE_{20}$ brought about by the addition of either octanol or octanoic acid, which is associated with an increase in the size of the mixed micelles solubilizing the oily material.

A simple method, the "talc test" suggested and applied by Reigsmond et al. [53], was reported sensitive in qualitatively measuring the surface viscoelasticity of surfactant/polymer solution [54].

H. Other Methods

1. Probe Method

Although dye solubilization, calorimetry, gel filtration, and small-angle neutron scattering (SANS) methods are commonly used in studying the interaction of mixture systems containing a surfactant, only the probe method is reviewed.

Mixed surfactant systems can be investigated by the use of certain probes. Ultraviolet-visible spectra can provide information on the state of mixed micelles, the cmc value, and whether the interaction between different components in mixed micelles is taking place. Meguro and co-workers pioneered the use of this method to study the surfactant mixture system. In their extensive study, probes, tetracyanoquindimethane [55–58], benzoylacetone [59], beznoylacetoanilide [60,61], sodium (p-sulfonatophenylazo-1-naphthol-2-sulfonate) [62,63], and 8-anilino-naphthalene sulfonic ammonium salt [64] were examined. Using the probe method, Meguro and co-workers [64] confirmed the existence of mixed micelles between hydrocarbon–fluorocarbon surfactant, which was a controversial question before.

2. Theoretical Methods

Blankschtein and co-workers [65] have done pioneer work through theoretical modeling, aided by the computer, to predict the properties of mixed surfactant systems. Also, based on the "necklace model" proposed by Shirahama et al. [67,68], they have proposed a molecular thermodynamic theory of the complexation of nonionic polymers and surfactants in diluted aqueous solutions [66]. Application of this method can help predict the interaction parameters for several nonionic polymer–surfactant mixtures.

III. INTERACTION OF AMINO ACID–BASED SURFACTANTS WITH OTHER MATERIALS

Like a conventional surfactant such as SDS, amino acid–based surfactants can interact with other surfactants, solvents, electrolytes, polymers, proteins, and membrane of cells to show specific behavior. In view of the unique structure of amino acids, where an amino group and a carboxyl group are combined in one molecule, the interaction of amino acid–based surfactants with other ingredients is of great interest to fundamental study and practical applications.

A. Inorganic Ions

It is well known that the presence of electrolytes can significantly influence properties of surfactant solutions, such as cmc and solubility, especially for ionic surfactants. Rosen and co-workers [69] reported that N-acyl-N-substituted glycinate showed much lower cmc and surface-tension values in hard water, which is composed of Ca^{2+} and Mg^{2+} ions, than in pure water. Even at high concentration these amino acid–based anionic surfactants were not observed to form any precipitates in hard water, indicating the possible chelation of amide and carboxyl groups with Ca^{2+} and Mg^{2+}. It is worth noting from the cmc, per molecule area (A_{min}), that in the case of the partially hydrophilic N-substituent, the surfactant molecule organized predominately in the cis conformation about the amide group in the hard water, as shown in Fig. 3. There are two possible reasons for this: One is the N-substituent's partial hydrophilicity, and another is the efficient interaction of the amide and carboxyl groups and oxygen atoms with Ca^{2+} and Mg^{2+} in hard water. These can make the partially hydrophilic N-substituent stand on micellar surface.

Takeshita et al. [70–72] made a systematic study of EDTA-derived surfactants. For ethylene alkyl EDTA chelate surfactant, it was found that the properties of the chelating surfactant were greatly influenced by the nature of the metal ions. Cu^{2+} chelate surfactant showed a lower specific surface tension

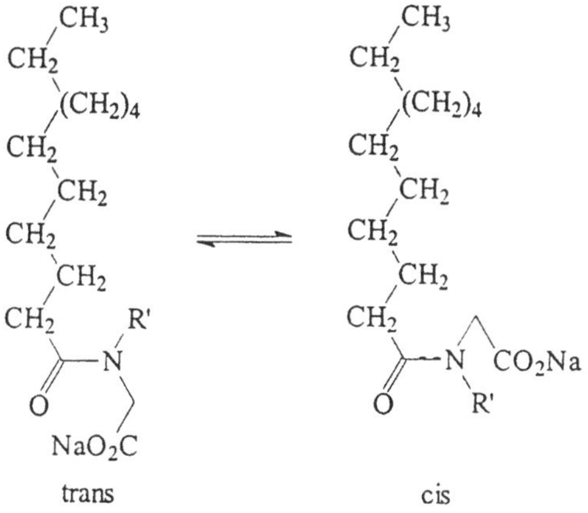

FIG. 3 Schematic model of cis and trans conformations of sodium N-acyl-N-substituted glycinate in a micellar solution.

and interfacial tension than Fe^{3+}, Zn^{2+}, Co^{2+}, Cd^{2+}, and Fe^{2+} chelate surfactants, and the iron chelate showed the excellent dispersing power of TiO_2. In the case of EDTA monoalkyl ester chelate surfactants [70], the water solubility is dependent on the metal ions. Al^{3+} and iron chelates had a higher solubility than the others, Ca^{2+}, Co^{2+}, Ni^{2+}, Zn^{2+}, Cd^{2+}, and Pb^{2+}. This is probably because these chelates are coordinated by hydrophilic hydroxo ions. Also, it is noteworthy that trivalent metal ions such as Al^{3+} and Fe^{3+} showed high surface activities as well as a high dispersing power.

Polymer surfactant, poly(Na-acrylamidoalkanoate)s [73] at a low concentration, was observed to increase its viscosity with dilution. Moreover, the intrinsic viscosity is strongly influenced by the presence of electrolytes. The viscosity of this polymerized surfactant decreased markedly with the addition of NaCl, indicating that the electrostatic repulsions between the charged head are suppressed, which causes a contraction of the polymer chain in the solution. The mixture of this surfactant with alum [74] was found to show coagulant properties in the same manner as cationic polyamines.

Martell et al. [75] reported on N-substituted ethylene diaminetriacetic acid that these compounds were efficient Ca^{2+} and Cupric chelating agents, and formed a stronger complex with cupric ions than with calcium ions. Their study showed that chelate stability was dominated by the structure of the N-substituent group. N-Benzylethylene diaminetriacetic acid formed a very stable complex with cupric ions (log K = 16.80) and calcium ions (log K = 6.70). It is very difficult to explain the affinity toward Cu^{2+} and Ca^{2+} by benzyl.

Presumably the greater tendency of an aromatic substituent group to release electrons compared to an alkyl group may account for the increased stability of the formed complex. The Martell group [76] also reported that N,N'-dicyclohexylethylene diaminediacetic acid did not form any complex with cupric ion, because the steric-hindrance effect of the dicyclohexyl group interferes with the formation of the structure of chelate rings.

Hidaka et al. [77] reported that amphoteric N-(2-hydroxyethyl)-N-(2-hydroxyalkyl)-β-alanines's cmc values greatly depend on the nature of the electrolytes added to its nearly neutral aqueous solution, and that the cmc value decreased in the following order: $NaCl > CaCl_2 > Na_2SO_4$. Also, their calcium stability is superior to that of N-dodecyl-β-alanines. The effect of pH on this amphoteric amino acid surfactant was studied in the presence of 0.1 M NaCl [78] and the results showed that the cmc value increased on the acidic side below the isoelectric point pI $=$ 6.8 and remained almost unchanged on the alkaline side. Examination of the configuration with molecular models indicates that the cationic ionization of the amino group on the acidic side probably takes place within the micelle, whereas under alkaline circumstances the anionic ionization of the carbonyl group occurs on the micellar surface. This makes the electrostatic potential for ionization different on the acidic and alkaline sides.

EDTD-monoalkylamide chelates were studied by Takeshita et al. [79] and the cmc values of several C_{12}-amide chelates are shown in Table 1. C_{18}-Amide chelate showed high dispersing ability toward TiO_2, and the nickel chelate was the best among them. These chelates exhibited excellent kerosene-emulsifying power as compared to SDS.

It is reported by Okahata [80] that glycine-based $2C_{16}$-gly-NH_3^+ bilayer-corked capsule membrane's permeability is pH responsive. This implies that the addition of HCl/NaOH causes a change in the molecular packing of $2C_{16}$-Gly-NH_3^+ in the bilayer. In DDS application, pH-sensitive liposomes, where dissociative bilayer-forming molecules are incorporated, would be useful for their clinical implications. They would release encapsulated drug when

TABLE 1 cmc Values of Several C_{12}-Amide Chelates

| | cmc $\times$ 10^{-5} (M) | | | | | |
	Al^{3+}	Fe^{3+}	Co^{2+}	Ni^{2+}	Cu^{2+}	Zn^{2+}
C_{12}-Amide	5.0	13.0	6.0	8.0	1.5	15.0

Source: Ref. 79.

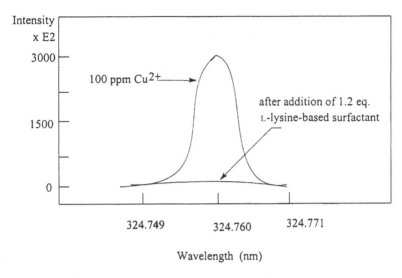

FIG. 4   Ability of N^{ε}-lauroyl-L-lysine to chelate Cu^{2+} ions.

passing around tumor cells that have a considerably lower pH than normal tissues [81]. Takehara [82] has reported that N^{ε}-lauroyl-L-lysine is a very strong cupric chelator, as shown in Fig. 4.

The effect of pH and electrolytes on the surface properties of betaine $C_{12}H_{25}N^+(CH_2C_6H_5)(CH_3)CH_2COO^-$ was studied by Rosen and Zhu [83]. At pH 5.65 and 3.0, the betaine interacts more strongly with Na_2SO_4 than with NaCl or $CaCl_2$, indicating a stronger electrostatic interaction with anions than with cations. With similar results, the greater effect of Na_2SO_4, compared to either NaCl or $CaCl_2$, on the cmc of $C_8H_{17}CH(COO^-)N^+(CH_3)_3$ was reported by Tori and Nakagawa [84], and for N-(2-hydroxyethyl)-N-(2-hydroxyalkyl)-β-alanines was reported by Takai et al. [77].

B. Interaction with Solvents Including Water

The behavior of solutions of either small or large molecules is considerably influenced by the nature of the solvent itself. For example, water plays a very important role in determining the properties of a surfactant, such as micelle formation and surface-tension reduction. It is well known that protein folding occurs in water in acquiring the native structure, and the final conformation is influenced by the solvent [85]. Similarly, the properties of amino acid–based surfactants are strongly dependent on the nature of the solvent. In water, surfactant can construct an organized solution to form micelles, lamellar phases, and liquid crystals. In organic solvent, it also can form reversed micelle, gel, and liquid crystals. Only lamellar-related issues in aqueous solution of PBS are briefly mentioned. The main topic is the interaction of amino acid–based surfactants with organic solvent.

Kunitake has done the pioneering work in the bilayer-related chemistry by preparing many kinds of bilayer-forming molecules from amino acid [86]. At the same time, stemming from the effect of hydrogen bonds, Marakami and co-workers [87] designed bilayer-forming molecules involving amino acid moieties, and they found that the formation and stability of bilayer are enhanced by the intermolecular hydrogen bond, the so-called "hydrogen bond belt." Kuwamura [88] has made an extensive study on long-chain N-acyl-α-amino acid POE monoester and found that the insertion of amino acid residue into POE nonionics can impart two important properties to the nonionics: hydrophilicity and crystallinity. The crystallinity can result in a high Krafft point but facilitate the formation of highly organized assemblies such as lamellar liquid crystals. Many kinds of nonionics, from lysine, aspartic, and glutamic acid, have been prepared by Infante's group as molecular mimics of lecithins in aqueous solution [89,90].

Imae et al. [91] have reported that gel-like solutions are obtained at medium pH for aqueous solutions of N-acyl-L-aspartic acid (C_nAsp, $n = 12$–18). The possible models of helical, fibrous assemblies are double stranded to form a twisted ribbon of a planar bilayer sheet. However, N-dodecyl-β-alanine formed cylindrical fibers that have no helical structure.

Hidaka et al. [92] reported that N-(2-hydroxydodecyl) amino acid formed twisted fibrous aggregates in organic solvents. Ihara and co-workers [93] reported that two L-glutamate derivatives having two N-dodecylamide groups formed a gel-like, highly oriented aggregate in benzene. This aggregate is believed to be stabilized by intermolecular hydrogen bonds among the L-glutamate-derivative molecules.

Voyer and Guerin reported [94] that cyclic peptidic receptors can recognize complex with aromatic amine by a combination of hydrogen bonding, electrostatic, and π–π interactions. Hanabusa and co-workers [95] have found that simple balaform amides derived from amino acids are excellent gelators and can gelate many solvents from polar to nonpolar solvent at a very low concentration. Some N-acylamino acid derivatives with more than one amide group in the molecule, shown in Table 2, have proven to gelate a variety of oils, as reported by Saito et al. [96]. Sakamoto et al. [97] studied the behavior of N-lauroyl-L-glutamic acid (LGA) in organic solvents and reported that this amino acid surfactant formed a cholesteric liquid phase in aromatic solvent, such as

TABLE 2 Gelatinizing Properties of N-Acyl-L-amino Acid Surfactants[a]

		Oil		
Compounds	Kerosene	Liquid paraffin	Toluene	Soybean Oil
N-C$_{12}$ Glutamic acid				
Dibutylamide	206	398	182	353
Distearylamide	41	35	0	None
N^{α}, N^{ε}-Dicaproyllysine				
C$_{12}$-Amide	183	137	5	105
C$_{12}$-Amine salt	85	None	153	42
C$_{12}$-Ester	18	32	None	122
N-C$_{12}$-Phenylalanine				
C$_{12}$-amide	4	8	None	52
12-Hydroxystearic acid[b]	249	379	109	266

[a] Concentration 1 wt%.
[b] Concentration 2 wt%.
Number indicates gel hardness (g/cm^2). *None* indicates no gel formation.
Source: Ref. 96.

benzene, xylene, chlorobenzene, methyl iodide, or n-hexane. The induced CD spectra support the assumption that this lyotropic liquid crystal has a supramolecular helical structure. On the other hand, when LGA was dissolved in EtOH, MeOH, no liquid crystals were observed. They concluded that an appropriate solvating ability to swell but not to dissolve LGA might be necessary to get such a liquid crystal phase.

β-Alanine betaines [R–N$^+$(CH$_3$)$_2$(CH$_2$)$_2$COO$^-$, R = C$_{12}$, C$_{16}$, C$_{18}$] were studied in formamide, water, and N-methylsydnone, an aprotic solvent [98]. The cmc value in formamide was almost double that in water for the same compound, and the cmc value in N-methylsydnone was higher than that in formamide. This presumably may be attributed to the dielectric constant and dipole moment of water, formamide and N-methylsydnone. During micellization, the energy required to dissolve these surfactant increases in the order of water < formamide < methylsydnone. A linear relationship between $-\log$ cmc and the number of carbon atoms in the alkyl chain was observed in the three solvents; the equations are shown in Table 3. It is clear that slopes of the plots are close to each other, but the intercepts b are a bit different. The intercept value is comparable in formamide and N-methylsydnone but higher than that in water. Since b is the contribution of the polar head group to micellization, the observed difference in b values can be explained in terms of the

TABLE 3 Values of Slopes a and Intercepts b
for β-Alaninebetaines in Water, Formamide,
and N-Methylsydnone

Solvent	log cmc = $aN + b$	
	a	b
Water	−0.32	0.9
Formamide	−0.33	3.58
N-Methylsydnone	−0.29	3.72

Source: Ref. 98.

interactions of the surfactant and solvent stemming from solvation through dipole–dipole interactions in formamide and N-methylsydnone.

   N-Acylamino acid derivatives with more than one amide group in the molecules have been found to be gelators for organic solvents [82]. This is associated with the formation of three-dimensional networks of molecules through intermolecular hydrogen bonding. Also, the effect of the chirality of these compounds on gelling power was found significant; a chiral compound formed a gel with sufficient hardness. In contrast, the corresponding racemic compound formed a gel but lacked strength.

C. Surfactant Mixtures Containing Protein-Based Surfactants

Studies on the interaction of betaines and other surfactants by Rosen [99] suggest that betaines show only weak interaction with nonionics, whereas betaines can show more strong interaction with cationics and anionics, depending on whether the betaine surfactant is capable of donating and accepting a proton in the solution. Harwigsson and Hellsten [100] studied mixtures of C_{16},C_{18} betaines with LAS. They found that a 0.1% solution of betaine/Na-LAS mixture at a molar ratio of 6/1 gave viscoelasticity, and the Krafft point of the mixture C_{18}-betaine/Na-LAS is 22°C, which is lower than the 30°C of C_{18}-betaine alone. Tsujii [101] reported the phenomenon of lowering the Krafft point in a mixture of anionic/betaine surfactants. In the diffusion study of mixtures of SDS and C_{12}-betaine across a collagen membrane, a minimum in the total monomer concentration was observed [102], which is due to the synergistic electrostatic interaction between the anionic SDS and the zwitterionic C_{12}-betaine head group. Since the monomer concentration in a surfactant solution is associated with irritation, a mixture of surfactants at a certain ratio can minimize the irritation. From a study of mixtures of sodium N-lauroylsarcosinate (SLS) and sodium N-caprinoyl-glycylsarcosinate (SCGS) with N-dodecylpyridinium chloride (DPC), Watzke et al. [103] reported results of the relaxation method (T_1, T_2), freeze fracture microscopy, and light scattering: the SLS/DPC mixture formed vesicle plus mixed micelles in the range of molar ratio x from 0.4 to 0.6, formed mixed micelles at low or high x value, and at $x = 0.5$ formed a lamellar phase only. However, the SCGS/DPC mixture formed only mixed micelles regardless of the composition. Furthermore, they [104] found that in the SLS/DPC mixture, the fraction of cis conformation of the SLS head group around the amide bond increased significantly upon vesicle formation, as shown in Fig. 5, indicating that the SLS head group experiences strong immobilization upon vesicle formation. This same phenomenon, in the presence of divalent ions Ca^{2+} and Mg^{2+}, was discussed by Zhu et al. [69].

Maza and Parra [105] systematically studied the solubilization of unilamellar liposomes by surfactants and found that dodecylbetaine had the lowest Re parameter among SDS, dodecyl betaine, and Triton-100. Re—namely, the effective surfactant/lipid molar ratio at which the intensity of light scattering starts to decrease—reaches 50% of original value ($Re_{50\%}$) and leads to total

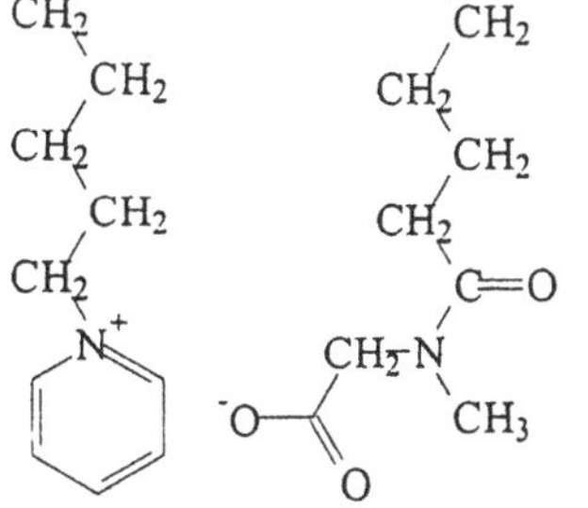

FIG. 5 Cis conformation with respect to the amide bond of sodium N-lauroylsarcosinate upon vesicle formation in an equimolar mixture with N-dodecylpyridinium chloride.

solubilization (R_{sat}) according to the definition of Lichtenberg [106]. For a mixed system of dodecyl betaine/SDS [107], when the molar ratio of betaine/SDS is 0.4, the cmc value reached a minimum, whereas their partition coefficient between the aqueous phase and lipid bilayer of liposomes exhibited a maximum. As a consequence, this causes a substantial change in the permeability of unilamellar liposomes. The study of the solubilization of liposomes by mixtures of SDS/betaines [107] reveals that the solubilization involves two essential transitional steps of the interaction: the saturation of bilayers by the surfactant system and the complete solubilization.

Rosen and Zhu [108] have reported on the interaction of N-dodecyl-N-benzyl-N-methylglycine with cationic $C_{12}H_{25}N^+(CH_3)_3$ Br^-, nonionic $C_{12}H_{25}EO_8$, and anionic $C_{12}SO_3Na$. The interaction with the anionics is considerably greater ($\beta = -5.7$) and increases with a decrease in the pH of the solution, whereas the degree of interaction with cationics yields a value of $\beta = -1.3$, which is very close to that with nonionics, having a value of $\beta = -0.6$. Sodium N-acyl-N-substituent glycinate/alcohol ethoxylates at pH 11 were found to have a medium interaction [109], and the interaction was influenced by the structure of the N-substituent group.

Hidaka and co-workers [110,111] studied the mixture of N-(2-hydroxyethyl)-N-(2-hydroxylalkyl)-β-alanines (HAA) and anionics. From the results of its mixtures with soaps, it is evident that the mixture of tallow soap with 10% of C_{12-14} NaHAA or more exhibited better solubility than tallow soap. Also, the mixtures at any ratios showed a synergistic effect and resulted in improving fabric detergency. The binary mixture of Na-HAA/soap showed a lower interfacial tension than the individual one, and the contact angle for the blend was higher than for soap alone. As a consequence, antisoil redeposition was enhanced in the Na–HAA/soap mixture, especially for polyester/cotton. The viscosity of the binary mixture of NaHAA with LAS, SDS, and LES (sodium laurylether sulfate) was studied at 1.0% total surfactant concentration; the solutions of mixture of Na-HAA/LAS and SDS were viscoelastic, and maximal viscosities were obtained at the ratio of Na-HAA/anionic surfactant 7/3 and 8/2 in weight for LAS and SDS, respectively. However, the Na-HAA/LES binary at any compositions did not show any change in viscosity. Moreover, the interaction between HAA and LAS was studied from the deviation of the molar extinction coefficient in UV spectra; a complex formation of HAA with LAS was proposed above and below the isoelectric point of HAA. For all of the binary systems, excellent detergency and other surface properties were obtained. In addition, binary systems showed less skin irritation than individual surfactants alone.

Takahara et al. [112,113] studied the effect of enantiomerism on the cmc of amino acid surfactant. The cmc of racemic compounds was slightly higher than for the corresponding optically pure isomer. This may result from the difference in the conformation of amino acid moieties in the micelles, due to intermolecular hydrogen bonding.

Nakama and co-workers [114] studied the interaction between amino acid–based anionics (Na N-lauroyl-N-methyl-β-alanine) and cationics (stearyltrimethyl ammonium chloride) in aqueous solution by the use of surface tension, conductivity, and the NMR method. A strong interaction between anionics and cationics was observed. By study of the phase diagram of mixtures, an equimolar mixture was found to show a cloud-point phenomenon at high temperature, at which the solution separated into two liquid phases. For this mixture, no remarkable rise in Krafft point with change in composition was observed, whereas this Krafft point increase was observed for other nonamino acid–based anionics/cationics. The cmc value of the mixture was much smaller than for the individual surfactants, and the specific conductance is found to be the smallest in an equimolar mixture. The NMR showed the signal of the carbon atom of the carbonyl group shifted to high field in this mixture. Taking account of these results, it has been concluded that the observed cloud-point phenomenon is caused by the decreased solute–solvent interaction, resulting in a breaking of the hydrogen bond between water and both the amide group and the carboxylic group in the ion pair of the mixture, as shown in Fig. 6. The protein (ovalbumin) denaturation rate was also found to show a much lower value around the composition of an equimolar mixture; this is ascribed to the minimum concentration of monomeric surfactants in an equimolar mixture.

Ogino et al. [115] studied the dissolution behavior of SDS/amphoteric surfactant N,N-dimethyl-N-lauroyl lysine (DMLL). Mixing the two surfactants resulted in a depression of the Krafft point of the two pure surfactants, to even less than 0°C at a certain composition. The effect of inorganic salt on the dissolution of mixed systems was significant, depending on the nature and concentration of the salt; the dissolution temperature of mixtures at a DMLL ratio less than 0.4 increases with an increase in the concentration of NaCl, while the dissolution temperature of the mixture remained almost unchanged at $X_{DMLL} > 0.8$. The effect of NaSCN is somewhat different; the dissolution temperature of a mixture at X_{DMLL} less than 0.4 increases when the concentration of NaSCN increases, while at X_{DMLL} more than 0.9, the dissolution temperature decreases. Tsujii et al. [116] attributed this phenomenon to the power of the anionic ion, corresponding to the lyotropic series, which results in de-

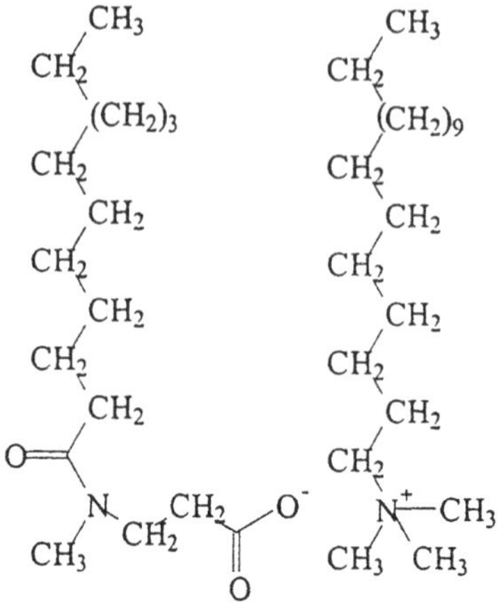

FIG. 6 Structure of the ion pair of stearyltrimethylammonium chloride and sodium
N-lauroyl-N-methyl-β-alanine mixture.

creasing the Krafft point. N-Acyl sarcosinates have been found to be very
effective at depressing the Krafft point of anionic surfactants and at increasing
the cloudy point of nonionic surfactants [117]. Addition of 0.5 wt% of Na N-
lauroyl sarcosinate to SDS depresses the Krafft point to 0°C. A mixture of K
salt of N-cocoylglycinate/K salt laurate was found to show a synergism in
foaming and foam stability in an equimolar mixture [118]. The addition of K
salt of myristate to K N-cocoylglycinate causes an increase in Bingham yield,
an index of lather viscoelasticity. Poly(N-methacryloyl-L-methionine) was
found to have a compact conformation in water. The addition of SDS has a
remarkable effect on its conformation [119], indicating the formation of a
cluster of SDS/this polymer surfactant.

In order to clarify whether the influence of the N-substituent group in an
amino acid surfactant on micellization is present in a mixture of N-acyl amino
acid surfactants, Miyagishi et al. [120] determined the cmc values of mixtures
of sodium salts of N-lauroyl amino acid derived from glycine, valine, and
phenylalanine. The results indicate that the steric hindrance of N-substituent
is not always significant in the formation of mixed micelle. The interaction
between the surfactants with and without a substituent group was very small.
The interaction parameter was zero for lauroyl phenylalanine/SDS. A strong
interaction was observed in a mixture of $C_{12}EO_6$ and N-lauroyl valinate.

The chiral separation of a racemic mixture is a very challenging task facing separation scientists as well as medical chemists and synthetic chemists, because the chirality of a compound has proven to be very significant to its biological activity and toxicological effects. In view of the availability of a chiral center, amino acid has been widely used in chiral recognition and enantiomer separation. Cohen et al. [121] first succeeded in chiral separation via micellar electrokinetic capillary chromatography (MEKC) by use of a mixed micelle chiral ligand consisting of N-dodecyl-L-alanine/SDS and Cu^{2+} ions. The development and application of MEKC was pioneered by Terabe's group [122]. They and other groups have made an extensive study of the use of various amino acid–based surfactants as a chiral selector for enantiomeric separation in MEKC [123–130]. Warner et al. [131] have reported that amino acid–based N-undecylenyl-L-valinate used as a chiral selector plays an important role in MEKC to achieve enantioseparation of neutral, acidic, and basic racemic compounds. In principle, the enantioselectivity of MEKC is achieved by the different interactions of analyte (here racemic mixtures) with the chiral micelles as well as the differences in electrophoretic mobilities of the micellar phase and aqueous phase. It was pointed out that hydrophobic and electrostatic interaction as well as hydrogen bonding appear to be critical for chiral recognition. Also, the same group reported [132] the effect of amino acid order in dipeptide surfactants on chiral separation by using polymerized Na N-undecylenyl-valine-L-leucine (L-SUVL) and Na N-undecylenyl-L-leucine-L-valine (L-SULV). They found that the order of amino acids has a dramatic effect on chiral recognition. Also, their results suggest that an increase in enantioselectivity appears to be due to some synergism of the dipeptide as compared with the nonpeptide amino acid surfactant, as well as the interaction of the chiral center of the analyte with the surfactants.

Radley and co-workers [133] reported that when chiral dopants—decyl esters of amino acids, serine, alanine, leucine, and methionine—were added to an aqueous solution of alkyl methyl ammonium bromides, mixed with decanol, amphiphilic cholesteric crystals were formed. They found that the sense and magnitude of the induced helix by decyl ester of amino acid hydrochlorides except decyl ester of alanine are dependent on the achiral cationic surfactant, alkyl methyl ammonium bromides. The formation of amphiphilic cholesteric crystals was interpreted in terms of the trans and cis rotamers of the chiral ester of amino acid, associated with the ester linkage. Also, they reported [134] that in the case of the ester hydrochloride of proline as a chiral dopant, a concentration-dependent reversal in helical twist was observed.

D. Interaction with Polymer, Bacteria, and Others

Vinardell and co-workers [135] studied the irritation of rabbit eye and skin from cationic amino acid–derived surfactants N-acyl-L-arginine methylesters. They found that amino acid–based surfactants are less irritant than a conventional cationic surfactant, cetrimide. Also, the degree of irritation depends on the chain length of the alkyl group, and a maximum of irritation was observed for C_{12} surfactant. The similar effect was reported for alkyl sulfates before [136], being ascribed to the high monomer concentration of surfactant.

Hocker et al. [137] synthesized anionic N-acylated L-amino acids, oligo-peptides, and protein hydrolysates and studied the adsorption on wool. Increased bath exhaustion and a more intense coloration of wool were obtained during dying of wool by the addition of the studied amino acid surfactants. These surfactants exhibited good biodegradabilities with 68–78% BOD_5/COD.

Takehara [138] reviewed the antimicrobial properties of amino acid–based surfactants. Long-chain N-acylamino acids have antimicrobial activities against gram-positive bacteria and some fungi. The maximal activities are obtained at C_{12-18} for neutral amino acid and C_{16-18} for acidic amino acid. In particular, it is worth noting that N-cocoylarginine ethyl ester showed a broad antimicrobial spectrum but was not very irritant to the human organism. Infante et al. [139] studied the antibacterial activity of acidic and basic N-lauroyl-arginine dipeptides, having an acidic amino acid, glutamic acid, or basic amino acid, lysine as the terminal amino acid. The cationic dipeptides are antibacterial agents, whereas the antimicrobial activity of amphoterics is lower. The cationic methyl ester of N-lauroyl-arginyl dipeptides containing glycine and phenylalanine [140] was found to have much higher antimicrobial activity than the methyl ester of N-lauroyl-L-arginyl-L-lysine dichloride. This can be ascribed to the ability of the guanidinium group to interact with polyanionic components of the cell surface due to their cationic nature, adsorbing and allowing the surfactant to penetrate the membrane, thereby inhibiting microbial growth. Arginine dipeptides from collagen [141] did not show very high antibacterial activity, but they can be used as preservative or protective substances. From the study of the chiral isomerism of N-alkyl amino acid [142], no effect of chiral isomerism on antimicrobial activity was found. On the contrary, the effect of hydrophobic alkyl chain was significant. Sugimoto and Toyoshima [143] studied the antimicrobial activity of N-cocoyl-L-arginine ethyl ester (CAE), and they found that CAE shows high antimicrobacterial activity, especially an inactivating effect on the hepatitis B surface antigen.

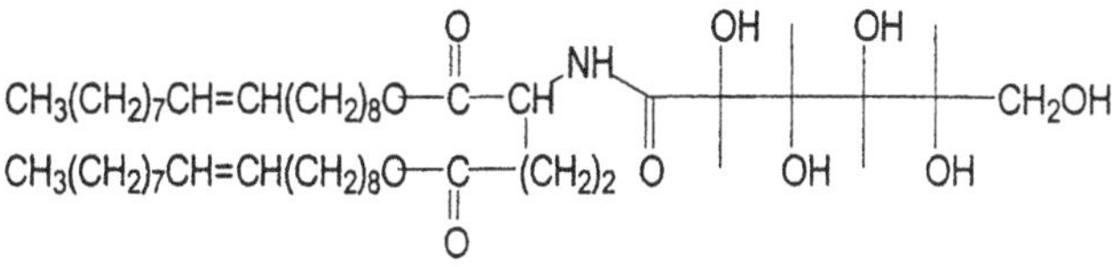

FIG. 7   Molecular structure of nonionic dioleyl L-glutamate ribitol amide.

The CAE caused inactivation at much lower concentrations than even sodium hypochlorite.

Goto et al. [145] studied the development of surfactant–chymotrypsin complex (CT-complex) as a novel biocatalyst in organic media. The CT-complex not only enhances enzymatic activity in reaction media but also reduces enzyme loss. They found that glutamic acid–based nonionic dioleyl-L-glutamate ritiol amide, shown in Fig. 7, significantly increased the enzymatic activity of the CT complex and created a very stable W/O emulsion, which encapsulates the enzyme in the organic media.

N-(1-Phenylalanine)-4-(1-pyrene)butyramide, shown in Fig. 8, has been found able to recognize and bind to specific sites of proteins and to cleave the protein backbone by laser flash photolysis, reported by Kumar et al. [144]. This molecule is composed of the hydrophilic phenylalanine carboxyl group and the hydrophobic pyrene group; as a consequence, this molecule appears to lodge only in the sites of a protein where a hydrophilic pocket is very close to a hydrophobic pocket.

Pulmonary surfactant protein C (SP-C) is a small hydrophobic peptide. The effect of acylation of SP-C on its structure and function was reported by Haagsman et al. [146] using the Whilelmy plate method and CD spectra measurement. Also, SP-C-induced bilayer interaction was studied via the lipid mixing method by the use of fluorescence spectra of pyrene-PC labeled vesi-

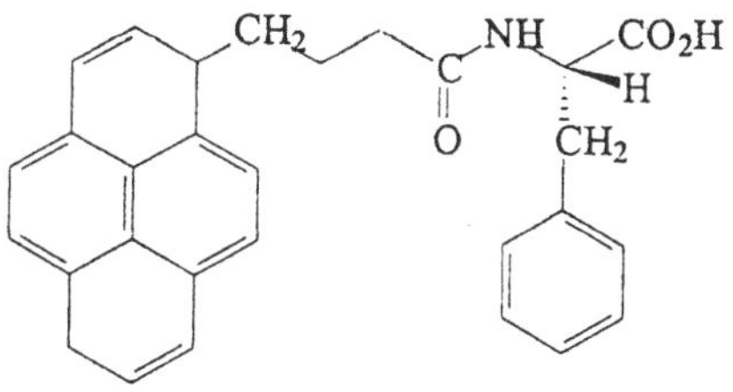

FIG. 8   Molecular structure of *N*-(1-phenylalanine)-4-(1-pyrene)butyramide.

cles consisting of DPPG/PG/Pyrene-PC (63/27/10, w/w/w). For vesicles (DPPG/PG, 7/3, w/w), the excimer ($\lambda = 475$ nm)/monomer ($\lambda = 377$ nm) (E/M) ratio showed no changes; even the concentration of SP-C added to the DPPG/PG bilayer is varied, indicating no lipid-mixing is taking place. However, in the case of DPPC/PC vesicles mixed with pyrene-labeled DPPC/PG, a decrease in the E/M ratios was observed when changing the SP-C concentration. This implies that an induced lipid-mixing occurs and the extent of lipid-mixing is SP-C dependent. No significant differences were found between the acylated and nonacylated SP-C in the ability to induce lipid-mixing, although the acylation of an SP-C significantly affected other properties and the structure of this protein surfactant. It appears that the electrostatic interaction of the head groups between SP-C and lipid membrane is critical in lipid mixing.

Polymerized anionic surfactants derived from amino acid have been developed as polyanion inhibitors of HIV and other viruses by Roque and coworkers [147]. They reported that all of the polyanions proved active against HIV-1 and HIV-2, their 50% inhibitory concentration (IC_{50}) ranges from 0.04 to 7.5 µg/mL), while they were not toxic to the host cell (CEM-4 or MT-4) at concentration as high as 100 µg/ml or higher. The HIV-inhibitory effect increased with the hydrophilic character of the amino acid moiety. This study suggests that polyanions with double-charged or multicharged anionic groups from amino acid may be more promising.

Tsujii and Tokiwa [148] concluded from the shift in the melting point of *calf thymus DNA* that lauroylprolylprolylglycine has a larger influence on the tertiary structure of the DNA than does SDS. The results showed that the amino acid sequence in polypeptide surfactants could play an important role in the interaction with the DNA. They concluded that the primary factor governing the interaction with DNA is the molecular structure of surfactants, rather than their surface activity or ionic nature.

REFERENCES

1. M. J. Rosen, Surfactants and Interfacial Phenomena, 2nd ed., Wiley, New York, 1989, Chapter 11.
2. J. F. Scamehorn (ed.), Phenomena in Mixed Surfactant Systems, ACS Symp. Ser. 311, American Chemical Society, Washington, DC, 1986.
3. K. Ogino and M. Abe (eds.), Mixed Surfactant Systems, Surfactant Science Series, Vol. 46, Marcel Dekker, New York, 1993.
4. E. D. Goddard and K. P. Ananthapadmanabhan (eds.), Interactions of Surfactants with Polymers and Proteins, CRC Press, Boca Raton, FL, 1993.
5. B. Jonson, B. Lindman, K. Holmberg, and B. Kronberg, Surfactants and Polymers in Aqueous Solutions, Wiley, New York, 1998.

6. J. Steinhardt and J. A. Reynolds, Multiple Equilibria in Proteins, Academic Press, New York, 1969.

7. J. C. Brackman and J. B. F. N. Engberts, Chem. Soc. Rev. 22:85 (1993).

8. R. Zana (ed.), Surfactant Solutions: New Methods of Investigation, Marcel Dekker, New York, 1987.

9. D. N. Rubingh, in: Solution Chemistry of Surfactants, K. L. Mittal (ed.), Plenum, New York, 1979, Vol. 1, p. 337.

10. M. J. Rosen and X. Y. Hua, J. Colloid Interface Sci. 86:164 (1982).

11. M. J. Rosen and D. S. Murphy, J. Colloid Interface Sci. 110:224 (1986).

12. M. J. Rosen and B. Gu, Colloids Surf. 23:119 (1987).

13. M. J. Rosen, Surfactants and Interfacial Phenomena, 2nd ed., Wiley, New York, 1989, pp. 394–395.

14. M. N. Jones, J. Colloid Interface Sci. 23:36 (1967).

15. M. J. Schwuger, J. Colloid Interface Sci. 43:491 (1973).

16. C. Davidson, P. Meares, and D. G. Hall, J. Membrane Sci. 36:511 (1988).

17. E. D. Goddard, Colloid Surf. 19:301 (1986).

18. R. Zana (ed.), Surfactant Solutions: New Methods of Investigation, Marcel Dekker, New York, 1987, pp. 463–467.

19. For example, K. Ogino, T. Kakihara, H. Uchiyama, and M. Abe, J. Am. Oil Chem. Soc. 65:405 (1988).

20. F. Tokiwa and N. Muriyama, J. Colloid Interface Sci. 30:338 (1969).

21. T. Gilanyl and E. Wolfram, Colloid Surf. 3:181 (1981).

22. J. Francois, J. Dayantis, and J. Sabbadin, Eur. Polym. J. 21:165 (1985).

23. R. Zana, J. Lang, and P. Lianos, Polym. Prepr. Am. Chem. Soc. Div. Polym. Chem., 23:39 (1982).

24. R. Zana (ed.), Surfactant Solutions: New Methods of Investigation, Marcel Dekker, New York, 1987, Chapter 5.

25. V. Ramamurthy (ed.), Photochemistry in Organized and Constrained Media, VCH, New York, 1991.

26. N. J. Turro, B. H. Baretz, and P.-L. Kuo, Macromolecules 17:1321 (1984).

27. A. Mallians, Adv. Colloid Interface Sci. 27:158 (1987).

28. M. Almgren and S. Swarap, J. Phys. Chem. 87:876 (1983).

29. H. Asano, K. Aki, and M. Ueo, Colloid Polym. Sci. 267:935 (1989).

30. N. J. Turro and A. Yekta, J. Am. Chem. Soc. 100:5951 (1978).

31. For example, F. M. Winnik, M. A. Winnik, and S. Tazuke, J. Phys. Chem. 91:594 (1987).

32. For example, M. D. Liu, D. H. Patterson, C. R. Jones, and C. R. Leidner, J. Phys. Chem. 95:1858 (1991) and references cited therein.

33. H. Yoshioka, J. Jpn. Oil Chem. Soc. 30:727 (1981), and references cited therein; L. J. Berliner, Spin Labeling Theory and Applications, Academic Press, New York, 1976; R. Baglioni, M. F. G. Ottaviani, and E. Ferrani, ESR Studies of Spin Labelled Micelles, in: Surfactants in Solution, K. L. Mittal and B. Lindman (eds.), Plenum Press, New York, 1984.

34. J. Martinic, J. Michon, and A. Rassat, J. Am. Chem. Soc. 97:1818 (1975).

35. C. Ramachandran, S. Vijayan, and D. O. Shah, J. Phys. Chem. 84:1561 (1980).
36. F. Zhao, M. J. Rosen, and N. L. Yang, Colloid Surf. 11:97 (1984).
37. K. Shimahara, M. Tohdo, and M. Murahashi, J. Colloid Interface Sci. 86:283 (1982).
38. F. M. Witte, P. Buwalda, and J. B. F. N. Engberts, Colloid Polym. Sci. 262:42 (1987).
39. F. M. Menger, Angew. Chem. Int. Ed. Engl. 30:1086 (1991).
40. F. Tokiwa and K. Tsujii, J. Phys. Chem. 75:3650 (1971).
41. M. I. Gjerde, W. Nerdal, and H. Hoiland, J. Colloid Interface Sci. 183:285 (1996).
42. B. Lindman, O. Soderman, and H. Wennerstorm, NMR Studies of Surfactant Systems, in: Surfactant Solutions: New Method of Investigation, R. Zana (ed.), Marcel Dekker, New York, 1987, Chapter 6, and Reference 3, p. 360.
43. H. Yamakawa, Modern Theory of Polymer Solutions, Harper & Row, New York, 1971.
44. P. Dubin, C. Chew, and L. Gan, J. Colloid Interface Sci. 128:566 (1989).
45. P. Dubin, T. D. McQuigg, C. Chew, and L. Gan, Langmuir 5:89 (1989).
46. Y. C. Chiu, Y. C. Han, and H. M. Cheng, in: Structure/Performance Relationships in Surfactants, M. J. Rosen (ed.), ACS Symp. Series 253, American Chemical Society, Washington DC, 1984, p. 89.
47. S. Saito, Kolloid Z. 154:19 (1957).
48. H. Lange, Kolloid Z. Z. Polym. 243:101 (1971).
49. T. Takagi, K. Tsujii, and K. Shirahama, J. Biochem. 77:939 (1975).
50. R. K. Prud'homme and J. T. Uhl, Soc. Pet. Eng. J. 24:431 (1984).
51. D. L. Chang and H. L. Rosano, in: Structure/Performance Relationships in Surfactants, M. J. Rosen (ed.), ACS Symp. Series 253, American Chemical Society, Washington DC, 1984, p. 129.
52. K. Ogino, H. Uchiyama, and M. Abe, in: Mixed Surfactant Systems, K. Ogino and M. Abe (eds.), Marcel Dekker, New York, 1992, p. 189.
53. S. Reigsmond, F. M. Winnik, and E. D. Goddard, Colloid Surf. A 119:221 (1996).
54. J. Tang and F. M. Winnik, Huaxue Tong Bao 3:54 (1998).
55. S. Muto, K. Deguchi, K. Kobayashi, E. Kaneko, K. Meguro, J. Colloid Interface Sci. 33:475 (1970).
56. K. Deguchi and K. Meguro, J. Colloid Interface. Sci. 38:596 (1972).
57. S. Muto, A. Aono, and K. Meguro, Bull. Chem. Soc. Jpn. 46:2872 (1973).
58. H. Akasu, A. Nishi, M. Ueno, and K. Meguro, J. Colloid Interface Sci. 54:278 (1976).
59. N. Shoji, M. Ueno, and K. Meguro, J. Am. Oil Chem. Soc. 53:165 (1976).
60. N. Shoji, M. Ueno, and K. Meguro, J. Am. Oil Chem. Soc. 55:297 (1978).
61. T. Suzuki, K. Esumi, and K. Meguro, J. Colloid Interface Sci. 93:205 (1983).
62. K. Meguro, Y. Tabata, N. Fujimoto, and K. Esumi, Bull. Chem. Soc. Jpn. 56: 627 (1983).

63. N. Kawashima, N. Fujimoto, and K. Meguro, J. Colloid Interface Sci. 103:459 (1985).
64. K. Meguro, Y. Muto, F. Sakurai, K. Esumi, ACS Symp. Series 311, 1986, p. 61.
65. N. J. Zoeller, A. Shiloach, and D. Blankschtein, Chem. Tech. 26:24 (1996) and references cited therein.
66. Y. J. Kikas and D. Blankschtein, Langmuir 10:3512 (1994).
67. K. Shimahara, K. Tsujii, and Takagi, J. Biochem. 75:309 (1974).
68. K. Shirahama and T. Tsujii, Hyomen 11:383 (1973).
69. Y-P. Zhu, M. J. Rosen, and S. Morrall, J. Surf. Deterg. 1:1 (1998).
70. T. Takeshita, N. Miyuuchi, S. Maeda, and T. Butsda, Yukagaku 25:354 (1976).
71. T. Takeshita, I. Wakabe, and S. Maeda, J. Am. Oil Chem. Soc. 57:430 (1980).
72. K. Yamashita, N. Miyauchi, S. Maeda, and T. Takeshita, Yukagaku 28:552 (1979).
73. L-M. Gan and C.-H. Chew, in: Organized Solutions, Stig E. Friberg and Lindman (eds.), Marcel Dekker, New York, 1992, p. 336.
74. L-M. Gan, K. W. Yeoh, C. H. Chew, L. L. Koh, and T. L. Tan, J. Appl. Polym. Sci. 42:225 (1991).
75. A. J. Bruno, S. Chaberek, and A. E. Martell, J. Am. Chem. Soc. 78:2723 (1956).
76. A. E. Frost Jr. and A. E. Martell, J. Am. Chem. Soc. 72:3743 (1950).
77. M. Takai, H. Hidaka, and M. Moriya, J. Am. Oil Chem. Soc. 56:537 (1979).
78. H. Hidaka, M. Moriya, and M. Takai, J. Am. Oil Chem. Soc. 56:914 (1979).
79. T. Takeshita, T. Shimohara, and S. Maeda, J. Am. Oil Chem. Soc. 59:104 (1982).
80. Y. Okahata, Acc. Chem. Res. 19:57 (1986).
81. J. Connor, M. B. Yatvin, L. Hunang, Proc. Natl. Acad. Sci. U.S.A. 81:1715 (1984).
82. M. Takehara, Colloid Surf. 38:149 (1989).
83. M. J. Rosen and B-Y. Zhu, in: Structure/Performance Relationships in Surfactants, M. J. Rosen (ed.), ACS Symp. Series 253, American Chemical Society, Washington DC, 1984, pp. 61–71.
84. K. Torì and T. Nakagawa, Kolloid Z. Z. Polym. 189:50 (1963).
85. W. Kauzmann, Adv. Protein Chem. 14:1 (1959).
86. T. Kunitake, Angew. Chem. Int. Ed. Engl. 31:70 (1992) and references cited therein.
87. Y. Murakami, A. Nakano, and H. Ikeda, J. Org. Chem. 47:2137 (1982); Y. Murakami, Kagaku to Kogyo 39:102 (1988).
88. T. Kuwamura, in: Structure/Performance Relationships in Surfactants, M. J. Rosen (ed.), ACS Symp. Series 253, American Chemical Society, Washington DC, 1984, pp. 27–47.
89. J. Seguer, C. Selve, M. Allouch, and M. R. Infante, J. Am. Oil Chem. Soc. 73: 79 (1996).
90. M. Allouch, M. R. Infante, J. Seguer, M. Stebe, and C. Selve, J. Am. Oil Chem. Soc. 73:86 (1996).
91. T. Imae, Y. Takahashi, and H. Muramatsu, J. Am. Chem. Soc. 114:3414 (1992).

92. H. Hidaka, M. Murata, and T. Onai, J. Chem. Soc. Chem. Comm. 562 (1984).

93. H. Ihara, H. Hachisako, C. Hirayama, and K. Yamada, J. Chem. Soc. Chem. Comm. 1244 (1992).

94. N. Voyer and B. Guerin, J. Chem. Soc. Chem. Comm. 1253 (1992).

95. K. Hanabusa, R. Tanaka, M. Suzuki, M. Kimura, and H. Shirai, Adv. Mater. 9:1095 (1997) and references cited therein.

96. T. Saito, M. Honma, M. Takesada, K. Honda, and A. Akamatsu, Proc. 36th Spring Annual Meeting of Chem. Soc. Jpn., The Chemical Society of Japan, Tokyo, 1977, 1C44.

97. K. Sakamoto, R. Yoshida, and M. Hatano, Chem. Lett. 1401 (1976), and J. Am. Chem. Soc. 100:6898 (1978).

98. I. Rico and A. Lattes, in: Organized Solutions, Stig E. Friberg and Lindman (eds.), Marcel Dekker, New York, 1992, pp. 115–124.

99. M. J. Rosen, Langmuir 7:885 (1991).

100. I. Harwigsson and M. Hellsten, J. Am. Oil Chem. Soc. 73:921 (1996).

101. K. Tsujii, K. Okahashi, and T. Takeuchi, J. Phys. Chem. 86:1437 (1982).

102. M. Garcia, I. Ribosa, J. Sanchez Leal, and F. Comelles, J. Am. Oil Chem. Soc. 69:25 (1992).

103. M. Ambiihl, F. Bangerter, P. L. Luisi, P. Skrabal, and H. J. Watzke, Progr. Colloid Polym. Sci. 93:183 (1993).

104. M. Ambiihl, F. Bangerter, P. L. Luisi, P. Skrabal, and H. J. Watzke, Langmuir 9:36 (1993).

105. A. dela Maza and J. L. Parra, J. Am. Oil Chem. Soc. 70:699 (1993).

106. D. Lichtenherg, Biochem. Biophys. Acta 821:470 (1985).

107. A. dela Maza and J. L. Parra, J. Am. Oil Chem. Soc. 70:685 (1993).

108. M. J. Rosen and B-Y. Zhu, J. Colloid Interface Sci. 99:427 (1984).

109. Y-P. Zhu and M. J. Rosen, unpublished data.

110. M. Takai, H. Hidaka, S. Ishika, M. Takada, and M. Moriya, J. Am. Oil Chem. Soc. 57:183 (1980).

111. M. Takai, H. Hidaka, S. Ishika, M. Takada, and M. Moriya, J. Am. Oil Chem. Soc. 57:382 (1980).

112. R. Yoshida, M. Takehara, and K. Sakamoto, Yukagaku 24:538 (1975).

113. M. Takehara, I. Yoshimura, and R. Yoshida, J. Am. Oil Chem. Soc. 51:419 (1974).

114. Y. Nakama, F. Harusawa, and I. Murotani, J. Am. Oil Chem. Soc. 67:717 (1990).

115. K. Ogino, K. Kato, and M. Abe, J. Am. Oil Chem. Soc. 65:272 (1988).

116. K. Tsujii and J. Mino, J. Phys. Chem. 82:1610 (1978).

117. J. J. Crudden, B. J. Lambert, and R. W. Kohl, in: Industrial Applications of Surfactants III, D. R. Karsa (ed.), Royal Society of Chemistry, London, 1992, pp. 95–119.

118. E. Shiojiri, K. Sano, M. Koyama, M. Ino, T. Hattori, H. Yoshihara, K. Iwasaki, and Y. Kawasaki, J. Soc. Cosmet. Chem. Jpn 30:410 (1996).

119. C. Methenitis, G. Pneumatikakis, M. Pitsikalis, and J. Morcellet, J. Polym. Sci. Part A: Polym. Chem. 33:2233 (1995).

120. S. Miyagishi, Y. Ichibai, T. Asakwa, and M. Nishida, J. Colloid Interface Sci. 103:164 (1985).
121. S. A. Cohen, A. Paulus, and B. L. Karger, Chromatographio 24:15 (1987).
122. S. Terabe, K. Otsuka, K. Ichikawa, A. Tsuchiya, T. Ando, Anal. Chem. 56:111 (1984) and subsequent references.
123. K. Otsuka and S. Terabe, J. Chromatogr. A 515:221 (1990).
124. K. Otsuka, S. Terabe, J. Chromatogr. A 559:209 (1991).
125. K. Otsuka, M. Kashrahara, Y. Kawaguchi, R. Koike, T. Hisamitsu, S. Terabe, J. Chromatogr. A 652:253 (1993).
126. A. Dobashi, T. Ono, S. Hara, and I. Yamaguchi, Anal. Chem. 61:1984 (1989).
127. A. Dobashi, T. Ono, S. Hara, and I. Yamaguchi, J. Chromatogr. A 480:413 (1989).
128. A. G. Peterson and J. P. Foley, J. Chromatogr. B 695:131 (1997).
129. A. G. Peterson, E. S. Ahuja, and J. P. Foley, J. Chromatogr. B 683:15 (1996).
130. J. Wang, and I. M. Warner, Chromatogr. A 711:297 (1995).
131. K. A. Agnew-Heard, M. S. Pena, S. A. Shamsi, and I. M. Warner, Anal. Chem. 69:958 (1997).
132. E. Billiot, J. Macossay, S. Thibodeaux, S. A. Shamsi, and I. M. Warner, Anal. Chem. 70:1375 (1998).
133. K. Radley, N. Mclay, and K. Gicquel, Mol. Cryst. Liq. Cryst. 303:249 (1997).
134. K. Radley and N. Mclay, J. Phys. Chem. 98:3071 (1994).
135. M. P. Vinardell, A. Vilchez, J. Seguer, and M. R. Infante, Rev. Toxicol. 13:8 (1996).
136. H. Schott, J. Pharm. Sci. 62:341 (1973).
137. K. Schafer, J. Wirsching, and H. Hocker, Fett/Lipid 99:217 (1997).
138. M. Takehara, Fragrance J. 2:52 (1989).
139. M. R. Infante, J. Molinero, and P. Erra, J. Am. Oil Chem. Soc. 69:647 (1992).
140. M. R. Infante, J. Molero, M. R. Julia, P. Bosch, and P. Erra, J. Am. Oil Chem. Soc. 66:1835 (1989).
141. J. Molinero, M. R. Julia, P. Erra, M. Robert, and M. R. Infante, J. Am. Oil Chem. Soc. 65:975 (1988).
142. S. Osanai, Y. Yoshida, K. Fukushima, and S. Yoshikawa, J. Jpn. Oil Chem. Soc. 38:39 (1989).
143. Y. Sugimoto and S. Toyoshima, Antimicrob. Agents Chemother. 16:329 (1979).
144. C. V. Kumar, A. Buranaprapuk, G. J. Opiteck, M. B. Moyer, S. Jockusch, and N. J. Turro, Proc. Natl. Acad. Sci. 95:10361 (1998).
145. K. Abe, M. Goto, and F. Nakshio, J. Ferment. Bioeng. 83:555 (1997).
146. L. A. Creuwels, R. A. Demel, L. van Golde, B. J. Benson, and H. P. Haagsman, J. Biochem. 268:26752 (1993).
147. A. Leydet, H. Hachemi, B. Boyer, G. Lamaty, J. P. Roque, D. Schols, R. Snoeck, G. Andrei, S. Ikeda, J. Neyts, D. Reymen, J. Este, M. Witvrouw, and E. De Clercq, J. Med. Chem. 39:1626 (1996).
148. K. Tsujii and F. Tokiwa, J. Am. Oil Chem. Soc. 54:585 (1977).

9
Potential Applications of Protein-Based Surfactants

IFENDU A. NNANNA   Proliant, Inc., Ames, Iowa

GUANG YU CHENG   Coty International Development Center, Morris Plains, New Jersey

JIDING XIA   Wuxi University of Light Industry, Wuxi, China

I. INTRODUCTION

Surfactants in general are applied to a broad spectrum of industrial products, where they fulfill emulsifying, detergency, wetting, solubilizing, foaming, or dispersing functions. With these functions, surfactants are considered indispensable in the industrial sector, and worldwide demand and supply are huge and growing. As evidenced from various chapters in this book, the last decade has experienced innovative approaches to surfactant development. This effort is driven in part by environmental concerns and legislative requirements for petrochemical-based surfactants and the desire by industry for multifunctional surfactants.

Because of these reasons and the fact that protein-based surfactants (PBS) are biodegradable and generally of low toxicity, the market place projection for PBS is encouraging. The trend is toward mild and biodegradable surfactant products. The biodegradability issue is expected to be a major consideration in new product and market development in the 21st century. The technological approaches presented in this book point to a trend toward new science and chemistry for developing mild, low-cost, biodegradable, and multifunctional surfactants.

The niche applications for PBS are envisioned especially for specific end uses, such as pharmaceutical formulations and personal care products, where a broad range of functionality is desired (e.g., safety, mildness to skin, high surface activity, antimicrobial activity, biodegradability) or for high-volume uses, such as in detergents, foods, and agriculture.

In this chapter, we provide information on the potential applications of PBS, with emphasis on the personal care, detergent, food, and agricultural sectors. Since the personal care and household detergent sectors represent the largest single use and market share of surfactants in general, the existing and potential applications of PBS in these two sectors will be extensively discussed.

II. PROTEIN-BASED SURFACTANTS AND PERSONAL CARE PRODUCTS

In the past few years, mildness, gentleness, and safety have become increasingly important for selecting surfactants in the personal care and pharmaceutical industries. Consumers are genuinely concerned about potential irritation, toxicity, and environmental issues. The perception is that peptide and protein derivatives such as PBS possess numerous properties desirable in personal care products. For example, some have film-forming properties, whereas acylated proteins exhibit high foaming in both soft and hard water. Amino acids can penetrate through the hair cuticle into the hair shaft, drawing in moisture. The PBS are generally known to have a long-chain acyl radical as a lipophilic group and aminocarboxylic acid as a hydrophilic group. It has been recognized that PBS generally have good foaming power, are mild to human skin, and provide a comfortable and soft feeling to the skin during and after use [1]. Because PBS are capable of binding and holding moisture (moisturizing action) and are highly biodegradable and rapidly decomposed in sewage, it has been likened to the natural moisturizing factor (NMF) present in the skin. The PBS may find application in formulating film-forming skin moisturizers and lubricants.

The physicochemical characteristics of surfactants, such as adsorption, micelle formation, foam properties, and in vitro skin mildness, as well as the PBS in personal care products will be discussed.

A. Selected Properties of Surfactants Important for Personal Care Product Applications

1. Interactions of Surfactant with the Skin and the Hair

Surfactants applied to the skin have to fulfill the following essential but different functions [2]. First, surfactants must care for the upper skin layer, the stratum corneum, by improving the elasticity of the skin surface through the absorption of fat or fatlike substances into the horny layer without disturbing skin functions. Second, surfactants must enhance skin permeability to bring active ingredients to the horny layer, which will improve its moisture-retention capacity and thereby simultaneously improve the appearance of the skin by a light tightening of the horny layer. Third, surfactants must promote a micro-distribution of substances at the surface of the skin and/or in the stratum corneum, to fulfill the following functions: protection from sunlight, tanning of the skin, bleaching of the skin, insect repulsion, deodorizing, retardation of perspiration, and odor releasing. Fourth, they must clean the skin surface. Preferably, only the surface film present on the skin should be removed, consisting of cutaneous excretions (sweat, fats), flakes, air-carried dirt, etc. This cleaning should neither result in an elution of the horny layer nor increase skin permeability to substances on the surface.

The keratin of the skin and of the hair has lots of similarities. Since the isoelectric point of hair is approximately 3.7 [3], its surface bears a net negative charge at the neutral pH of most shampoos. The binding of surfactants to hair keratin is determined by the charge of the surfactants, surfactant molecular structure, the isoelectric point of hair, the pH of the surroundings, and the electrolyte in the formulation [3]. Anionic surfactants bind via hydrophobic interaction at the surface of the protein, causing it to become overall negatively charged and to exhibit a tendency for overall solubilization. The potential for solubilization decreases with increasing ethoxylation in alkyl ether sulfates because of increasingly greater head group mobility accompanied by increasing repulsion between the head groups that limits the amount of surfactant that can be adsorbed at the protein surface. Increased chain branching of anionic surfactants increases the overall propensity of each molecule for hydrophobic binding to the protein but reduces the net surface charge density. In the case of cationic surfactants, the positively charged head group binds to the negatively charged groups at the surface of the protein associated with aspartyl or glutamyl sidechains or sulfonated sites to make the protein surface hydrophobic. Hydrophobic interaction between this protein at the surfaces inhibits overall swelling by water.

2. Solubility–Krafft Point Relationship

The Krafft point is the temperature at which the solubility of hydrated surfactant crystals increases sharply with increasing temperature and forming micelles. This increase is so sharp that the solid hydrate dissolution temperature is essentially independent of concentration above the critical micelle concentration (cmc) and is therefore often called the Krafft point without specifying the surfactant concentration. The steep increase in solubility above the sharp bend is caused by micelle formation. Micelles exist only at the temperature designated as the Krafft point. This is a triple point at which surfactant molecules coexist as monomers, micelles, and hydrated solid [2]. Below the Krafft point, the surfactant exists in molecular form up to the solubility limit. At higher concentrations, the surfactant molecules are in equilibrium with crystals. Above the Krafft point, the crystals are in equilibrium with micelle and monomers. It was found, as shown in Table 1, that the Krafft point for surfactants in a homologous surfactant series rises with increasing hydrophobic chain length [4]. The Krafft point of surfactants depends greatly on the counterion.

The Krafft points of monosodium salts of *N*-acylglutamic acids were determined by the electric conductivity method. The Krafft points were a few degrees below the clear points, the temperatures at which the surfactant dissolved completely [4]. The monosodium salt of *N*-oleoyl-L-glutamic acid hardly hydrolyzed because it had an affinity with water and the Krafft point is lower. Triethanolamine salts of *N*-acylglutamic acid are generally more soluble in water and less irritating on the skin and have a lower Krafft point, compared with the corresponding sodium salts.

In general, a surfactant can satisfactorily show surface activity when used at a temperature higher than its Krafft point. Anionic surfactants have very high Krafft points. Because anionic surfactants are major ingredients in personal wash products, limitation is placed on the condition of their use. Especially when divalent or higher-valence metallic ions such as calcium ions are present, the Krafft point sharply increases via the occurrence of ion pair exchange of a surfactant. The surfactant precipitates in water within a practical temperature range as a hydrated solid, such as a calcium salt of the surfactant, or deposits on an article being washed, thus losing the surface activity. Accordingly, to improve hard-water resistance, the Krafft point of the calcium salt of a surfactant has to be lowered to a temperature beneath the temperature of use. As a consequence, good foaming and good detergency in hard water are ensured. A particularly effective reduction of the Krafft point results from the introduction of an *N*-hydroxyalkyl group into *N*-acyl-glycine [6]. This will

TABLE 1 Krafft Point of Monosodium Salts
of *N*-Acylglutamic Acids

Acyl group	Krafft point (°C)	Clear point (°C)
Lauroyl	38.1	42.0
Myristoyl	45.8	50.5
Palmitoyl	54.2	58.3
Stearoyl	62.5	66.0
Oleoyl	10.5	14.2

produce strong steric hindrance. As a result, the crystallinity of the hydrated solid of the surfactant lowers with a drastic reduction of the Krafft point. In addition, if the surfactant forms divalent metallic salts, such as a calcium salt, it dissolves in a micellar form because of the imparting of hydrophilicity with the hydroxyl group, showing satisfactorily the surface activity in hard water. This is considered the reason why good foaming and high detergency are obtained.

3. In Vitro Skin Mildness

Surfactants are major ingredients in personal wash products, such as body wash, facial wash, and hair shampoo. Personal wash products often contain irritating ingredients, especially when applied to human skin that is sensitive. Therefore, it is not surprising that one of the most desirable claims for personal wash products is that it is "mild."

There are numerous in vivo and in vitro test methods for evaluating "mildness." The in vivo irritation tests, such as Flex Wash and the Frosch-Kligman chamber, have been widely used in the personal care area. The Flex Wash test is currently the preferred in vivo method for the measurement of surfactant harshness. It has proved to be a highly sensitive, reproducible measure of surfactant irritation. The test provides a measure of the irritation of both harsh and mild surfactants. The concentrations used and the type of washing involved are particularly relevant to skin- and hair-washing products. The benefits of the in vivo test are that it enables the evaluation of an entire living system's reaction. The drawbacks of the in vivo test are low reproducibility, unsuitability for clarification of the toxic mechanism, and the legal issues in some countries. Therefore many attempts have been made to find simple in vitro tests to predict surfactant skin irritation without having to resort to in vivo tests involving people or animals. The most successful methods are Zein solubilization, acid phosphatase activity, and collagen film swelling and callous breakdown. The most used screening test is the Zein test, a measure of the solubilization of the water-insoluble protein "Zein" by anionic surfactants. The Zein method is highly reproducible and gives excellent compatibility with the ranking obtained from the in vivo Flex Wash method. These methods involve the surfactant–protein interaction and seem to correlate well with the action of surfactants on in vivo skin. The in vitro testing methods yield very consistent results, involve controlled test conditions, and are suitable for clarifying the toxic mechanism.

Protein-based surfactants are composed of two naturally occurring moieties, amino acid and fatty acid such as acylglutamate. Acylglutamates by themselves are not ultramild surfactants by any means. Their in vitro Zein values are higher than those of alkyl polyglycosides, alkylamphoacetate, amphoteric betaine, sodium cocoyl isethionate, etc. [6]. The pH of acylglutamate solution is approximately 5.5, nearly equal that of normal human skin. One major advantage for the acylglutamates is their foam enhancement at acidic pH [6]; their irritation potential is also reduced. Therefore at pH around 6, acylglutamate is able to deliver a good lather but will possess a high degree of mildness compared to other surfactants. It has been used as a primary surfactant for mild cleansing products. Kanari et al. [7] developed an in vitro cell toxicity test that best characterized various toxic mechanisms and mild surfactants, such as sodium monolaurylphosphate, sodium acylmethyltaurate, sodium acylisethionate, and sodium lauryl POE (3) sulfate, in various cytotoxicity tests. These evaluations showed that acylglutamate had the lowest irritation scores. They also conducted dose-response tests on five binary combinations of 1% surfactant solution, 100/0, 75/25, 50/50, 25/75, 0/100 (acylglutamate/other anionic surfactant), determined the anti-irritating effect of acylglutamate, and showed that increasing the acylglutamate level decreased the irritation potential of the other anionic surfactant in cytotoxicity tests. Lee et al. [8] investigated the irritant contact dermatitis potential and the possible anti-irritating potential of sodium lauroyl glutamate in a mixed surfactant system on human skin by using visual scores and the measurement of transepidermal water loss (TEWL). They found that increasing the sodium lauroyl glutamate concentration decreased the visual scores and TEWL values.

III. COMMERCIALLY AVAILABLE PROTEIN-BASED SURFACTANTS FOR PERSONAL CARE PRODUCTS

A. N-Acyl Amino Acids and Their Salts

Personal wash products are required not only to have good detergent properties, such as detergency and foaming power, but also to provide a low stimulus to the skin and the eye, especially when they are used in direct contact with the skin and the scalp. N-Acyl amino acids and their salts are known as mild surfactants carrying both amino groups and carboxylic acid groups on the same molecule. In modern usage, the term *amino acid* is sometimes restricted to the principal products of the hydrolysis of proteins. In this sense, amino acids are the monomeric building blocks of peptides and proteins. Most of the amino acids from natural sources, with the exception of glycine, possess one or more asymmetrical centers and, therefore, may exist in at least two optically active forms. Most of the natural amino acids are of the L configuration. Since amino acids carry two types of salt-forming functions, their behavior in aqueous solutions is dependent on pH. At low pH, the amino group is protonated, and amino acids act as cations. At high pH, the carboxyl groups are ionized, and these acids behave as anions. At intermediate pH, the amino group may partially neutralize the carboxyl group. In this case, the amino acid exists as a zwitterionic molecule possessing both cationic and anionic charges on the same molecule. Commercially available products such as, in particular, *N*-acyl glutamic acid salts, *N*-acyl sarcosine salts, *N*-acyl β-alanine salts, *N*-acyl methyl taurine salts, and the like, are broadly used in personal wash products. In the following, an overview of *N*-acyl amino acid salts recommended for the preparation of personal care products is presented.

1. Glutamates

Glutamate surfactants are amides derived from L-glutamic acid and natural higher fatty acids. The commercially available glutamate surfactants include: cocoyl glutamic acid, disodium cocoyl glutamate, disodium hydrogenated tallow glutamate, disodium stearoyl glutamate, hydrogenated tallowoyl glutamate acid, lauroyl glutamic acid, myristoyl glutamic acid, palmitoyl glutamic acid, potassium cocoyl glutamate, potassium lauroyl glutamate, potassium myristoyl glutamate, sodium cocoyl glutamate, sodium hydrogenated tallowoyl glutamate, sodium lauroyl glutamate, sodium myristoyl glutamate, sodium stearoyl glutamate, stearoyl glutamic acid, TEA-cocoyl glutamate, TEA-hydrogenated tallowoyl glutamate, and TEA-lauroyl glutamate. They are promoted mainly as mild surfactants resistant to hard water and also are claimed not to defat the skin to the same extent as other surfactants. Some of the features of glutamate surfactants include the following [4,10]:

Moderate foaming power and good detergency. The monosodium salt has lower surface tension and better foaming properties than the disodium salt. The optimum acyl glutamate chain length lies between lauroyl and myristoyl.

In the aqueous solution the monosodium salts show a pH of 5–6.5 and have the buffer capacity of an unneutralized carboxylic acid; the disodium salts show a pH of 8–9, which was the same value as for a sodium salt of carboxylic acid.

Comfortable and soft feeling to the skin during and after use.

The monosodium salts are less soluble in water and become pasty at higher concentration. Triethanolamine salts are highly soluble in water, and no crystals precipitate at −5°C.

The monosodium salts have better wetting and emulsifying properties than the disodium salts.

Remarkably low irritation potential and anti-irritating effect.

Acyl glutamate surfactants are carboxylates, and as such their properties will depend heavily upon the pH of the medium. The optimum acyl group chain length would appear to be cocoyl, with optimum pH for this species appearing to be approximately pH 6. In terms of foaming power, monotriethanolamine N-myrestoyl-L-glutamate has excellent foaming power and stability [10]. N-Oleoyl glutamic acid salt also has good foaming power. The physical properties of this surfactant, such as low Krafft point and cmc [4], can be useful in cleansing and skin care applications. The solubilities of glutamate surfactants follow a similar order to those of soaps. Lower-chain-length fatty acids are more soluble. Blends—for example, Coco—are more soluble than their components. Finally, counterions follow the order sodium < potassium < triethanolamine. The mildness and anti-irritant effect of glutamate surfactants can be dramatically improved at acidic pH by adding coactives [11]. Acyl glutamates are widely used by Japanese companies in skin cleansing formulations. One of their present Japanese products (Page One, from Lion) uses acyl glutamate as the major active with other coactives, such as acyl sarcosinate and soap. Current patent activity is mainly in Japan and is centered around generally mild anionics, especially acylated amino acid surfactants in combination with other cosmetic ingredients. Kao patented the following detergent composition containing N-acyl amino acids and their salts to improve foaming power and give a good skin feel on use [12]:

Component	Wt%
N-Lauroyl-β-alanine TEA salt	30.0
N-(N'-Lauroyl-β-alanyl)-β-alanine TEA salt	0.6
Lauric acid TEA salt	5.0
Lauryldimethylamine oxide	2.0
Perfume	0.5
Dibutylhydroxytoluene	0.2
Ethanol	3.0
Water	Up to 100%

Although glutamate surfactants are effective and mild cleansing agents, there are only a few commercially available products. The major reason is price, for they are relatively expensive (generally 5–10 times the price of SLES). Clariant has developed a new process for increasing the production of acyl glutamates. The process is expected to reduce the price from the current $4.50–5/lb to about $2.50/lb [13].

2. Sarcosinates

Sarcosinate surfactants are derived from natural fatty acids and an amino acid, sarcosine (*N*-methyl glycine). Typically the sarcosinate is used in the form of its sodium, potassium, or ammonium salt solution. *N*-Acyl sarcosinates are produced commercially by the Schotten–Baumann reaction of the sodium salt of sarcosine with the appropriate fatty acid chloride under carefully controlled conditions. The commercially available sarcosinate surfactants are cocoyl sarcosine, lauroyl sarcosine, myristoyl sarcosine, oleoyl sarcosine, stearoyl sarcosine, sodium cocoyl sarcosinate, sodium lauroyl sarcosinate, sodium myristoyl sarcosinate, ammonium lauroyl sarcosinate, TEA lauroyl sarcosinate, ammonium cocoyl sarcosinate, monoethanolamine cocoyl sarcosinate, and TEA oleoyl cocoyl sarcosinate.

The sarcosinates are modified, or "interrupted," soaps that, while being very soaplike, possess many of the more desirable properties of synthetic surfactants [14]:

Outstanding lather building and resistance to sebum delathering.
Excellent foaming and wetting properties; good detergency.
Compatibility with quaternary ammonium germicides and conditioners.
Imparting a good feel, softness, and lubricity to skin and hair.
Mildness to skin and eyes; improving of moisture retention in the skin.
Low cloud point; good solubility in wide pH range.

All these characteristics form a composite of functional benefits that give sarcosinate surfactants their unique identity as anionic conditioning surfactants. Sarcosinate surfactants represent one of the few classes of anionic surfactants that are compatible with quaternary compounds. Because they can adsorb onto proteinaceous and cellulosic surfaces, they are also highly substantive. Interestingly, sarcosinates are capable of inhibiting corrosion as well as the growth of certain enzymes and bacteria. The ability of sarcosinates to display these inhibitory tendencies, coupled with their inherent nature as conditioning surfactants, gives rise to a number of different applications.

Sarcosinate surfactants have been widely used in personal wash, skin care, and hair care formulations as coactives. They are particularly useful in providing a rich, stable lather, especially in the presence of sebum. Lathers produced from other anionic and amphoteric surfactants are remarkably enhanced by the addition of sarcosinates. Maximum foaming and detergency are developed in the pH range 4–8. The adsorption of sarcosinates onto the hair results in manageability and reduced static buildup. Sarcosinate surfactants are easily adsorbed on the surface of skin, especially within the mildly acidic pH range, and this adsorbed material has been shown to retard moisture loss from the skin.

Sarcosinate surfactant has been used mainly as a major toothpaste ingredient, since it is nontoxic, strongly foaming, and enzyme inhibiting. Recently, it has been used in skin cleansing formulations as a coactive, especially for the low-temperature-storage condition. Recently, Woodbury and his co-workers patented a new process for preparing sodium N-acyl sarcosinates [15]. This process eliminates the use of phosphorus trichloride or thionyl chloride and carboxylic acid chlorides by reacting the alkali metal N-acyl amino acid directly with a fatty acid at elevated temperatures, with constant removal of water generated in the reaction. It would simplify the process for manufacturing N-acyl sarcosinates and eliminate the environmental drawbacks of the conventional processes. Generally sarcosinates are more expensive than most nonionics and anionics, but less so than amphoterics such as betaines.

3. Taurates

Taurate surfactants contain a strongly electronegative sulfonic group and a cationic amine group. It is known that the dissociation constant of a strongly electronegative sulfonic group greatly exceeds that of the cationic amine group. Taurates consequently demonstrate strong anionic tendencies and minimal cationic properties. The commercially available taurate surfactants include sodium methyl cocoyl taurate, sodium methyl lauroyl taurate, sodium methyl myristoyl taurate, sodium methyl oleoyl taurate, sodium methyl palmitoyl taurate, and sodium methyl stearoyl taurate.

Taurate surfactants are similar to those of the corresponding fatty acid soaps in soft water, but these materials are effective in both hard and soft water, are not sensitive to lower pH, and are better wetting agents. They show good stability to hydrolysis by acids and alkali, good skin compatibility, and good lime soap–dispersing power [16]. This amide structure of the N-acyltaurine surfactants stabilizes these derivatives of fatty acids against hydrolysis under acid and alkaline conditions, even at high temperature [17].

4. Selected Related Patents

Ajimonoto patented an excellent detergent composition comprising an N-acyl amino acid salt, which is highly safe and shows quick, effective, and creamy lathering and reduced slimy touch to the skin [18]. They conducted intensive studies by jointly using a salt of a specified N-acyl amino acid and a salt of a specified higher fatty acid, by blending them with a specified higher alcohol or polyhydric alcohol. The N-acyl amino acid salts are cocoyl glycine TEA, cocoyl alanine TEA, cocoyl glycine potassium, lauroyl β-alanine potassium, etc. The higher fatty acid salts are soap, sodium laurate, potassium myristate, etc. The higher alcohol molecules are lauryl alcohol, cetyl alcohol, etc. Such a combination improved lathering, bubble retention, and a refreshed feeling significantly.

Yuichi and co-workers [19] patented a dermal drug for external use that had improved stability of urea and a moisture-imparting effect to the skin. The product contained N-acylglutamic acid salt and N-acylaspartic acid salt and urea as essential components and was claimed to produce no skin stimulation and to have excellent frothing property. Lynn and co-workers patented a lipstick composition containing N-acylglutamic acid amide as a gellant. They claimed that this composition facilitates the retention of emollient oils, particularly under high humidity and temperature and provides good moisturization of the lip [20]. Kazutami and others [21] patented a hair care composition containing N-long-chain-acyl dipeptide or salt that exhibits low irritation, an excellent resistance to hard water, and freedom from turbidity and odor. It has an excellent feeling upon use, without any creaking feeling of the hair or stretching feeling of the skin [21]. Ei [22] patented an oil/water type of emulsion of high stability by using an oligomeric condensate from fatty acids, glycerol, and dicarboxylic acid as well as N-acyl acidic amino acids or their salt. This emulsion shows good spreadability on skin and gives a freshened feeling after being flown off with water. Koji [23] patented a water/oil emulsified composition containing an oil-soluble N-long-chain acyl acidic amino acid mono- or diester. The emulsion possessed excellent storage stability over a wide temperature range.

Hideo and his co-workers [24] used an N-long-chain acyl acidic amino acid mono salt with an amphipathic substance in water to obtain a composition capable of thickening and gelling an aqueous phase. A makeup-removal composition was made [25] by blending an N-caproyl-L-arginine methyl ester hydrochloride, sugar beet pectin, low-viscosity dimethyl polysiloxane, and water as well as an antibacterial and antifungal agent.

N-Acyl amino acids have been used in various applications, including a composition or formulation for improving a hair-restoration-stimulating effect, moisture-retaining activity, and subcutaneous blood-flow-stimulating activity [26], a treated powder cosmetic or powder-rich foundations [27–29]. The powder cosmetic [29] consisted of a fine-particle ultraviolet-ray-blocking agent whose surface is treated with an N-mono-long-chain acyl basic amino acid, which has transparency in a high visible light region and ultraviolet-ray-blocking effect and is capable of suppressing surface activity.

B. Acylated Protein Hydrolysates

1. Anionics

This surfactant category was one of the oldest synthetic detergents and was originally introduced during the early 1930s in Germany as the lamepons [17]. The protein fatty acid condensates are produced from hydrolyzed protein from scrap leather and other waste protein, such as gelatin, and the condensation process is by the Schotten–Baumann reaction [30].

They are available commercially as cocoyl hydrolyzed collagen, cocoyl hydrolyzed keratin, cocoyl hydrolyzed soy protein, isostearoyl hydrolyzed collagen, lauroyl hydrolyzed collagen, lauroyl hydrolyzed elastin, myristoyl hydrolyzed collagen, palmitoyl hydrolyzed collagen, palmitoyl hydrolyzed milk protein, palmitoyl hydrolyzed wheat protein, potassium cocoyl hydrolyzed casein, potassium cocoyl hydrolyzed collagen, potassium cocoyl hydrolyzed corn protein, potassium cocoyl hydrolyzed keratin, potassium cocoyl hydrolyzed potato protein, potassium cocoyl hydrolyzed rice protein, potassium cocoyl hydrolyzed soy protein, potassium cocoyl hydrolyzed wheat protein, potassium lauroyl hydrolyzed collagen, potassium lauroyl hydrolyzed soy protein, potassium myristoyl hydrolyzed collagen, potassium oeroyl hydrolyzed collagen, potassium stearoyl hydrolyzed collagen, potassium undecylenoyl hydrolyzed collagen, potassium undecylenoyl hydrolyzed soy protein, sodium cocoyl hydrolyzed collagen, sodium cocoyl hydrolyzed keratin, sodium cocoyl hydrolyzed soy protein, sodium myristoyl hydrolyzed collagen, sodium oleoyl hydrolyzed collagen, sodium stearoyl hydrolyzed collagen, sodium stearoyl hydrolyzed soy protein, sodium stearoyl oat protein, sodium/TEA-lauroyl hydrolyzed collagen, sodium/TEA-undecylenoyl hydrolyzed soy protein, TEA-cocoyl hydrolyzed collagen, TEA-cocoyl hydrolyzed soy protein, TEA-isostearoyl hydrolyzed collagen, TEA-lauroyl hydrolyzed collagen, TEA-oleoyl hydrolyzed collagen, and TEA-undecylenoyl hydrolyzed collagen [31]. These products were prepared from collagenaceous wastes and were available as 30–40% active solutions.

The condensate of a hydrolyzed collagen protein can be used as a hair growth stimulant [32]. Masato and others [33] patented a shampoo composition containing a quaternary ammonium derivative of a peptide derived from animal protein and an amino acid type of anionic surfactant, and claimed that this composition has an excellent conditioning effect on hair and results in little damage to the hair on repeated shampooing. Su [34] patented a mild skin cleansing composition; the prototype formula was described as follows:

Component	Wt%
Acylated hydrolyzed animal protein (collagen) (30 wt%)	10.0
Lauric acid	1.5
Cocoamphodiacetate (50%)	10.0
Acylglutamate (30%)	1.0
Miracare 2MCT	3.75
Sodium tridecth sulfate	4.5
Hexylene glycol	2.25
Isopropanol	0.5
Triethanolamine	1.85
Benzophenone 6	0.03
Dimethicone copolyol	0.5
Aculyn polymer 22 (30%)	7.0
Deionized water	up to 100%
Fragrance	0.3
DMDM hydantoin	0.3
Tetrasodium EDTA	0.3
Colorant (0.01%)	0.3

2. Cationics

This surfactant category represents the quaternized acyl protein hydrolysates, combining the care effects of the cationic substances with the positive dermatological effects of the protein derivatives and making them valuable for personal care products.

The commercially available surfactants include: cocamidopropyldimonium hydroxypropyl hydrolyzed collagen, cocodimonium hydroxypropyl hydrolyzed collagen, cocodimonium hydroxypropyl hydrolyzed keratin, cocodimonium hydroxypropyl hydrolyzed soy protein, laurdimonium hydroxypropyl hydrolyzed soy protein, laurdimonium hydroxypropyl hydrolyzed wheat protein, lauryldimonium hydroxypropyl hydrolyzed collagen, lauryldimonium hydroxypropyl hydrolyzed keratin, lauryldimonium hydroxypropyl hydrolyzed soy protein, oleamidopropyldimonium hydroxypropyl hydrolyzed collagen, steardimonium hydroxypropyl hydrolyzed collagen, steardimonium hydroxypropyl hydrolyzed keratin, steardimonium hydropropyl hydrolyzed silk, and steartrimonium hydroxyethyl hydrolyzed collagen. The unique properties [35,36] exhibited by this class of acylated protein hydrolysates are as follows:

Are strongly cationic
Are compatible with anionics, cationics, amphoterics, and nonionics
Are foaming proteins
Remain soluble at acidic pH
Have outstanding dermatological characteristics
Exhibit bactericidal properties
Have high tolerance to alcohol
Have a broad application profile

The substantiveness that quaternized acyl protein hydrolysates gives to virgin and permed hair has been investigated [36]. An N-acyl neutral amino acid–based cationic surfactant was made by Hiroyiki and Yoshiaki [37] and was reported to have good foaming power and stability and to be soft on the hair and compatible with an anionic surfactant. The substantiveness of protein hydrolysates increases with concentration up to about 5% concentration and then approaches a plateau. It is this dependence upon concentration that has led to the belief that to impart significant benefits to hair, more than a token amount of protein should be present. For the quaternized acyl protein hydrolysates, hair attains its maximum plateau value of substantiveness at very low active concentrations (0.5–1%) compared with unmodified protein. Results have shown a better wet compatibility and feel in dry hair by combining quaternized protein and alkyl polyglucoside [35]. Healthy hair is hydrophobic and distinguished by a negative wetting tension. Hydrophilicity increases with increasing damage of the hair and yields a higher, positive wetting-tension value. The wetting-tension value can be lowered by adjusting the proportion of cationic protein. Due to their unique physical and dermatological characteristics, quaternized protein hydrolysates have been widely used in hair treatments, shampoo, skin cleanser, and facial cleanser. Masato and his co-workers [38] have patented an acidic dyeing agent containing quaternized cationic protein that causes hardly any hair damage but imparts good gloss and moistness to hair after dyeing. A hair-restoring agent was developed [39] to change hair structure, restore and strengthen damaged hair, and suppress the fuzzing or peeling of cuticle or broken hair. The composition was prepared by reacting an animal protein–derived polypeptide and a vegetable protein–derived polypeptide with a cationizing agent. A film-forming polymer, such as an amphoteric polymer or a cationic polymer, was formulated with those protein derivatives to provide a hair-strengthening agent.

3. Ester of Glycine Betaine

Most of the synthetic cationic surfactants used in hair care products, such as cetyl trimethylammonium chloride, tend to be rather toxic, irritating, and incompletely biodegraded. A natural source of quaternary ammonium groups is available, trimethyl glycine, obtained as a by-product of the processing of sugar beet. Long-chain fatty esters of this material are cationic surfactants that have been studied as potential alternatives to synthetic surfactants. The use of long-chain fatty esters of trimethylglycine in hair treatment formulation has been patented by Wella [40]. So far, trimethylglycine is the only commercially available natural source of the quaternary ammonium group, sold under the trade name "Natural Extract AP" from Witco.

C. N-Acyl Glutamic Acid Cholesteryl Ester

N-Acyl glutamic acid cholesteryl esters are cholesteryl derivatives synthesized from *N*-acyl glutamic acid and mixed alcohol, including cholesterol. Such esters also include: cholesteryl/behenyl/octyldodecyl lauroyl glutamate, cholesteryl/octyldodecyl lauroyl glutamate, diethyl palmitoyl aspartate, dihexyldecyl lauroyl glutamate, dioctyldodeceth-2 lauroyl glutamate, dioctyldodecyl lauroyl glutamate, dioctyldodecyl stearoyl glutamate, disteareth-2 lauroyl glutamate, and disteareth-2 lauroyl glutamate.

Sakamoto and coworkers have investigated the recovery effect of lipo-amino acid cholesteryl derivatives for damaged skin, and whether lipo-amino acid cholesteryl derivatives could be used as a substitute for ceramide [41]. The recovery effect was evaluated in damaged skin using dermal scores, water-holding capacity (conductance), and transepidermal water loss (TEWL). Cholesteryl esters showed a better recovery effect upon visual and instrumental assessment than petrolatum and cholesteryl hydroxy stearate. These results indicate that cholesteryl esters applied on damaged skin may be incorporated into the lipid component in the stratum corneum to reconstruct damaged lipid layers. *N*-Acyl glutamic acid cholesteryl esters are derivatives of an amino acid bearing an amide group and having a structural resemblance to ceramide. Ceramides have been reported to be the main component in stratum corneum lipids that form the lamellar structure. The results showed the models with ceramide and cholesteryl ester both produced virtually the same recovery effect in damaged skin. Cholesteryl esters contributed to the formation of the lamellar structure in the same way as ceramide, and this structure promotes the recovery by holding water between the intercellular lipid bilayers.

Ajinomoto [42–44] patented cosmetics and pharmaceutical preparations containing *N*-long-chain acyl amino acid esters, which are novel compounds with excellent emulsifying and hydrating performance. They found that when the cholesteryl esters of *N*-acyl amino acid are incorporated into skin cosmetics such as face wash cream, cleansing cream, massage cream, moisture cream, and baby skin protectors, they exhibit extremely excellent performance as an oily base and an emulsifier and impart an affinity to the skin.

D. Amphoteric Amino Carboxylic Acids and Their Salts

Amphoteric surfactants are characterized by a molecular structure containing two different functional groups, with anionic and cationic character, respectively. An amphoteric surface-active amino acid derivative should contain opposing acidic and basic groups with pKa and pKb values of about 2–4 to minimize the predominance of one group over the other. Some factors that can influence the neutralization and zwitterionic properties of amphoterics are the proximity of the acidic and basic functions to the hydrophilic group and the positioning of such group on a sidechain. The ampholytic materials discussed are pH-sensitive zwitterionic. The pH where these ionic groups neutralize each other is called the *isoelectric point* or *isoelectric area*. These materials show the properties of anionics above the isoelectric area and those of cationics below the isoelectric area. In the vicinity of their isoelectric points they exist mainly as zwitterionics and show minimum solubility in water and minimum foaming, wetting, and detergency [16]. Examples include derivatives of amino acids and their salts. The cationic functionality of aminocarboxylic acids is generally a protonated secondary or tertiary amine group (in so-called *weak nitrogen amphoterics*). The weak nitrogen amphoterics are protonated only at lower pH levels [45]. These compounds make up a unique family of amphoteric surfactants. They may be described as N-substituted amino acid derivatives (β-alanines). They are derived by the condensation of fatty primary amines and acrylic monomers. This reaction can be controlled to give two types of the commercially available products: *N*-fatty aminopropionates and *N*-fatty iminodipropionates [46].

    N-Fatty β-aminopropionate has an equal number of anionic and cationic groups. A number of *N*-fatty β-aminopropionates, under the trade name Deriphat (Henkel products), has been used in shampoo formulations. Deriphats, as shown structurally in Table 2 [45], either singly or in combination, have shown promise in shampoo formulations. A shampoo with good cleaning and foaming properties has been obtained by combining *N*-dodecyl-β-alanine and a cationic surfactant (dioctadecyl dimethylammonium chloride). The presence of Deriphats in a shampoo formulation could reduce eye irritation and sting

TABLE 2 Chemical Structures of Deriphats

Trade name	Chemical name	Structure
Deriphat 151	Sodium N-coco-β-aminopropionate	$RNHCH_2CH_2COONa$
Deriphat 151C	N-Coco-β-aminopropionic acid	$RNHCH_2CH_2CHCOOH$
Deriphat 170C	N-Lauryl/myristyl-β-aminopropionic acid	$RNHCH_2CH_2CHCOOH$
Deriphat 154	Disodium N-tallow-β-iminodipropionate	$RN(CH_2CH_2CHCOONa)_2$
Deriphat 160	Disodium N-lauryl-β-iminodipropionate	$RN(CH_2CH_2CHCOONa)_2$
Deriphat 160C	Partial sodium salt of N-lauryl-β-iminodipropionic acid	$CH_2CH_2CHCOOH$ / RN \ $CH_2CH_2CHCOONa$

Source: Ref. 45.

caused by sodium lauryl sulfate or ether sulfate. Deriphats could also counter-act the dry feel of hair, often attributed to electrostatic charge on each hair strand.

Deriphat 151C is another Henkel product that functions effectively as a wetting agent. The isoelectric range for Deriphat 151C is 2.9–4.5 [46]. It is very soluble in aqueous solutions of strong acids and alkalis at different tem-peratures and an excellent wetting agent. Solubility is low in most organic solvents, including ethanol and isopropyl alcohol. The product exhibits good emulsifying action for semipolar and slightly polar materials.

As for N-fatty β-iminodipropionates, they have an unequal number of an-ionic and cationic groups. The commercially available products from Henkel are Deriphat 154, 160, and 160C. The isoelectric ranges of these surfactants are from 1.3 to 4.2 [46]. They are more soluble in water than those monopropi-onic acid derivatives. They show a very low order of skin and eye irritation. They may be removed from substrates onto which they have adsorbed at pH below their isoelectric points by raising the pH [16].

Aminocarboxylic ampholytes have been used extensively in hair care and other cosmetic formulations. These materials are substantive to hair and have also been claimed to possess antistatic and bacteriostatic properties [45]. Many amphoteric surfactants have poor foaming properties and are blended with foaming anionics. The triethanolamine salt of Deriphat 170C shows excep-tional foam volume. Optimum foaming properties of the N-alkyl-β-aminopropionates and dipropionates are at pH value above the isoelectric point of 4.3–4.7. Although their best foaming properties are demonstrated in a weakly alkaline pH range, the amphoteric compounded shampoos exert their optimum effect on hair manageability at an acid pH. Therefore, amphoteric shampoos formulated specifically to improve hair manageability will inevitably show a decrease in foaming.

E.   *N*-Acyl Amino Acid Amides

N-Acyl amino acid derivatives used as a gelling agent include *n*-acyl amino acid amides and *n*-acyl amino acid esters, which include *n*-lauroylglutamic acid diethylamide, *n*-lauroylglutamic acid dibutylamide, *n*-lauroylglutamic acid dihexylamide, *n*-lauroylglutamic acid dioctylamide, *n*-lauroylglutamic acid didecylamide, *n*-lauroylglutamic acid didodecylamide, *n*-lauroylglutamic acid distearylamide, *n*-stearoylglutamic acid dibutylamide, *n*-stearoylglutamic acid dihexylamide, *n*-stearoylglutamic acid diheptylamide, *n*-stearoylglutamic acid dioctylamide, *n*-stearoylglutamic acid didecylamide, *n*-stearoylglutamic acid didodecylamide, *n*-stearoylglutamic acid ditetradecylamide, and *n*-stearoylglutamic acid distearylamide.

The gelling agent, commonly referred to as GP-1, is *N*-acyl glutamic acid diamide, from Ajinomoto [47]. GP-1 has *N*-acyl amino acid in it as a basic skeleton, which is a condensate of amino acid and fatty acid. Ajinomoto markets it as an oil-structuring agent [47]. GP-1 was found to form self-supporting gels with a wide range of the solvents commonly used in cosmetic products.

There are a number of antiperspirant stick patents from Procter & Gamble containing dibutyl lauroyl glutamide and 12-hydroxystearic acid as the gelling agent [48–50]. The primary structurants are the dibutyl lauroyl glutamide and 12-hydroxystearic acid. The combination of these two gelling agents is very effective because the conventional sticks use a total of 18% wax as structurant, whereas the stick containing these two gelling agents uses only 8% to achieve a similar hardness. The 12-hydroxystearic acid is a natural product derived from castor oil and used for thickening various oils.

Atsuko and his co-workers [51–56] patented antimicrobial and low-irritant cosmetic compositions containing *N*-long-chain acyl basic amino acid derivatives, plant extracts, and other ingredients. They can be widely used as a skin cosmetic, a makeup cosmetic, or a hair care formulation.

F.   Section Summary

It is clear from the foregoing that *N*-acyl amino acids and their salts have played a dominant role in the personal care sector. Although the preceding overview was not intended to be exhaustive, it provides insights into various applications and the huge market potential for PBS in the personal care industry. The patent literature is replete with numerous PBS applications, which are growing. The suppliers of PBS worldwide are also growing, as shown in Table 3. The increasing popularity of mild and environmentally friendly products is expected to boost the research and development of novel PBS with unique applications.

IV. PROTEIN-BASED SURFACTANT APPLICATIONS IN THE DETERGENT AND RELATED INDUSTRIES

Major manufacturers of consumer goods (i.e., laundry detergents and other home cleaning products) are paying close attention to new products and technological improvements of existing products. It has been estimated that United States demand for detergent additives will increase 5.5% annually to nearly $4 billion in 2001, propelled by more expensive specialty additives. The PBS fit into the so-called specialty additives and can potentially improve detergent efficiency by improving water properties and/or function as an enzyme stabilizer.

As with the personal care industry, the detergent industry is aware of the environmental concerns and is looking into product development innovation as a way to grow the business. This implies that innovative additives such as PBS that can impart new or improved properties to cleaning products would be particularly attractive. A number of areas in the detergent or related industries offer interesting prospects for PBS, and some selected areas are discussed in this section.

A. Household Detergents

A surfactant with ampholytic and chelating properties derived from aspartic acid is an effective detergent [57]. *N*-Alkyl-β-amino-ethoxy acids blended with anionic detergents are employed in reducing skin-irritating effects in detergent applications. Liquid detergent containing *N*-coco-β-aminopropionate (Deriphat 151) is claimed to be an effective cleanser for ''burnt-on'' grease spots on metallic surfaces [58]. Aminocarboxylic surfactants, such as $C_{14}CHOHCH_2NHCH_2$–COONa are used for laundering textiles, cleaning carpets, and washing hair, food, and glass [59].

It is reported that 2-hydroxy-3-aminopropionic-*N,N*-diacetic acid derivatives are used as good complexing agents, bleaching agent stabilizers, and

TABLE 3 Supplier Information

Surfactants	Trade name	Supplier
Acyl alaninates	Nikkol alaninate	Nikko
Acyl aspartates	Monosodium *N*-lauroyl-L-aspartate	Mitsubishi Petrochemical
Acyl glutamates	Amisoft	Ajinomoto
	Hostapon	Clariant
Acyl sarcosinates	Amilite	Ajinomoto
	Crodasinic	Croda
	Hamposyl	Hampshire
	Maprosyl	Stepan
	Nikkol sarcosinate	Nikko
	Oramix	SEPPIC
	Sarkosine	Hoechst
	Soypon	Kawaken
	Vanseal	Vanderbilt
Acyl taurates	Adinol	Croda
	Geropon	Rhodia
	Hostapon	Hoechst
	Jeepon	Jeen
	Nikko	Nikko
	Somepon	SEPPIC
	Tauranol	Finetex
Acyl hydrolyzed protein (anionic surfactant)	Aminofoam	Croda
	Foam-Coll, Foam-Keratin	Brooks
	Lamepon	Henkel
	Maypon	Inolex
	May-Tein	Maybrook
	Monteine	SEPPIC
	Nikkol	Nikko
	Peptein	Hormel
	Texatein	Lanaetex
	Promois	Seiwa Kasei
Acyl hydrolyzed protein (cationic surfactant)	Croquat	Croda
	Lamequat	Henkel
	Promois	Seiwa Kasei
	Quat-Coll, Quat-Soy	Brooks
Glycine betaine amino-carboxylic acids	Natural Extract AP	Witco
	Deriphat	Henkel
	Miranol	Rhodia
	Mirataine	Rhodia
	Unitex	UPI
Surfactin	Surfactin	Wako chemical
Amide of acyl aminoacid	GP-1	Ajinomoto
Cholesteryl/octyldodecyl lauroyl glutamate	Eldew	Ajinomoto

builders in detergents. The calcium-binding power and iron-binding power of such complexing compounds are given in Table 4 [60,61].

An N-(N'-long-chain acyl-β-alanyl)-β-alanine is used in detergent compositions to impart good foaming power, foaming stability, foam breakage, and good feel after washing without showing slimy feel, different from conventional products [62]. A typical composition is exemplified in Table 5.

Sano and Hattori [63] patented detergent compositions that comprised various acyl amino acid salts. The compositions were all rated favorably on various quality characteristics, as shown in Table 6. Kao Co. [64] developed a detergent composition having a combination of low skin irritation with high detergency and high hard-water resistance, which comprised N-acyl-1-N-hydroxyalkyl-β-alanine, including alkali metal, ammonium, alkanolamine, and alkyl-substituted ammonium. Ajinomoto Company [65] has offered amino acid derivatives, such as L-glutamic acid, L-lysine, and L-arginine derivatives, as functional ingredients for detergent products, shampoos, and cosmetics, claiming special features such as biodegradability and low irritation, moderate foaming, and good detergency properties: The specification and detergency of some N-acyl glutamates are shown in Table 7.

The polyaspartic acids α,β-poly[(2-hydroxyethyl)-d-l-aspartamide] with good biodegradability are used in detergents and cleaners. They are prepared by polycondensation of aspartic acid in the presence of phosphoric acid in a molar ratio of from 1:0.5 to 1:10 at a temperature of at least 120°C to give polysuccinimide. Subsequent hydrolysis of the polysuccinimide with bases is

TABLE 4 Metallic Iron–Binding Power in Detergent Solution

	Ca-binding power, mg $CaCO_3$/g active		Iron-binding power, mg Fe^{3+}/g active substances	Perborate in (%) detergent formulation
	RT/pH 11	80°C/pH 10		
2-Hydroxyl-3-aminopropionic N,N-diacetic acid	275	200	113	43.3
NTA/Na₃	350	250	54	24.5
Na Triphosphate	215	150	—	—
EDTA/Na₄	275	240	44	20

Source: Ref. 60.

TABLE 5 Detergent Composition Containing Amino Acid
Surfactants

Composition	% by weight
N-Lauroyl-β-alanine triethanolamine (TEA)	30
N-(N'-Lauroyl-β-alanyl)-alanine TEA salt	0.6
Lauric acid TEA salt	5
Lauryldimethylamine oxide	2
Perfume	0.5
Dibutylhydroxytoluene	0.2
Ethanol	3
Water	The balance

Characteristic	
Foam volume (mL)	
Immediately	240
After 2 min	225
Feel at use	
Foam breakage in rinsing	Very good
Smoothness in rinsing	Very good
Smoothness after rinsing	Good
Smoothness after drying	Good

performed to give at least partially neutralized polyaspartic acid, as an additive to detergents and cleaners in amounts of from 0.1 to 10% by weight based on the detergents [66].

N-Acylamino acid surfactants, including those based on glutamic acid, methylglycine, alanine, aspartic acid, and serine, have been used as aqueous liquid cleansing agents [67]. The derivatives showed no increase in viscosity, even at a low temperature. Thus they can be pushed out easily from a foamer container to give uniform foam at a low temperature [67].

B. Detergents Containing Enzymes

Enzyme demand in detergent formulations is rising worldwide. The enzyme types used in detergents are proteases, lipases, amylases, and cellulases. New enzymes with unique properties are already being applied to replace the cleaning power of chlorine, while others are being developed for bleaching applications. These enzymes, as protein molecules, would be subjected to harsh con-

TABLE 6 Detergent Composition Containing Glycine or Alanine Salt

Acyl amino acid salt	Sample, % by weight				
	1	2	3	4	5
Lauroyl glycine TEA	90				
Lauroyl glycine Na		92			
Myristoyl glycine K			90		
Lauroyl alanine TEA				90	
Lauroyl β-alanine K					90
Results:					
Bubble volume (mL)	300	310	320	315	300
Creaminess of bubbles	v.c.	v.c.	v.c.	v.c.	v.c.
Retention of bubbles	98	95	99	97	00
Touch of bubbles	Good	Good	v.g.	v.g.	v.g.
Jarring feeling	v.g.	v.g.	v.g.	v.g.	v.g.
Sliminess	Good	Good	Good	Good	Good

v.c. = very creamy; v.g. = very good.
Source: Ref. 63.

ditions that may inactivate them. Nonionic PBS can potentially promote enzyme activity by either providing a protective effect or enhancing substrate adsorption.

Protease-containing liquid aqueous detergents are well known, especially in the context of laundry washing, where they attack protein-based stains. A commonly encountered problem in such protease-containing liquid aqueous detergents is the degradation second enzymes (e.g., lipase, amylase, and cellulase, or protease itself) by the proteolytic enzyme in the composition. As a

TABLE 7 General Properties of the *N*-Acyl Glutamates

Fatty acid moiety	Salt	Form	Solubility, 40°C (1% active matter)	Purity (%)
Lauroyl	Mono Na	White	Soluble	>93
Myristoyl	Mono Na	White	—	>93
Cocoyl/tallowyl	Mono Na	White	Soluble hazy	>93
Stearoyl	Mono Na	White	Soluble with precipitate	>93

Source: Ref. 65.

result, the stability of the second enzyme or the protease itself in the detergent composition is affected and the detergent composition consequently performs less well. In response to this problem, it has been proposed to use various protease inhibitors or stabilizers, such as benzamidine hydrochloride, lower aliphatic alcohols, a mixture of a polyol, a boron compound, or aromatic borate ester.

Labeque (Procter & Gamble Co.) et al. [68,69] proposed the use of a low level of peptide trifluoromethyl ketones as reversible protease inhibitors in aqueous liquid detergent composition. This approach is attractive and particularly critical in the formulation of concentrated liquid detergent compositions.

C. Wool Finishing

Anionic N-acylated L-amino acids, oligopeptides, and protein hydrolysates can be used in the wool finishing process. An increased bath exhaustion and a more intense coloration of wool were obtained during the dyeing of wool by the addition of N-acyl-L-amino acids or commercially available protein–fatty acid condensates [70].

D. Drug Delivery

The potential for the use of novel PBS in drug delivery is interesting and needs research attention. Recently, several glutamic acid dialkyl amides with varying alkyl chain length bearing a variety of peptides (1–4 amino acids) {peptide-glu-$(NHC_nH_{2n+1}, n = 12, 14, 16)$} have been rapidly and efficiently converted into high-axial-ratio microstructures (HARMs) in aqueous buffer at physiological pH and ionic strength or in buffer containing methanol or ethanol. Helical ribbons and tubular HARMs were produced that were stable for as long as 6 months below the phase-transition temperatures of the compounds. Such lipopeptide particles can retain their morphology long enough in vivo to be useful as drug delivery vehicle axial-ratio microstructures from peptides modified with glutamic acid dialkyl amides [71].

E. Antimicrobial Agents

Many natural substances possess antibacterial and/or indiscriminate cytotoxic properties. For example, these include PGLa (frog skin), defesins (human phagocytes), cecropins (silkmoth pupae or pig intestine), apidaccins (honeybee lymph), melittin (bee venom), bombinin (toad skin), and the magainins (frog skin), which consist primarily of protein and a system of cellular immunity in the producing organism. Knowledge of these natural protein-based antimicrobial agents and an understanding of their modes of action have been and will continue to be useful in developing new antimicrobial PBS.

Previous chapters have amply demonstrated that several amino acid surfactants and peptide surfactants have good antimicrobial activities. Understandably, several of the antimicrobial surfactants are still at the experimental stage, but it is important to note that many commercial PBS have been widely used as efficient and safe antimicrobial agents. The biological activity of the PBS antimicrobial activity can range from being specifically bacteriocidal or fungicidal.

Some studies have been devoted to understanding the structural basis of the antimicrobial activity of certain PBS. This subject is discussed in some detail in previous chapters, and the reader is referred to sections in the various chapters. Here, we point to perspectives based on amino acid sequences. Lee et al. [72,73] described the preparation of basic model synthetic peptides containing highly conserved amino acid sequences (leu, ala, lys, and arg) possessing antimicrobial activity especially against gram-positive bacteria. The antimicrobial activity of these products is parallel to their alpha helix content and correlated with the position of the basic and hydrophobic groups in the sequence. Thus, high alpha-helix content corresponds to high biological activity and peptides, where the hydrophobic groups and the hydrophilic cationic groups are segregated on the face of the helix exhibit the most potent biological activity. Houghten and Blondelle [89] describe that there is no antimicrobial activity when the length of the peptide exceeds 25 amino acids or contain fewer than 8 amino acids and that the most active sequences contain hydrophilic content of between 50 and 60%. Furthermore, when the charge content of the sequence exceeds 54%, the oligopeptide loses potent activity against gram-negative bacteria even though an amphiphilic secondary structure is maintained. Haynie [72] provided a series of non-natural oligopeptides that share a common amino acid sequence, referred to as the *core oligopeptide*, that have activity against gram-negative and gram-positive bacteria and against yeast but that are not indiscriminately cytotoxic. Here, the term *core oligopeptide* refers to the peptide given by the formula (leu-lys-lys-leu-leu-lys-leu)$_n$, wherein the amino acid sequence occurs from left to right as shown or from right to left and where $n = 1$, 2, or 3. The core oligopeptide may be as short as 7 amino acids and as long as 21, where a length of at least 14 amino acids is preferred. It is found that the N-addition analogues have greater antimicrobial activity than the core oligopeptides that have an amphiphilic helical secondary structure, resembling naturally occurring antimicrobial peptides. The experimental results showed that the minimal inhibition concentration (μg/mL) of core oligopeptide was 15.6 for *E. coli* and 31.2 for *S. aureus*, while that of N-addition analogues was 7.8 for *E. coli* and 15.6 for *S. aureus* [74].

V. PROTEIN-BASED SURFACTANT PROSPECTS IN FOOD APPLICATIONS

Many foods contain mixtures of oil-soluble and water-soluble components; such foods invariably require surface-active agents. In a broad sense, special ingredients like surface-active agents, also called emulsifiers or stabilizers, are needed to ensure uniform quality and shelf life stability of food emulsions and foams. Generally, the function of surface-active materials in food systems can be one or more of the following: (1) to promote emulsion stability, control agglomeration of fat globules, and stabilize aerated systems; (2) to improve texture and shelf life of starch-containing products by complex formation with starch components; (3) to modify the rheological properties of wheat doughs via interactions with gluten proteins; (4) to improve the consistency and texture of fat-based products by controlling polymorphism and the crystal structure of fats [75].

Food-grade emulsifiers are esters of edible fatty acids originating from animal or vegetable sources and polyvalent alcohols like glycerol, propylene glycol, sorbitol, and sucrose [75]. These products can be modified by making derivatives with ethylene oxide or by esterification with acetic acid, diacetyl tartaric acid, succinic acid, citric acid, or lactic acid, which makes it possible to tailor-make surface-active materials with specific properties [75].

As amply discussed in Chapter 2 [Secs. III, IV], proteins from different sources can be modified to produce a wide variety of PBS. Some of the techniques used in protein modification have been indicated in Chapter 5 [Secs. II–VII], with reference to PBS produced via a degradation process such as proteolysis and deamidation, in addition to the cases of the surfactants produced by covalent attachment of nonprotein moieties. Enzymatic modification of proteins with specific reference to food application has been discussed in Chapter 1, Section II.B. Briefly, it was hypothesized that by reacting a hydrophilic protein as the substrate with highly hydrophobic amino acid ester as the nucleophile, a product with amphiphilic properties would result from the localized regions of hydrophobicity [76]. To obtain adequately hydrophilic proteins as substrates, succinylation could be used to modify the proteins prior to their use as substrates for the one-step process [76]. To obtain adequately hydrophobic nucleophiles, amino acid esters with specific functional properties were produced [76]. For example, application of the papain-catalyzed one-step process for L-norleucine n-dodecyl ester attachment to succinylate α-s_1-casein yielded a surface-active 20-kDa product with increased emulsifying activity compared to α-s_1-casein or succinylated α-s_1-casein [77]. The one-step process has been applied to prepare PBS with different hydrophile–lipophile-balance (HLB) values by using as nucleophiles L-leucine, n-alkyl esters of varying chain length and commercially available proteins as hydrophilic substrates [76]. The commercial proteins used in the experiments were gelatin, fish protein concentrate, soy protein isolate, casein, and ovalbumin. Generally, the succinylation (acylation) of these proteins caused an improvement in whippability, foam stability, and emulsifying activity [76]. Furthermore, the attachment of leucine alkyl ester of short chain length (C_4) to the succinylated proteins was more effective in enhancing their functional properties than was longer chain length (C_{12}) [76].

As seen in Chapter 4 [Sec. II], extensive research has been done throughout the past century in the area of PBS, especially leading to the commercial development of N-acyl amino acids. The salts of long-chain N^α-acyl peptides have been finding increasing utility in detergents, cosmetics, pharmaceuticals, and foods, as cited in Chapter 6 [Sec. I]. A number of N-acylated peptides with different hydrophobic contents are manufactured commercially from hydrolyzed animal protein (collagen and keratin) or from plant proteins (soy, wheat).

The foregoing is intended to provide some insight into the potential research and application opportunities of PBS in the food sector. The surface activity of food proteins is regarded as the most important property, for it influences the quality and utility of emulsions and foams. Accordingly, enhancement of the surface activity of food proteins is very important and can be achieved by hydrophobic modification. The contribution of amino acids or peptides per se to the overall hydrophobicity of food proteins is limited, compared to the contribution of a hydrophobic tail in classic surfactants, such as ethoxylated fatty acids [78]. Therefore, it can be expected that the incorporation of long hydrocarbon chains will affect significantly the surface activity of proteins.

It is our opinion that progress has been slow in developing commercial PBS for food applications. Acylation processes (acetylation and succinylation) have been applied successfully to some extent to improve the functionality of proteins or their hydrolysates. What seem to be needed now are novel approaches (chemical or enzymatic) to develop new surfactants with enhanced surface activity and utility for food applications. The trend toward safer emulsifiers will justify more research and development in enzyme technology for the preparation of new surfactants.

VI. POTENTIAL APPLICATIONS OF PROTEIN-BASED SURFACTANTS IN AGRICULTURAL SECTORS

The prospects for PBS in agricultural sectors are seemingly huge, especially in crop protection via pesticide application, biomass conversion, and animal-feed-utilization improvement.

Currently, the active ingredients and adjuvants of the products used for crop protection in agriculture are nonrenewable (generally of petrochemical origin) and present industrial and environmental risks [79]. The role of surfactants in modifying pesticide behavior has been reviewed on a number of occasions over the past 25 years [References in Ref. 80]. Many published reports have shown that the incorporation of surfactants into pesticide sprays improved efficacy [80]. Since most surfactants used in pesticide formulations are petroleum based, it is conceivable that the growing propensity toward products that are environmentally friendly will work in favor of PBS as potential replacements.

On the other hand, the application of surfactants in biomass conversion and feed-utilization improvement is mostly at the stage of research and development. Reports on surfactant utility in biomass conversion [81–85] and feed utilization [86] are relatively scanty but growing.

The following are discussions on the current and/or potential roles of surfactants in pesticide formulation and crop protection, biomass conversion, and feed utilization.

A. Surfactant Roles in Pesticide Formulation and Crop Protection

Surfactants are often incorporated into pesticide formulations to function as emulsifiers, wetting agents, and dispersing agents, especially to facilitate the dispersion and distribution of insoluble active ingredients during spraying. The activity of compounds used in spray applications critically depends on the deposition and adhesion of the active ingredients involved [87]. In nearly all cases, active ingredients are dispersed or suspended in an aqueous medium before being sprayed onto the plant surface [87]. Due to the hydrophobic character of the plant surfaces involved, these aqueous solutions produce drops that have a tendency to "rebound" (bounce) off the surface, which leads to poor adhesion and delivery of the active ingredient [86]. Often, less than 50% of the initial spray is retained by the plant surface (e.g., foliage), which leads to both increased costs and contamination risks [87]. For this reason, surfactants are added to the spray tank to aid droplet adhesion and the wetting of foliage [80]. The main effect of the surfactant will be to reduce the contact angle, thereby encouraging larger droplets to adhere.

A number of commercial formulations, which contain substantial quantities of surfactant for enhancement purposes, are now available [80]. Most are aqueous solutions of herbicides, e.g., paraquat and glyphosate; but there are examples of fungicides, which have limited solubility in water, e.g., ethirimol and diclobutrazol. The most commonly used surfactants are nonionics [80], alkylphenol ethoxylates, alkanol ethoxylates, and alkylamine ethoxylates. Another example is the Roundup™ formulation, which contains glyphosate as an active ingredient and surfactant, tallow amine, $C_{18}N[(CH_2CH_2O)_{15}-H]_2$. The concentrated product contains 5–15% of surfactant and is diluted to ~0.1% upon application. A solution containing tallow amine (e.g., Ethomeen T25, a cationic) and glyphosate is sprayed onto the plant (leaves). The surfactant may initially dissolve the surface wax and penetrate through the cuticles and membranes to the cell walls. Evidence indicates that tallow amine causes chemical abrasion of the membrane and possibly of the cell walls, resulting in a porous environment. This facilitates the effective delivery of glyphosate to the cells and causes the inhibition of the targeted enzyme(s) and ultimate plant death.

B. Surfactant Roles in Biomass Conversion

Surfactants have been reported to enhance enzyme stability and hydrolysis. Certain surfactants have been shown to increase the rate of hydrolysis as well as prolong the life of the enzyme [85]. Many investigators have studied the effective utilization of cellulosic materials, with an emphasis on enzymatic conversions of cellulosic materials to glucose [84]. Cellulosic materials (e.g., plant biomass) are annually renewable resources [82]. Utilization of the glucose stored in cellulosic substrates for the production of industrial chemicals and microbial protein requires the previous hydrolysis of the polysaccharides by acid or enzymatic treatment [82]. Limitations in acid treatment have favored enzymatic hydrolysis. However, cost analysis revealed that enzyme production accounts for about 60% of the process cost of manufacturing sugar from waste cellulosic materials [82]. Therefore, recycling and optimizing enzyme usage during hydrolysis would substantially improve its economics. This has prompted several studies [81–85] aimed at exploring the potential of surfactants in enhancing enzyme hydrolysis and stability during cellulose saccharification. Ooshima et al. [84] showed that nonionic (Emulgen 147, Tween 20, Tween 81), amphoteric (Anhitole 20BS), and cationic (Q-86W) surfactants enhanced the saccharification, while anionic (Neopelex F-25) surfactant did not. Further, it was noted that cationic (Q-86W, 0.008%) and anionic (Neopelex F-25, 0.001%) denatured cellulase. Tween 20 was the most effective in enhancing saccharification.

Other studies [85] have shown that cationic bacitracin, anionic AOT, and nonionic Tween 80 all enhanced cellulose hydrolysis, implying that the charge of the surfactant was not an important consideration. In fact, Tween 80 (0.1%) increased the rate and extent of saccharification by up to 40% [82,85]. The structure of the hydrophilic head group of the surfactant also had little significance [85]. Those with a sugar group, sophorolipid and rhamnolipid, worked well, as did bacitracin, which has a peptide hydrophilic group [85].

To rationalize the mechanism by which surfactants influence cellulose hydrolysis, several hypotheses have been proposed. Because hydrolysis involves the transport of enzymes and soluble sugars between the solid substrate and the bulk reaction fluid, it follows that modification of the interfacial energy may have some impact on this transfer [85]. The enzymatic hydrolysis of cellulose is a heterogeneous reaction, with soluble cellulolytic enzymes converting solid cellulose into soluble sugars [85]. The first step in this reaction is adsorption of cellulolytic enzymes from the reaction medium onto the surface of the cellulosic substrate [85]. It has been reasoned that perhaps certain surfactants may facilitate mass transfer and the cellulase adsorption–desorption process. Another school of thought has implicated the hydrophobic–hydrophilic nature of cellulose, suggesting a hydrophobic interaction between surfactant and cellulose. Cellulose is known to have hydrophobic regions, which may attract surfactants by hydrophobic interaction, resulting in alignment of the polar head groups in the surrounding aqueous medium and an increase in hydrophilicity of the substrate [85].

According to Helle et al. [85], surfactants probably have two effects on cellulose hydrolysis. The first is the disruption of cellulose, making the cellulose more accessible to the enzymes. The second is prevention of the adsorbed enzyme from becoming inactive, probably by facilitating its desorption before inactivation occurs.

As indicated previously, the cost of cellulase enzymes account for a large proportion (60%) of the total cost of cellulosic material conversion [82]; the use of surfactants will result in substantial savings [85]. This is an area that awaits PBS application.

C.   Surfactant Roles in Feed-Utilization Improvement

Feed remains the most important cost of animal production; as a consequence, strategies are needed that improve feed efficiency and reduce the pollution associated with manure. During the last two decades, the feed efficiency of ruminants has remained unchanged, while that of pigs and chickens has been dramatically improved by improved feeding strategy and breeding.

The normal diet of the ruminant animal is forage [88], which provides fiber for the maintenance of healthy rumen function, energy, and protein for milk production. Forage includes grasses, legumes, and cellulolytic by-products of agricultural production [88]. These are fed fresh as either pasture or green chop; in a dry form as hay; or in a preserved state as silage. The ability to utilize these materials as sources of nutrients is possible only as a result of pregastric bacterial fermentation in the rumen, the nonfundic portion of the animal's stomach [88]. Here, bacterial action reduces the complex structural carbohydrates, cellulose, hemicellulose, and lignin, and the associated non-structural carbohydrates, pectin, starches and sugars, to either fatty acids or more chemically simplistic carbohydrate forms, which are subjected to gastric action in the fundic stomach and small intestine [88]. Forage utilization is limited, due to the slow rates of rumen fermentation. The slow rate has been attributed in part to the limited accessibility of enzymes to the substrates. The limited accessibility is presumed to be due to the steric hindrance imposed by the structure and association of the cell wall polysaccharides.

The PBS can potentially enhance forage digestibility as feed additives or forage conditioner to provide the following functions: (1) facilitate substrate accessibility by rumen bacteria via modification of the interfacial energy; (2) increase the wettability and dispersibility of substrates; (3) provide enzyme stability in cases where enzyme supplementation is practiced.

REFERENCES

1. M. Takehara, I. Yoshimura, K. Takizawa, and R. Yoshida, J. Am. Oil Chem. Soc. 49:157 (1972).
2. J. Falbe, in: Surfactants in Consumer Products, Springer-Verlag, New York, 1987.
3. C. Robbins, in: Chemical and Physical Behavior of Hair, Van Nostrand Reinhold, New York, 1979, Chap. 5.
4. M. Takehara, H. Moriyuki, I. Yoshimura, and R. Yoshida, J. Am. Oil Chem. Soc. 49:143 (1972).

5. Y. Kazuyuki, T. Kaoru, H. Hajime, and M. Yoshihisa, U.S. Patent 4,824,604 to Kao Soap Co. (1989).

6. Mild Anionic Surfactants-Acyl Glutamate, Technical Bulletin, Clariant, 1999.

7. M. Kanari, Y. Kawasaki, and K. Sakamoto, Abstract of 17th IFSCC International Congress, Yokohama, 1992, p. 462.

8. C. H. Lee, Y. Kawasaki, and H. I. Maibach, Contact Dermatitis. 30:205 (1994).

9. M. Takehara, H. Moriyuki, A. Arakawa, I. Yoshimura, and R. Yoshida, J. Am. Oil Chem. Soc. 50:227 (1973).

10. M. Takehara, I. Yoshimura, and R. Yoshida, J. Am. Oil Chem. Soc. 51:419 (1974).

11. M. Kanari, Y. Kawasaki, and K. Sakamoto, Abstract of 17th IFSCC International Congress, Yokohama 1992, p. 462.

12. O. Tachizawa, M. Kawaguchi, and K. Sotoya, U.S. Patent 5,284,602 to Kao Soap Co. (1994).

13. Mild Anionic Surfactants-Acyl Glutamate, Technical Bulletin, Clariant, 1999.

14. Hamposyl Surfactants, Technical Bulletin, Hampshire Chemical Corp., 1982.

15. R. Woodbury, R. Gaudette, and F. Wood, U.S. Patent 5,710,295 to Hampshire Chemical Corp. (1998).

16. M. J. Rosen, in: Surfactants and Interfacial Phenomena, Wiley, New York, 1978.

17. M. M. Rieger, ed., Surfactants in Cosmetics, Marcel Dekker, New York, 1985, Chaps. 9, 11.

18. K. Sano and T. Hattori, U.S. Patent 5,529,712 to Ajinomoto Co. (1996).

19. T. Yuichi and others, Japan Patent 4,001,129 to Kanebo (1992).

20. E. Magdna, V. Lee, W. William, L. Campbell, and P. Lynn, World Patent 95,11000 to Procter & Gamble (1995).

21. S. Kazutami, O. Eiko, H. Tatsuya, and Y. Hideki, European Patent 826,766 to Ajinomoto (1998).

22. Y. Ei, Japan Patent 10,036,221 to Noevir Co. (1998).

23. A. Koji, Japan Patent 8,020,529 to Shiseido Co. (1996).

24. A. Shinichi, O. Toru, and N. Hideo, Japan Patent 9,301,846 to Shiseido Co. (1997).

25. T. Kazutoshi, Japan Patent 9,188,605 to Noevir Co. (1997).

26. O. Kimihiko, Y. Hisashi, and T. Yoshihiro, Japan Patent 8,337,515 (1996).

27. T. Takehito, Japan Patent 8,217,989 (1996).

28. O. Katsumoto and N. Masayuki, Japan Patent 8,067,609 to Shiseido Co. (1996).

29. N. Norimoto and others, Japan Patent 7,277,937 to Ajinomoto Co. (1995).

30. The Lamepon S-Range-Protein Fatty Acid Condensates, Technical Bulletin, Henkel, 1986.

31. C. Fox, ed., Emulsions and Emulsion Technology, Marcel Dekker, New York, 1974.

32. N. Osamu and O. Katsuyoshi, World Patent 93,24100 to Fujisawa Pharma (1993).

33. Y. Masato et al., Japan Patent 5,078,224 (1993).

34. D. Su, U.S. Patent 5,693,604 to Colgate-Palmolive (1997).

35. Cationic Protein Derivatives, Technical Bulletin, Croda, 1985.

36. Lamequat L, Technical Bulletin, Henkel, 1994.
37. S. Hiroyuki and K. Yoshiaki, Japan Patent 9,012,522 to Ajinomoto Co. (1997).
38. Y. Masato and others, Japan Patent 10,175,830 to Seiwa Kasei (1998).
39. T. Masaru and others, Japan Patent 10,175,824 to Lion Corp. (1998).
40. Wella, DE Patent 3,527,974 to Wella AG.
41. H. Ishii, N. Mikami, and K. Sakamoto, J. Soc. Cos. Chem. 47:351 (1996).
42. I. Tomomichi, F. Shigetoshi, and K. Tohru, European Patent 443,592 to Ajino-moto Co. (1991).
43. I. Tomomichi, F. Shigetoshi, and K. Tohru, U.S. Patent 5,474,778 to Ajinomoto Co. (1995).
44. H. Tatsuya and M. Naoko, European Patent 538,764 to Ajinomoto Co. (1993).
45. B. Bluestein and C. Hilton, ed., Amphoteric Surfactants, Marcel Dekker, New York, 1982.
46. Deriphat, Technical Bulletin, Henkel, 1996.
47. Amino Acid Gelatinization Agent, Technical Bulletin, Ajinomoto, 1988.
48. C. Motley, World Patent 96,04886 to Procter & Gamble (1996).
49. P. Sawin and J. Luebbe, World Patent 94,24997 to Procter & Gamble (1994).
50. G. Guskey and T. Orr, U.S. Patent 5,744,130 to Procter & Gamble (1998).
51. T. Kazutoshi and A. Akari, Japan Patent 10,114,618 to Nevi Co. (1998).
52. M. Hidden and I. Atsuko, Japan Patent 10,182,334 to Nevi Co. (1998).
53. M. Hidden and I. Atsuko, Japan Patent 10,182,335 to Nevi Co. (1998).
54. Y. Yasuyuki and I. Atsuko, Japan Patent 10,109,923 to Noevir Co. (1998).
55. T. Shin, Japan Patent 10,045,563 to Noevir Co. (1998).
56. I. Atsuko, Japan Patent 10,045,557 to Neovir Co. (1998).
57. P. Bert and R. Barbier, 5th Int. Cong. Det. 1:175 (1969).
58. J. L. Periman and R. D. Norden, U.S. Patent 3,031,408 (1962).
59. H. Marumo, German Patent 2,222,899 (1972).
60. R. Baur et al. (BASF), U.S. Patent 4,827,014 (1989).
61. Becke, U.S. Patent 3,962,319 (1976).
62. O. Tachizaka and M. Kawaguchi, U.S. Patent 5,284,602 to Kao Co. (1994).
63. K. Sano and T. Hattori, U.S. Patent 5,529,712 to Ajinomoto Co. (1996).
64. Y. Kazuyuki, Japan Patent 1,090,295 to Kao Co. (1989).
65. M. Schmidt, Local Project No. 346,862 (1992).
66. K. Makoto, S. Koshiro, and M. Takashi, Japan Patent 4,253,942 to KaoCorp. (1992); U.S. Patent 5,770,553 (1998).
67. M. Moriyama, H. Tanabe, and K. Yasushi, U.S. Patent 5,712,232 to Kao Corp. (1998).
68. R. Labeque and M. Melver, U.S. Patent 5,582,762 to Procter & Gamble (1996).
69. R. Labeque, M. Melver, et al., U.S. Patent 5,840,678 (1998).
70. K. Schafer, J. Wirsching, and H. Hocker, Fett. Lipid 99(6):217 (1997).
71. K. C. Lee, A. Lukyandy, M. Beld, and P. Yager, Biochimica Biophysica Acta Biochembranes 1371(2):168 (1998).
72. S. L. Haynie, U.S. Patent 5,847,047 (1998).
73. K. C. Lee et al. Biochem. Biophys. Acta 862,211 (1986).

74. O. P. Guptaand and C. Manjula, Surf. Technol. 22(2):175 (1984).
75. N. Krog, J. Am. Oil Chem. Soc. 54:124 (1977).
76. S. Nakai and E. Li-Chan, in: Hydrophobic Interactions in Food Systems, CRC Press, Boca Raton, FL, 1988, pp. 145–151.
77. S. Arai and M. Watanabe, Agric. Biol. Chem. (Japan) 44:1979 (1980).
78. S. Magdassi and O. Toledano, in: Surface Activity of Proteins: Chemical and Physicochemical Modifications (S. Magdassi, ed.), Marcel Dekker, New York, 1996, p. 40.
79. S. Palmieri and G. Venturi, Agro-Food-Industry Hi-Tech. 5:51 (1999).
80. D. Seaman, in: Solution Behavior of Surfactants: Theoretical and Applied Aspects (K. L. Mittal and E. J. Fendler, eds.), Plenum Press, New York, 1982, pp. 1365–1380.
81. E. T. Reese, J. Appl. Biochem. 2:36 (1980).
82. M. Castonon and C. R. Wilke, Biotechnol. Bioeng. 23:1365 (1981).
83. M. H. Kim, S. B. Lee, D. D. Y. Ryu, and E. T. Reese, Enz. Microb. Technol. 4:99 (1982).
84. H. Ooshima, M. Sakata, and Y. Harano, Biotechnol. Bioeng. 28:1727 (1986).
85. S. S. Helle, J. B. D. Sheldon, and D. G. Cooper, Biotechnol. Bioeng. 42:611 (1993).
86. J. F. Tobey, J. S. McGee, C. W. Cobb, and W. Cortner, U.S. Patent 5,662,901 (1997).
87. V. Bergeron, J-Y. Martin, and L. Vovelle, Agro-Food-Industry Hi-Tech. 5:21 (1999).
88. W. E. Julien, U.S. Patent 5,709,894 (1998).
89. R. A. Houghten and S. Blondelle, WO Patent 9,201,462 (1992).

10

Current Market Developments and Trends in Amino Acid- and Protein-Based Surfactants

KAZUTAMI SAKAMOTO   Ajinomoto Company, Inc., Kanagawa, Japan

I. GENERAL FEATURES OF THE SURFACTANT MARKET

Surfactants are molecules with distinguishing structural and functional characteristics, which can be described as amphiphilic structure and property. They have been commonly explained as the molecules with dual affinities, to oil (hydrophobicity) and to water (hydrophilicity); thus surfactants tend to accumulate at the interfaces of oil and water, to change the physicochemical properties of a system. These properties, i.e., surface activities, are exploited in our daily life and in industrial processes in the form of agents that foam, emulsify, create dispersion, clean, wet, etc. With these functions, surfactants are indispensable, and worldwide demand and supply are huge and growing.

Surfactants are consumed in various industrial applications directly or indirectly connected to our daily lives. Market shares of surfactants in the United States in 1997 are shown in Figure 1. Consumption of household cleansing products, personal care products, and industrial and institutional (I&I) cleansing products exceeds more than 60% of the 5.14 billion pounds of surfactants produced in the United States in 1997 [1]. In Japan, I&I consumption exceeds over 50% of total production [2]. Anionic surfactants are the most common type utilized for every application because of their highly effective detergency and relatively lower cost. Thus anionic surfactants dominate household cleaning products.

Nonionics are expanding their share, especially in I&I cleansing products, where the weakness of nonionics in foaming power is not a determinant prop-

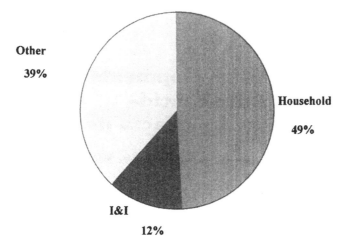

Other

39%

Household

49%

I&I

12%

FIG. 1 Surfactant consumption in the United States (1997), estimated from SRI data. Personal care excludes soaps and syndet bar.

erty or a disadvantage for the actual application. In Europe, nonionics are gaining share rather rapidly (5% per year annual growth) and are even estimated to exceed the share of anionics by 2002. Amphoterics show an even faster annual growth, up to 6%, although the current market share for amphoterics is much smaller than for anionics and nonionics, that is, only 14 million pounds out of 3 billion pounds consumed for household and I&I use in the United States [1]. In the personal care market, such trends are more evident, as seen in Figure 2, where the market share of amphoterics is almost half of nonionics in western Europe. The development of new surfactants and applications are focused mainly on functionality and cost reduction. Intensive research and development under competitive circumstances established today's huge surfactant market.

Recently, it has been noticed that excessive use of surfactants sometimes causes problems to humans and the environment. This concern opened the door for the nonionics and amphoterics to penetrate into the anionics market due to their better functionality and lower adverse impact on humans and the environment.

Therefore, environmental concerns and low-toxicity requirements are providing new incentives and opportunities for the development of specialty surfactants. Amino acid– and protein-based surfactants (AAS, PBS) are the candidates most likely to fulfill these requirements because of their structural and

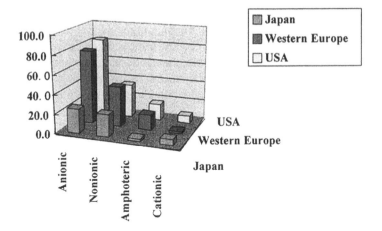

FIG. 2 Surfactant consumption (thousand metric tons) for personal care products (1997), estimated from SRI data, excluding soaps and syndet bar.

functional likeness to the natural system. Although the current production scales of amino acid– or protein-based surfactants are too small to be reflected in the statistics, market demand and growth are remarkable. To provide a better understanding and to encourage further developmental effort, the characteristics of and prospects for amino acid–related surfactants are explained in this chapter.

II. MILDNESS AS A MEASURE OF CONCERN IN PERSONAL CARE PRODUCTS

The history of surfactant applications in personal care products can be traced back to the ancient Egyptians. Soap was the only available surfactant in those early centuries. It was not until the Second World War that synthetic surfactants were developed and formulated into daily personal care products. Diversification and expansion of synthetic surfactants based on the technological and economical advances of the oleochemical and petrochemical industries made surfactants indispensable for daily life in industrialized countries such as the United States, western Europe, and Japan, and the same trends are expanding throughout the world. We now face a controversial problem: cleansing ability as the primary function of surfactants versus gentleness to humans and environmental impact.

Synthetic surfactants such as sulfates, sulfonates, and ethoxylates were developed to be superior in surface activity, along with cost effectiveness. An abundance of consumer products with high-potency cleansers and excessive shampooing and cleansing often cause skin troubles, especially for people with sensitive skin. As a result, mildness has become a measure of concern, and even of market opportunity, in the personal care industry. For this reason, product formulators are looking for surfactants with superb mildness. Amino acid- and protein-based surfactants are the molecules most likely to satisfy such market demand.

III. HISTORY OF AND TRENDS IN MILD SURFACTANTS

Acylproteins have been used in personal care formulations for many years, mainly because of their perceived attractiveness on the ingredient label. The acylproteins are usually derived in part from hydrolyzed protein or peptide. The acylproteins are rather expensive, because they are manufactured on a small scale. Another drawback related to the use of acylproteins is that they always contain mixtures of various reaction products and residual raw materials. For these reasons, extensive study of their mildness was required to warrant the application of acylproteins in personal care products.

Acylsarcosinates were the first commercial surfactants applied in personal care products at a meaningful concentration. The main application of acylsarcosinate was as a foaming additive for toothpaste but it was never a leading surfactant to create mildness.

In the late 1960s, extensive research and development was conducted on acylglutamate as a mild anionic surfactant. The evaluation included a primary skin and eye irritation test, a maximization test, and an acute and even chronic toxicity test for animals. This was followed by clinical studies, including patients with skin troubles, as shown in Table 1 [3–6]. Ajinomoto Co. Inc. launched acylglutamate in Japan in 1972 and applied it to the solid bar cleanser, which was launched by Yamanouchi Pharmaceutical Co. through pharmacy channels and was purchased mainly by patients with a doctor's recommendation. The bar cleanser has three significant mildness features: (1) it is nonalkaline (actually weakly acidic, the same as skin pH); (2) it is nonallergenic (hypoallergenic) and nonfragrant (probably the first product involving the concept of hypoallergenic; (3) it is nonirritating [7].

Several years later, as a line extension, paste-type or cream-type facial cleansers were launched with the same idea, which helped expand the mild-cleanser market in Japan. At the same time there were epidemic problems of hyperpigmentation by some cosmetic products in Japan. Nakayama et al. [8,9] proposed an allergen-controlled system (ACS) to treat such patients, and

TABLE 1 Clinical Safety Evaluation for Acylglutamate

Test	Object	Material	Result	
Closed patch test	40 healthy males	AGS 0.1%, 0.5%, 1%, 3%	0.1%, same as water 5%, Extremely weak primary irritation	
Percutaneous sensitization Phototoxicity and photo-sensitization			No abnormality found No abnormality found	
Closed patch test	40 healthy males	CS-11, LS-11, GS-11 CT-12, LT-12, SLS and Soap (0.5% & 1.5% aq.)	*Extent of irritation* Almost none (0–1.0) Slight (0.26–1.0)	*Material* Blank (no treatment) LS-11, CT-12, LT-12 Soap (all 1.5%) Distilled water
			Light (1.1–2.0)	LS-11, CT-12, LT-12 (all 0.5%) CS-11, GS-11 (both 0.5 & 1.5%)
			Moderate (2.1–5.0) Strong (>5.0)	Soap (0.5%), SLS (0.5%) None SLS (1.5%)
Closed patch test	47 Patients with eczema and derma-titis	AGS, SLS, Soap (1.5% solution)	Irritation order Water = AGS = soap < SLS	
Use test Hand, facial washing, and bathing	48 healthy males Patients with eczema and dermatitis as their stage toward re-mission	AGS solid bar AGS solid bar	No abnormality found Safely applicable	
Use test Hand, facial washing, and bathing	Patients with eczema and derma-titis	AGS solid bar	Safely applicable	

LS-11: monosodium lauroylglutamate; CS-11: monosodium cocoylglutamate; LT-11: monotriethylammonium lauroylglutamate; CT-12: monotri-ethylammonium cocoylglutamate; GS-11: monosodium cocoylglutamate; AGS: monosodium cocoyl/stearoylglutamate; AGS solid bar: AGS/cetylalcohol.

cleansers based on acylglutamate played a key role in the system due to its hypoallergenic and mild properties. This phenomenal result led other pharmaceutical companies to launch similar mild cleansing products. Professional hair care manufacturers and cosmetic companies followed the trend by establishing a mild-cleanser market. In 1980 the mild-cleanser market reached 50 billion yen, or about 5% of the total soap market, which was a significant part of the personal care cleanser market at that time [10].

This trend was further boosted in the early 1980s by Kao's launch of a facial cleanser formulated with monoalkylphosphate, with a similar mildness claim. Shiseido expanded the mildness concept to the hair care market in the late 1980s, with acylmethyltaurate as a primary surfactant. It is noteworthy that facial cleansers in solid bar and paste products by Yamanouchi Pharmaceutical Co. contained acylglutamate as a primary surfactant and were marketed for over a quarter of a century and merely had claims for their package but almost none for adverse skin troubles, and they continue to hold a significant market share in the mild-cleanser category sold through pharmacies and drug stores in Japan.

In the United States and Europe, the mildness concept is led by isethionate for a cleansing bar by Lever Brothers (Unilever Affiliate), followed by an amphoteric surfactant and nonionic alkylglycocide, which are really secondary surfactants to reduce the potential for irritation with a combination of typical anionics. The utilization of acylamino acid surfactants has just started in Western markets. In Europe, Biersdorf was the first company to use acylglutamate in a product for the mass market, although several upscale cosmetic cleansers pioneered its use on a small scale.

In the United States, Germain Monteil first applied acylglutamate to facial cleansers in the mid-1980s, but the market was not yet ready to acknowledge such mild products at that time. Recently, a major launch has been achieved by Bath & Body Works for their bath product with acylglutamate in the formula.

A. Substantiation of Mildness

Cleansing products, by physicochemical action, are used to wash and remove materials deposited from outside or excreted from the skin. This helps to keep the skin clean and facilitate normal physiological and homeostatic skin functions. Without proper cleansing, residual materials on the skin might decompose or be denatured by oxygen, UV radiation, or microbial transformation to the undesirable material that interferes with skin metabolism or even damages skin, especially sensitive skin, such as atopic or dry skin. On the other hand, surfactant, as a major component of the cleansers, has side effects such as potential irritation because of its nature to strip precious constituents, such as water-soluble natural moisturizing factor (NMF) and skin lipids, which provide a natural emollient to reduce the barrier function. Surfactants cause a further undesirable effect as an irritant when they penetrate the skin. Thus, skin irritation might be an outcome of successive and complicated physical and physiological responses caused by the surfactant as a foreign substance [11,12].

There are many in vivo and in vitro approaches to assessing the irritation potential of surfactants. For overall results, in vivo applications and observations are the most useful and realistic measures. The Draize method has been utilized for many years for this purpose and been found useful for assessing irritation potential. Recently, it has become controversial to utilize the Draize method with animals because of concerns for animal welfare, despite its purpose of isolating caustic materials in order to avoid human exposure. Many in vitro methods have been developed recently that are useful alternatives to the animal tests and even as functional tools for analyzing the cause of irritation [13,14].

Extensive studies have been conducted comparing acylglutamate with other anionic surfactants, along with isolating the possible cause from outside to inside the skin. As shown in Figs. 3a and b [15], acylglutamate has sufficient detergency to remove the surface lipid, e.g., squalene, but does not to strip the constitutive lipid in the stratum corneum, e.g., cholesterol. Thus selective detergency for squalene is higher for acylglutamate compared to other anionics. Also notice that the stripping of NMF was relatively insignificant, as shown in Fig. 4 [16]. From these results, it can be assumed that acylglutamate caused less damage to the skin barrier function. In vitro measurement, by the ESR method, of the mobility and orientation of intercellular lipid bilayers for the human stratum corneum further supported the mildness of acylglutamate (Fig. 5). This preservation of the barrier function is confirmed by in vivo human test by means of transepidermal water loss (TEWL) measurement, which is a well-accepted parameter corresponding to the skin barrier function (Fig. 6) [18].

Skin penetration by acylglutamate has been found rather small compared to that of soap or alkylsulfate. The cytotoxicity of acylglutamate for human keratinocyte was significantly lower than for other anionics, including acyl-methyltaurate, as shown in Fig. 7.

Thus acylglutamate has been demonstrated by several tests to show extreme mildness when tested in vivo with animals and humans or when the actual product is used. Acylmethyltaurate also showed a relatively lower irritation potential by these test methods.

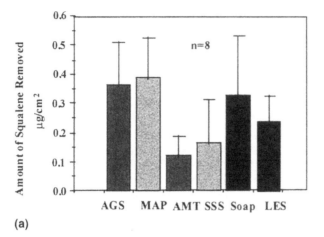

(a)

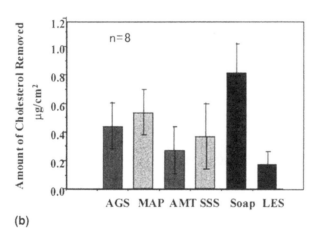

(b)

FIG. 3 (a) Loss of skin surface lipid: The amount of squalene removed from the skin by a 1% surfactant solution represents lipids secreted from the sebaceous glands. (b) Loss of intercellular lipid: The amount of cholesterol removed from the skin by a 1% surfactant solution represents lipids. AGS: acyl glutamate; MAP: monoalkyl phosphate; AMT: acylmethyl taurate; SSS: sulfosuccinate; LES: lauryl ether sulfate. (From Ref. 11.)

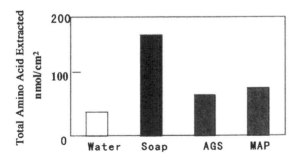

FIG. 4 Loss of NMF: The amount of amino acid extracted from the skin by surfactant. (From Ref. 16.)

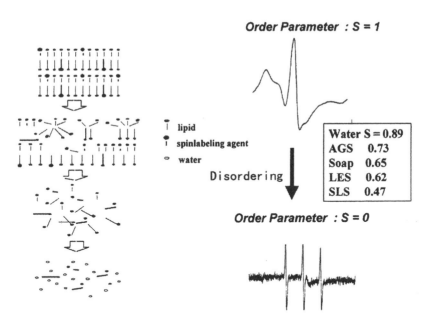

FIG. 5 Disordering of intercellular lipid layers in stratum corneum: Ordered structures of intercellular lipid bilayers in the stratum corneum are measured in terms of the order parameter (*S*) by the in vitro ESR spin-probe method. Intact stratum corneum showed a higher *S* value. With SLS treatment, structural deformation is evidenced by a smaller *S* value. (From Ref. 17.)

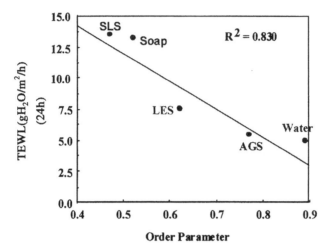

FIG. 6 Destruction of the barrier function by surfactant as measured by in vivo TEWL and ESR (in vitro). Correlation between the order parameter (*S*) and TEWL. (From Ref. 18.)

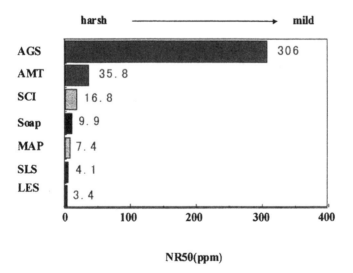

FIG. 7 In vitro safety assessment: Cytotoxicity for human keratinocyte. AGS: acyl glutamate; AMT: acylmethyl taurate; SCI: cocoyl isethionate; MAP: monoalkyl phosphate; SLS: lauryl sulfate; LES: lauryl ether sulfate. (From Ref. 11.)

IV. PRODUCT DEVELOPMENT

A. Anionic Surfactant

Commercially available amino acid– and protein-related surfactants are listed in Table 2. Anionic surfactants include most varieties, and glutamic acid and sarcosine are two major amino acids utilized. Acylglutamates successfully created a mild-surfactant market in Japan, as mentioned before. The reason is not only their extreme mildness but also, and more importantly, their versatile process adaptability for making products with a suitable selection and combination of acyl radicals and type of base and degree of neutralization. Thus, most types of mild cleansing products with a high mildness can be prepared with acylglutamate as a primary surfactant. Regular anionic surfactants do not have this flexibility, because their anionic group is either too water soluble, such as sulfate or sulfonate, or poorly water soluble, such as carboxylate, to accommodate versatile formulations, from clear solution to solid bar, in a single anionic moiety without losing the appropriate cleansing efficiency. Another distinguishing property of acylglutamate is that it leaves skin feeling soft and moist after use, which makes the skin supple.

Sometimes there are difficulties in combining acylglutamate as cosurfactant to the traditional anionic surfactant system in terms of the viscosity and lathering of shampoo or bath cleansers. Acylsarcosinate and acylmethylalaninate can be used as a tertiary cosurfactant or as an alternative to acylglutamate for this purpose [19]. Acylmethyltaurate has also been used as a primary surfactant for mild-shampoo formulation [20,21].

Recently, further developments have enabled better choices for formulators in selecting amino acid–related surfactants. Acylglycinate and acylalaninate make up the first class of this new anionic surfactant; both give a creamy and resiliant foam as a primary surfactant and even as cosurfactant with other traditional anionics. They exhibit synergistic effects with soap for lathering volume and retention. After the use of acylglycinate and acylalaninate the skin feels clean and fresh without the tautness that soap produces. Various types of formulations, i.e., shampoo, facial cleanser, and body wash, can be prepared with the appropriate selection and combination of amino acid residue, fatty acid residue, and counterion [22].

Another class is the amino acid salt of fatty acid. In this case, a basic amino acid, such as lysine or arginine, can be used as a base to neutralize fatty acid instead of a regular organic base, such as ethanolamines. Soaps with these basic amino acids have better foaming properties over a wider pH range than soaps with triethanolamine (TEA). Additionally, these amino acids have suit-

TABLE 2 Amino Acid–based Anionic Surfactants on the Market

INCI or chemical name	Amino acid	Trade name	Supplier	Reference no.[a]
TEA-cocoyl alanate	Alanine	Amilite ACT-12	Ajinomoto	1 (p. 14)
Arginine cocoate	Arginine	Aminosoap AR-12	Ajinomoto	1 (p. 14); 2 (p. 1821)
TEA-palmitoyl aspartate	Aspartic acid			3 (p. 238)
Cocoyl glutamic acid	Glutamic acid	Amisoft CA	Ajinomoto	1 (p. 14); 2 (p. 1821)
Disodium cocoyl glutamate		Amisoft ECS-22	Ajinomoto	1 (p. 14); 2 (p. 1821)
Disodium hydrogenated tallowyl glutamate		Amisoft HS-21	Ajinomoto	1 (p. 14); 2 (p. 1821)
Hydrogenated tallowyl glutamic acid		Amisoft HA	Ajinomoto	1 (p. 14); 2 (p. 1821)
Lauroyl glutamic acid		Amisoft LA	Ajinomoto	2 (p. 1821)
Potassium lauroyl glutamate		Amisoft LK-11	Ajinomoto	1 (p. 14); 2 (p. 1821)
Potassium cocoyl glutamate		Amisoft CK-11	Ajinomoto	1 (p. 14); 2 (p. 1821)
Sodium cocoyl glutamate		Amisoft CS-11	Ajinomoto	1 (p. 14); 2 (p. 1821)
		Hostapon KCG	Hoechst Celanese	2 (p. 2050)
Sodium lauroyl glutamate		Amisoft LS-11	Ajinomoto	1 (p. 14); 2 (p. 1821)
Sodium myristoyl glutamate		Amisoft MS-11	Ajinomoto	1 (p. 14); 2 (p. 1821)
Sodium stearoyl glutamate		Amisoft HS-11P	Ajinomoto	1 (p. 14); 2 (p. 1821)
TEA-cocoyl glutamate		Amisoft CT-12	Ajinomoto	1 (p. 14); 2 (p. 1821)
TEA-lauroyl glutamate		Amisoft LT-12	Ajinomoto	1 (p. 14); 2 (p. 1821)
Caplyroyl glycine	Glycine			2 (p. 1874)
Potassium cocoyl glycinate		Amilite GCK-12	Ajinomoto	
Sodium cocoyl glycinate		Amilite GCS-11	Ajinomoto	
TEA-cocoyl glycinate		Amilite GCT-12	Ajinomoto	
Lysine cocoate	Lysine	Aminosoap LYC-12	Ajinomoto	1 (p. 14); 2 (p. 1821)
Lauroyl methyl alanine	Methyl alanine			
Sodium cocoyl methyl alaninate		Alanon	Kawaken Fine Chem.	3 (p. 359)

Sodium methyl cocoyl taurate	Methyl taurine	Neoscope CN-30	Touhou Chem.	3 (p. 360)
		Nikkol CMT-30	Nikkol Chem.	3 (p. 360)
		Hostapon CT	Hoechst Celanese	2 (p. 2050)
Sodium palmitoyl methyl taurate		Nikkol PMT	Nikkol Chem.	3 (p. 238)
Caplyroyl collagen amino acids	Protein			2 (p. 1874)
Capryloyl hydrolyzed animal keratin				2 (p. 1874)
Capryloyl hydrolyzed collagen				2 (p. 1874)
Capryloyl hydrolyzed keratin				2 (p. 1874)
Capryloyl keratin amino acids				2 (p. 1874)
Capryloyl pea amino acids				2 (p. 1874)
Capryloyl quinoa amino acids				2 (p. 1874)
Capryloyl silk amino acids				2 (p. 1874)
Lauroyl hydrolyzed collagen				2 (p. 2093)
Lauroyl hydrolyzed elastin				2 (p. 2093)
Lauroyl silk amino acid				2 (p. 2093)
Potassium cocoyl hydrolyzed animal protein		Lamepon S	Henkel	1 (p. 107)
Potassium cocoyl hydrolyzed collagen		Maypon 4C	Inolex Chemical	1 (p. 124)
		May-Tein C	Maybrook, Inc.	1 (p. 124)
		Lamepon PA-K	Grunau	2 (p. 2088)
		NIKKOL CCK-40	Nikko Chem.	3 (p. 355)
		Promois ECP	Seiwa Chem.	3 (p. 355)
Potassium cocoyl hydrolyzed soy protein		Promois ESCP	Seiwa Chem.	3 (p. 355)
Sodium TEA hydrolyzed keratin		May-Tein KTS	Maybrook, Inc.	1 (p. 124)
TEA cocoyl hydrolyzed soy protein		May-Tein SY	Maybrook, Inc.	1 (p. 124)
TEA cocoyl hydrolyzed collagen		Lamepon	Grunau	2 (p. 2088)
			Henkel	2 (p. 2088)
		May-Tein CT	Maybrook, Inc.	1 (p. 124)
TEA hydrolyzed animal protein		Promois ECT	Seiwa Chem.	3 (p. 356)

TABLE 2 Continued

INCI or chemical name	Amino acid	Trade name	Supplier	Reference no.[a]
TEA-abietoyl hydrolyzed animal protein		Lamepon PA-TR	Henkel	1 (p. 107)
TEA-cocoyl collagen amino acids		AminoFoam C	Croda Inc.	1 (p. 14)
Ammonium lauroyl sarcosinate	Sarcosine	Hampsyl AL-30	Hampshire Chem. Co.	1 (p. 90)
Cocoyl sarcosine		Hampsyl C	Hampshire Chem. Co.	1 (p. 90)
		Soypon SCA	Kawaken Fine Chem.	3 (p. 357)
		Vanseal CS	R.T. Vanderbilt Co.	1 (p. 207)
Lauroyl sarcosine		Hampsyl L	Hampshire Chem. Co.	1 (p. 90)
		Vanseal LS	R.T. Vanderbilt Co.	1 (p. 207)
Myrisroyl sarcosine		Hampsyl M	Hampshire Chem. Co.	1 (p. 90)
		Vanseal MS	R.T. Vanderbilt Co.	1 (p. 207)
Oleoyl sarcosine		Vanseal OS	R.T. Vanderbilt Co.	1 (p. 207)
Sodium cocoyl sarcosinate		Hampsyl C-30	Hampshire Chem. Co.	1 (p. 90)
		Soypon SC	Kawaken Fine Chem.	3 (p. 357)
		Vanseal NACS-30	R.T. Vanderbilt Co.	1 (p. 207)
Sodium lauroyl sarcosinate		Maprosyl 30	BASF Co.	1 (p. 120)
		Hampsyl L-30	Hampshire Chem. Co.	1 (p. 90)
		Crodasinic LS-30	Croda Inc.	1 (p. 49)
		Closyl LA	Boehme Filatex	1 (p. 48)
		Vanseal NALS-30	R.T. Vanderbilt Co.	2 (p. 2050)
Sodium myristoyl sarcosinate		Hampsyl M-30	Hampshire Chem. Co.	1 (p. 90)
Stearoyl sarcosine		Hampsyl S	Hampshire Chem. Co.	1 (p. 90)
TEA-lauroyl sarcosinate		Soypon SCTA	Kawaken Fine Chem.	3 (p. 357)
		Hampsyl TL-40	Hampshire Chem. Co.	1 (p. 90)
		Hostapon KTW	Hoechst Celanese	2 (p. 2050)
Sodium methyl oleyl taurate	Taurine	Hostapon T	Hoechst Celanese	2 (p. 2050)

[a]Ref. 1: McCutheon's Emulsifiers & Detergents, 1998 North American edition; Ref. 2: International Cosmetic Ingredient Dictionary and Handbook, 7th ed. 1997, CTFA; Ref. 3: JCID/Japan Cosmetic Ingredients Dictionary.

able solubility for liquid-product formulation. The soaps with basic amino acids have improved mildness and biodegradability compared to TEA soap, and they give the skin a supple feeling [23].

B. Cationic Surfactant

Cocoyl arginine ethylester (CAE) as PCA salt is the only amino acid–based cationic surfactant commercially available, as shown in Table 3. Cocoyl arginine ethylester has unique properties, such as disinfecting and antimicrobial action [34], antistatic and conditioning effect for hair, excellent mildness to skin and eyes, and high biodegradability [25]. With these characteristics, CAE has been utilized in personal care products as a cationic surfactant and a germicidal agent. Examples are hand sanitizer, mouthwash, hair rinse and conditioner, shampoo, and skin cream and lotion [26]. Quaternary acylproteins are another example of cationic surfactants based on hydrolyzed proteins, as listed in Table 3. These materials are used mainly in hair care products as a conditioning agent.

C. Amphoteric Surfactant

Only two amphoteric surfactants derived from amino acid are known, as shown in Table 4. Lauroyl lysine (LL) is a condensate of lysine and lauric acid, as explained in Chapter 4. Because of a strong intramolecular neutralization, LL is insoluble in almost all kinds of solvents except strong acidic or alkaline solutions in spite of its amphoteric structure. And LL has high lubricity and a low frictional coefficient as an organic powder and excellent adhesive properties to hydrophilic surfaces to impart hydrophobicity to the substance. Thus LL has been widely used as a surface modifier in makeup products and as a lubricant in skin creams and hair conditioners [27].

Alkyloxy(2-hydroxypropyl) arginine (AA) is another amino acid–based amphoteric surfactant recently developed (Table 4). As an amphoteric molecule, AA changes its ionic structures in different pH. Because of a strong cationic moiety guanidine group in arginine residue, AA shows cationic properties under neutral to acidic conditions. Thus AA has a substantive conditioning effect for hair. And AA exhibits extreme mildness and biodegradability when compared to traditional cationic surfactants commonly used in hair care preparations, such as alkyl quaternary ammonium salts. As such, AA with its functionality and gentleness to humans and the environment is expected to be utilized widely in personal care products as an alternative to quaternary cationic surfactants.

TABLE 3 Amino Acid–Based Cationic Surfactants on the Market

INCI or chemical name	Amino acid	Trade name	Supplier	Ref. no.[a]
PCA salt of cocoyl arginine ethylester	Arginine	CAE	Ajinomoto	1 (p. 29)
Cocodimonium hydroxypropyl hydrolyzed keratin protein	Protein	Promois WK-HC	Seiwa Chem.	3 (p. 51)
Cocodimonium hydroxypropyl hydrolyzed silk protein		Promois S-CAQ	Seiwa Chem.	3 (p. 53)
Hexadecyl ester of hydrolyzed animal protein		Promois AH	Seiwa Chem.	3 (p. 82)
Lauryldimonium hydroxypropyl hydrolyzed casein				
Lauryldimonium hydroxypropyl hydrolyzed collagen		Lamequat	Henkel	1 (p. 107); 2 (p. 2088)
Lauryldimonium hydroxypropyl hydrolyzed keratin		Lamequat	Grunau	2 (p. 2088)
Lauryldimonium hydroxypropyl hydrolyzed silk				
Lauryldimonium hydroxypropyl hydrolyzed soy protein				
Stearyldimonium hydroxypropyl hydrolyzed collagen		Promois W-42SA	Seiwa Chem.	3 (p. 53)
Stearyldimonium hydroxypropyl hydrolyzed keratin		Promois W-HSA	Seiwa Chem.	3 (p. 53)
Stearyldimonium hydroxypropyl hydrolyzed silk protein		Promois S-SAQ	Seiwa Chem.	3 (p. 53)

[a] Ref. 1: McCutheon's Emulsifiers & Detergents, 1998 North American edition; Ref. 2: International Cosmetic Ingredient Dictionary and Handbook, 7th ed., 1997, CTFA; Ref. 3: JCID/Japan Cosmetic Ingredients Dictionary.

TABLE 4 Amino Acid–Based Amphoteric Surfactants on the Market

INCI or chemical name	Amino acid	Trade name	Supplier	Ref.
Alkylhydroxypropyl arginine	Arginine	Amisafe	Ajinomoto	
Lauroyl lysine	Lysine	Amihope LL	Ajinomoto	CTFA p. 1820

Ref.: International Cosmetic Ingredient Dictionary and Handbook, 7th ed., 1997, CTFA.

D. Nonionic Surfactant

Amino acid–based nonionic surfactants are listed in Table 5. Most of the materials are used either as a coemulsifier or as an emollient in skin care and hair care products [28–30]. Acylglutamate cholesteryl ester (AGCE) promotes recovery of damaged skin, which is similar to the effect found for ceramides, key constituents in the lipid membrane of the stratum corneum [31].

V. FUTURE PROSPECTS

Because of increasing concern for the effect of industrialization on human life and the environment, every chemical material or process is scrutinized for safety and compatibility to minimize impact. Surfactants are functional chemicals in principle, so many synthetic surfactants are developed to improve their cost effectiveness.

In nature, there are many physiological structures and functions supported by amphiphilic molecules (namely, surfactants), such as lecithin, fatty acid, lipoamino acids, and lipopeptides. From this viewpoint, amino acid– and protein-based surfactants are logically one of the most important materials to be reviewed and developed.

In this chapter, the development and usefulness of many amino acid– and protein-based surfactants has been discussed. The cost of amino acids and proteins and complicated manufacturing processes were the prohibiting factors for the broad utilization of this class of surfactants. Large-scale production and market consumption of certain amino acids, such as glutamic acid and lysine, have helped to alleviate this limiting factor.

Increased market application of acylglutamate and acyllysine, still in the niche area though, helped to increase awareness of these surfactants and will drive expansion of their applications, which eventually will lead to a larger cost reduction. The combination of synthetic and biological processes will be another prospective driving force to help reduce the production cost of amino

TABLE 5 Amino Acid–Based Nonionic Surfactants on the Market

INCI or chemical name	Amino acid	Trade name	Supplier	Ref. no.[a]
PCA salt of cocoyl arginine ethylester	Arginine	CAE	Ajinomoto	1 (p. 29)
Diethyl palmitoyl aspartate	Aspartic acid	ASP	Nippon Chem.	3 (p. 238)
Di(cholesteryl,behenyl,octyldodecyl)N-lauroyl glutamic acid ester	Glutamic acid	Eldew CL-301	Ajinomoto	1 (p. 158)
Dioxyethylene octyldodecyl-diester of lauroyl glutamic acid ester		Amiter LGOD-2	Ajinomoto	1 (p. 15)
Dioxyethylene stearyl-diester of lauroyl glutamic acid ester		Amiter LGS-2	Ajinomoto	1 (p. 15)
Octyldodecyl-diester of lauroyl glutamic acid		Amiter LGOD	Ajinomoto	1 (p. 15)
Aminopropyl laurylglutamine	Glutamine			2 (p. 1821)
2-Ethylhexyl PCA	PCA			3 (p. 425)
Lauryl PCA				2 (p. 2093)
PCA monoolein ester		Amifat P-30	Ajinomoto	1 (p. 14)
POE(25) glycerin monoPCA monoisostearic diester		Pyroter GPI-25	Ajinomoto USA	1 (p. 155)
POE(40) hydrogenated castor oil monoPCA monostearic diester		Pyroter CPI-40	Ajinomoto USA	1 (p. 155)

[a]Ref. 1: McCutheon's Emulsifiers & Detergents, 1998 North American edition; Ref. 2: International Cosmetic Ingredient Dictionary and Handbook, 7th ed., 1997, CTFA; Ref. 3: JCID/Japan Cosmetic Ingredients Dictionary.

acid– and protein-based surfactants. With all of this improvement, the food and industrial applications of amino acid– and protein-based surfactants will be promoted even further.

REFERENCES

1. P. M. Morese, Chem. Eng. News, Feb. 1, 1999, p. 35.
2. Fine Chemical, 24(15):13 (1995).
3. R. Yoshida and M. Takehara, Yukigouseikagaku kyoukaishi (J. Org. Synthesis) 33(9):671 (1975).
4. R. Yoshida, M. Takehara, A. Shibuya, and Y. Usuba, Yukagaku (J. Oil Chem. Soc. Jpn) 26:747 (1977).
5. M. Takehara, R. Yohida, and M. Yoshikawa, Cosmet. Toilet 94:31 (1979).
6. H. Nakayama, A. Oshiro, I. Yoshimura, R. Reiko, and M. Miyashiro, Fragrance J. 5:15 (1977).
7. T. Saitou, Fragrance J. 10:47 (1981).
8. H. Nakayama, H. Hanaoka, and A. Oshiro, Allergen Controlled System, Kanehara Shuppan Co., Tokyo, 1974.
9. H. Nakayama, J. Derm. Soc. Jpn. 84:697 (1974).
10. T. Saitou, Cosmet. Toilet 98:111 (1983).
11. K. Sakamoto, J. Japanese Soc. Cosmet. Sci. 21:125 (1997).
12. K. J. Carpenter, Cosmet. Toilet 110:31 (1995).
13. M. Kanari, Y. Kawasaki, and K. Sakamoto, J. Soc. Cosmet. Chem. Jpn. 27:498 (1993).
14. A. Domsch, B. Irrgang, and C. Moller, SOFW J. 122:353 (1996).
15. K. Sakamoto, J. Japanese Soc. Cosmet. Sci. 21:125 (1997).
16. M. Kawai, G. Imokawa, and A. Okamoto, Hihuka Shinnryou (Diagnostic Dermatology) 11:430 (1989).
17. Y. Kawasaki, D. Quan, K. Sakamoto, and H. I. Maibach, Dermatology 194:238 (1997).
18. Y. Kawasaki, K. Sakamoto, and H. I. Maibach, J. Soc. Cosmet. Chem. Jpn. 29: 252 (1995).
19. H. Yoshihara and D. Kaneko, Fragrance J.
20. K. Miyazawa, U. Tamura, Y. Katumura, K. Utikawa, T. Sakamoto, and K. Tomita, Yukagaku 38:297 (1989).
21. K. Miyazawa, J. Soc. Cosmet. Chem. Jpn. 29:95 (1995).
22. E. Shiojiri, K. Sano, M. Koyama, M. Inou, T. Hattori, H. Yoshihara, K. Iwasaki, and Y. Kawasaki, J. Soc. Cosmet. Chem. Jpn. 30:410 (1996).
23. H. Abe, N. Nakanishi, and N. Mikami, J. Soc. Cosmet. Chem. Jpn. 30:396 (1996).
24. R. Yoshida, K. Baba, T. Saito, and I. Yoshimura, Yukagaku 25:404 (1976).
25. M. Takehara, Colloids Surf. 38:149 (1989).
26. K. Sagawa and M. Takehara, Yukagaku 17:52 (1989).
27. K. Sagawa, M. Takehara, and H. Yokota, Fragrance J.

28. N. Mikami, T. Hattori, K. Sakamoto, S. Fukami, and T. Ichikawa, J. Soc. Cosmet. Chem. Jpn. 27:474 (1993).
29. N. Mikami, Bio Ind. 10:351 (1993).
30. M. Koyama, Fragrance J.
31. H. Ishii, N. Mikami, and K. Sakamoto, J. Soc. Cosmet. Chem. 47:351 (1996).

Index

T - #0040 - 111024 - C0 - 234/156/18 - PB - 9780367397296 - Gloss Lamination